ITAL

INS

INBETWEEN: LIVING + DYING

"TWO VERY BEAUTIFUL NAKED GIRLS ARE CROUCHED FACING EACH OTHER. THEY TOUCH EACH OTHER SENSUALLY, THEY KISS EACH OTHER'S BREASTS LIGHTLY, WITH THE TIP OF THE TONGUE. THEY ARE ENCLOSED IN A KIND OF CYLINDER OF TRANSPARENT PLASTIC. EVEN SOMEONE WHO IS NOT A PROFESSIONAL VOYEUR IS TEMPTED TO CIRCLE THE CYLINDER IN ORDER TO SEE THE GIRLS FROM BEHIND, IN PROFILE, FROM THE OTHER SIDE. THE NEXT TEMPTATION IS TO APPROACH THE CYLINDER, WHICH STANDS ON A LITTLE COLUMN AND IS ONLY A FEW INCHES IN DIAMETER, IN ORDER TO LOOK DOWN FROM ABOVE, BUT THE GIRLS ARE NO LONGER THERE. THIS WAS ONE OF THE MANY WORKS DISPLAYED IN NEW YORK BY THE SCHOOL OF HOLOGRAPHY. HOLOGRAPHY, THE LATEST TECHNICAL MIRACLE OF LASER RAYS, WAS INVENTED BACK IN THE '50S BY DENNIS GABOR; IT ACHIEVES A FULL COLOR PHOTOGRAPHIC REPRESENTATION THAT IS MORE THAN THREE-DIMENSIONAL. YOU LOOK INTO A MAGIC BOX AND A MINIATURE TRAIN OR HORSE APPEARS; AS YOU SHIFT YOUR GAZE YOU CAN SEE THOSE PARTS OF THE OBJECT THAT YOU WERE PREVENTED FROM GLIMPSING BY THE LAWS OF PERSPECTIVE."

UMBERTO ECO, 1967[1]

REALITY ISN'T WHAT IT USED TO BE!

Throughout history humanity has been moving further and further away from reality at an increasing speed. Every new technology since the Agricultural Revolution began 12,000 years ago has tried to enhance our way of life. But it wasn't until this last century, when technology became so ingrained in society, that humans began exploring ways to give life to alternate realities free from the confines of the physical world.

This increasing use of technology over shorter periods of time has initiated the largest technological and social human experiment in history. We must now learn how to adapt to a new evolutionary way of communicating with each other and with machines by reorganizing our cultural, work, entertainment, and social lives. It is no longer an exception to see people sitting at a table eating or in a business meeting simply pick up the phone and try to multi-task, or walking down the street with a headset talking on the phone while appearing to be talking to ourselves.

It is only in the last century with the exponential disruption of the internet, genetic engineering, the Metaverse, and artificial intelligence that humans have been so easily exposed to living with and augmenting multiple realities at the same time.

The architecture of this reality must now not only accommodate the material gravity of the real but, must also expand to embrace multiple realities at the same time as humans move toward the virtual world. It is inevitable that society and its future environments must learn to live seamlessly INbetween the real and the emerging virtual realities. In the twenty-first century the reality of truth has merely become a matter of perspective.

R

AR

VR

REALITY: INBETWEEN THE REAL + THE VIRTUAL

Reality is no longer confined by physics, or borders. The impossible becomes possible in this new frontier and is only limited by the imagination. The children being born today have entered a world where time has become compressed with instantaneous news, knowledge is an internet search away, communication is broadcast to millions at their fingertips, and social interaction has begun moving to interactive sites that hover between the real and the virtual. Researchers have found that young children cannot distinquish between a virtual, immersive expeience and one from a memory that is lived in a real life.[2] Each of these steps are moving us further away from the physical reality of the moment and into a virtual world. It's undeniable. The Metaverse is here and

REALITY

A UNIVERSAL TRUTH UNTAINTED BY HUMAN PERCEPTION

headed toward mass adoption. The question is how will we navigate this new world? How does classic, static architecture with a physical reality adapt to living and moving between these emerging multiple realities? How should a new language of architecture be developed that orients people to its own time the way triangular pediments with sculptures atop buildings told stories to the public

AUGMENTED REALITY

A VIEW OF REALITY ENHANCED BY TECHNOLOGY OR THE MIND

in architecture 2,000 years ago. Now that the private world of the home has instantaneous access to the public realm through the global reach of the computer, how can architecture evolve to enhance life INbetween the physical and the virtual world? Through a deep dive into how humans have learned a new language to talk with machines, how we have just begun to live in digital twins

VIRTUAL REALITY

EXISTS IN THE MIND, EXISTS IN ESSENCE, BUT NOT IN FACT

of the world in real time, the philosophy and fragility of truth, the disruption and dawn of artificial intelligence and the Metaverse, and how deciphering the genetic code has changed the materials of architecture and are all revealed by unpacking this new narrative in the twenty-first century. In the same way that agriculture changed the world 12,000 years ago, technology is once again creating a paradigm shift and this time it's happening at lightning speed. The climate crisis, social injustices, and an unsustainable economic system are creating a perfect storm, but architecture could embrace technology and become the North Star showing the world what it means to be living and dying INbetween the real and the virtual world.

REAL ENVIRONMENT

AUGMENTED REALITY

REAL I N B E

The real and the unreal are not a duality, they exist on a spectrum, and most of our lives are spent navigating INbetween these worlds.

In the beginning, there was reality—the sun, the moon, the stars, the earth and its inhabitants. But, as humans created new technologies, our reality evolved and we started spending our time in an increasingly unreal world. This spectrum of realities now exists in nearly every aspect of our lives.

Augmented reality (AR) is a term that became popular with contemporary technologies like Pokémon GO, but humans actually started augmenting their reality long before that. In Pokémon GO, AR creates an interactive experience on a device in which the real world is altered by computer-generated images, sounds, and other sensory experiences: The use of devices expands what's possible in the real world by creating a new reality INbetween the real and the virtual. While we may not realize it, we've been doing this for millennia.People used to travel to hunt and gather food, but by enhancing the real world with agriculture, humans expanded what was possible, and the need for foraging was virtually eliminated. Oceans and mountains once created natural borders, but by augmenting our reality with new modes of transportation, we can now travel anywhere on earth and beyond. In healthcare, we can even alter our bodies with stents, transplants, pacemakers, or plastic surgery. But AR has limits because, in the real world, you can't completely change nature, you can only enhance it.

In 1993, Neal Stephenson wrote in *Snow Crash*, "He is not seeing real people, of course. This is all a part of the moving illustration drawn by his computer according to specifications coming down

REALITY SPECTRUM: BEING HERE VS. BEING THERE

AUGMENTED VIRTUAL

VIRTUAL REALITY

T W E E N VIRTUAL

the fiber-optic cable. The people are pieces of software called avatars. They are the audiovisual bodies that people use to communicate with each other in the Metaverse."[3] Stephenson predicted a virtual world that is only just now becoming a reality.

This new, infinite world is limitless, free of the physical constraints of the real and the social constructs that have previously defined what it means to be human. Even death is no longer a certainty in the virtual world, due to artificial intelligence, avatars, and machine learning. *The Matrix*[4], once seen as science fiction, is now becoming a reality with the rise of the Metaverse. These new and exciting possibilities also present challenges with a real-life impact.

By taking a deep dive into the spectrum of evolving technologies, the role of architecture can be explored in this new world of the INbetween. Artificial intelligence and machine learning can transform the architectural profession as it is practiced today. These new tools can be used to augment design and push the envelope, redefining what architecture can become in the twenty-first century. In the real world, "smart homes" will be able to truly be smart, instead of just being a home with smart devices, when they are designed to provide more than just shelter. We can then innovate and operate INbetween the real and the virtual.

As Web3 (a decentralized version of the internet that utilizes blockchain technology) and the Metaverse become more accessible, we will also have the ability to create new worlds with endless possibilities, a transformative platform of new architectural dimensions that will redefine what it means to be human in the twenty-first century.

With the introduction of the Metaverse, humanity is on the cusp of a new era in its evolution. By documenting the reality spectrum, from real to virtual, it allows humankind to take a step back and be intentional when deciding which path society should take in the future.

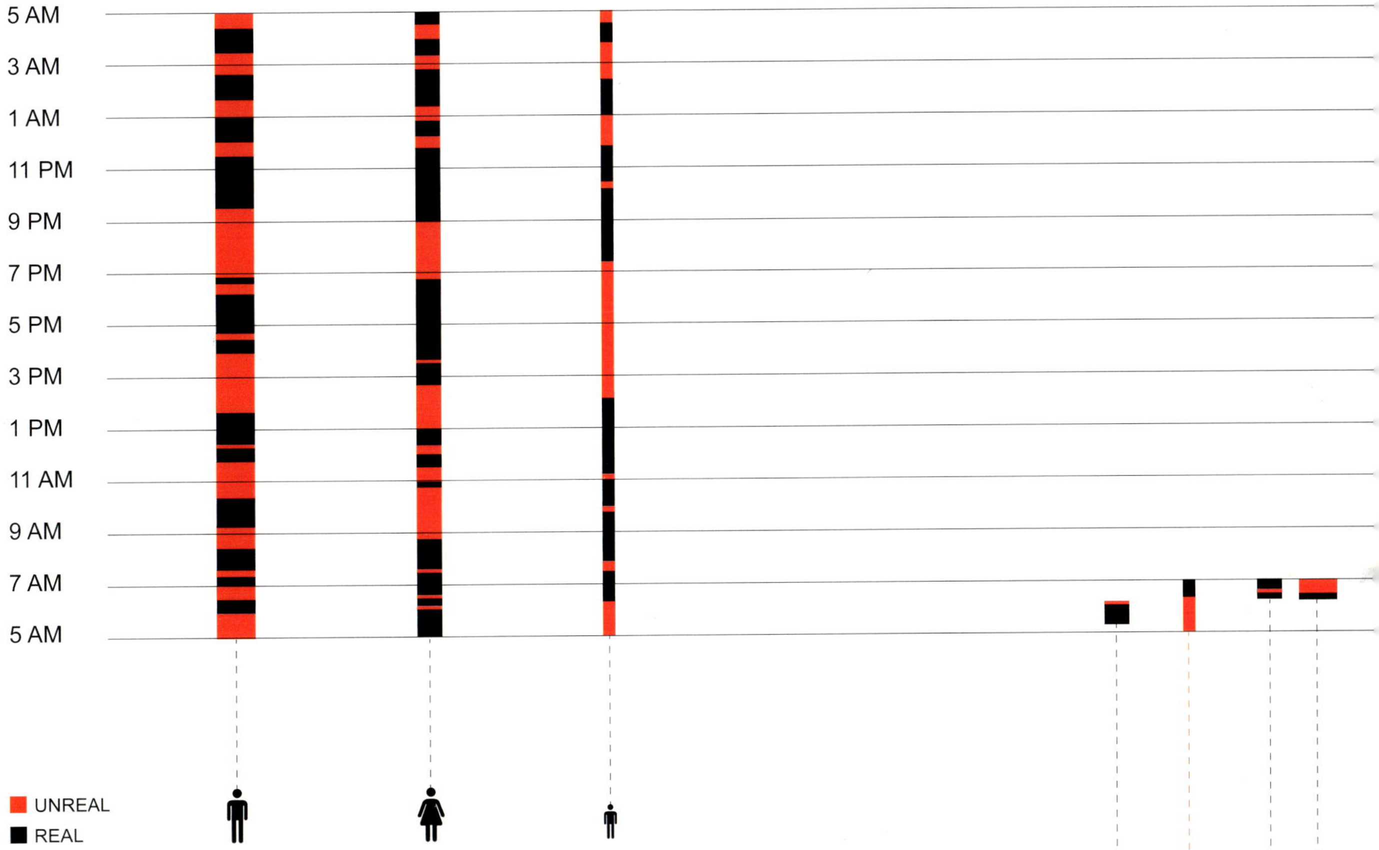

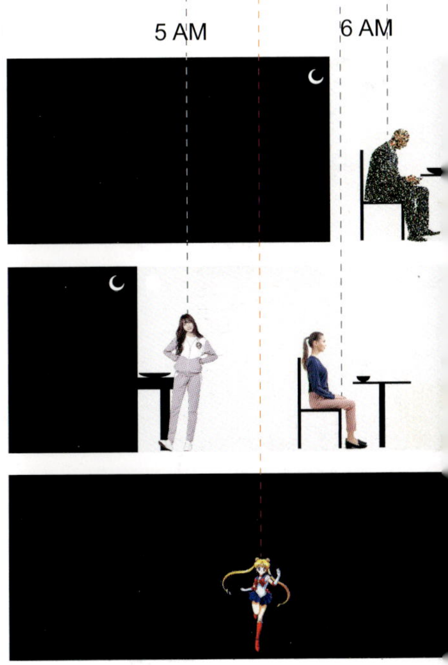

MAN

+ 38 YRS
+ USA
+ GEN X
+ MARRIED
+ WHITE
+ OFFICE
+ $55,291

WOMAN

+ 38 YRS
+ USA
+ GEN X
+ MARRIED
+ WHITE
+ OFFICE
+ $45,097

CHILD

+ 8 YRS
+ USA
+ GEN ALPHA
+ 3RD GRADE
+ WHITE

REAL + VIRTUAL: 24 HOUR CYCLE[5]

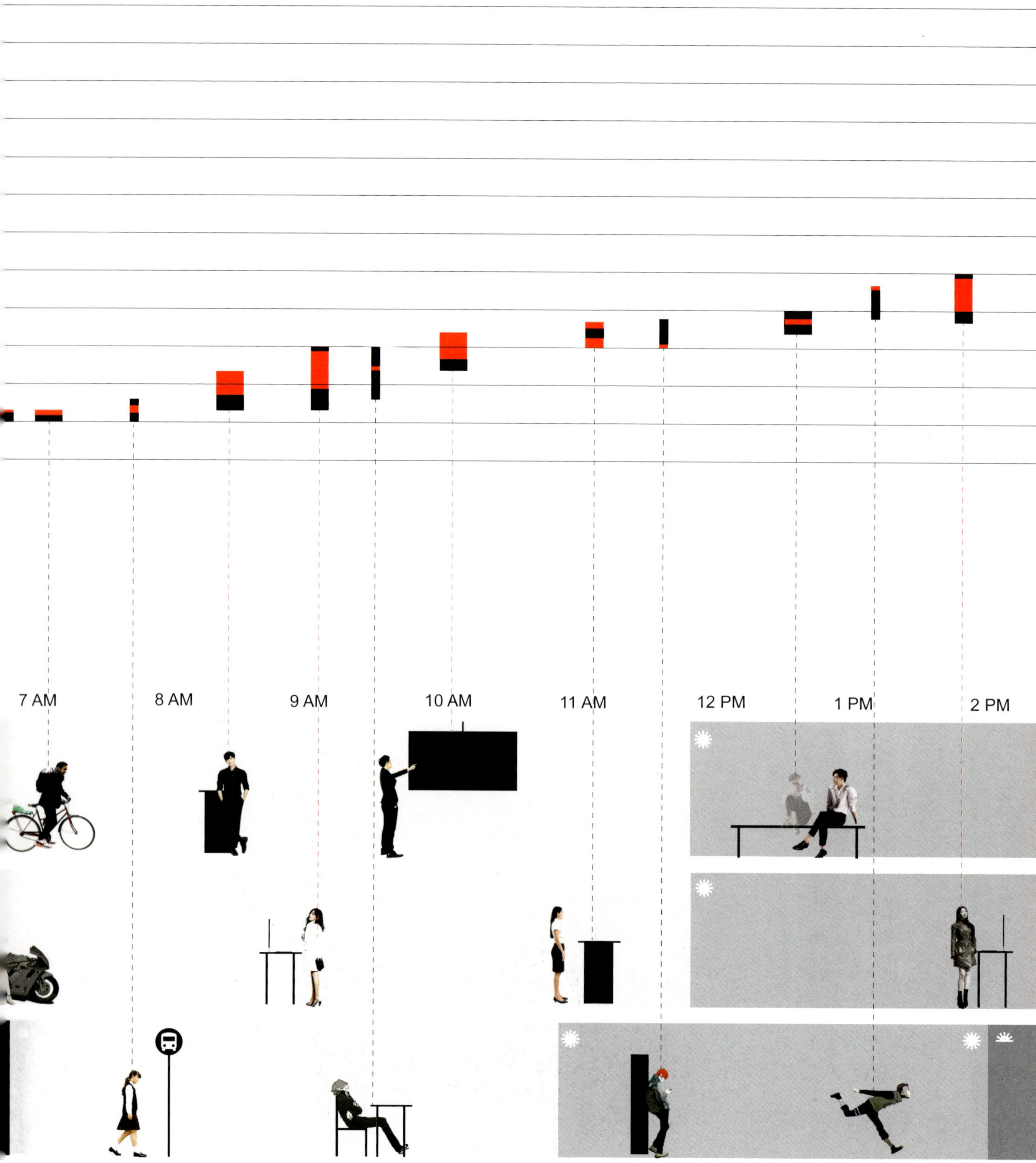

CONTINUE NEXT SPREAD

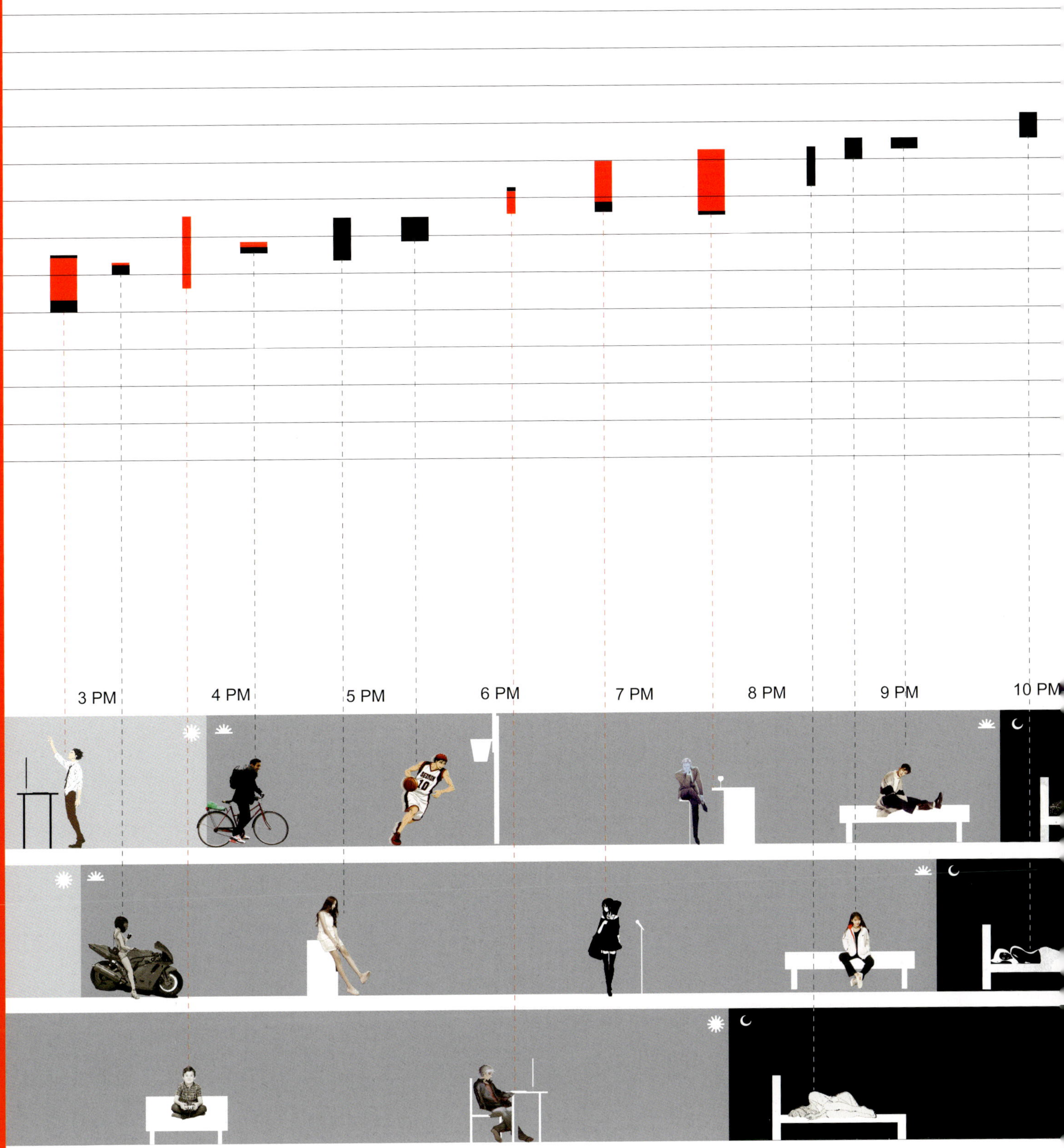
3 PM
4 PM
5 PM
6 PM
7 PM
8 PM
9 PM
10 PM

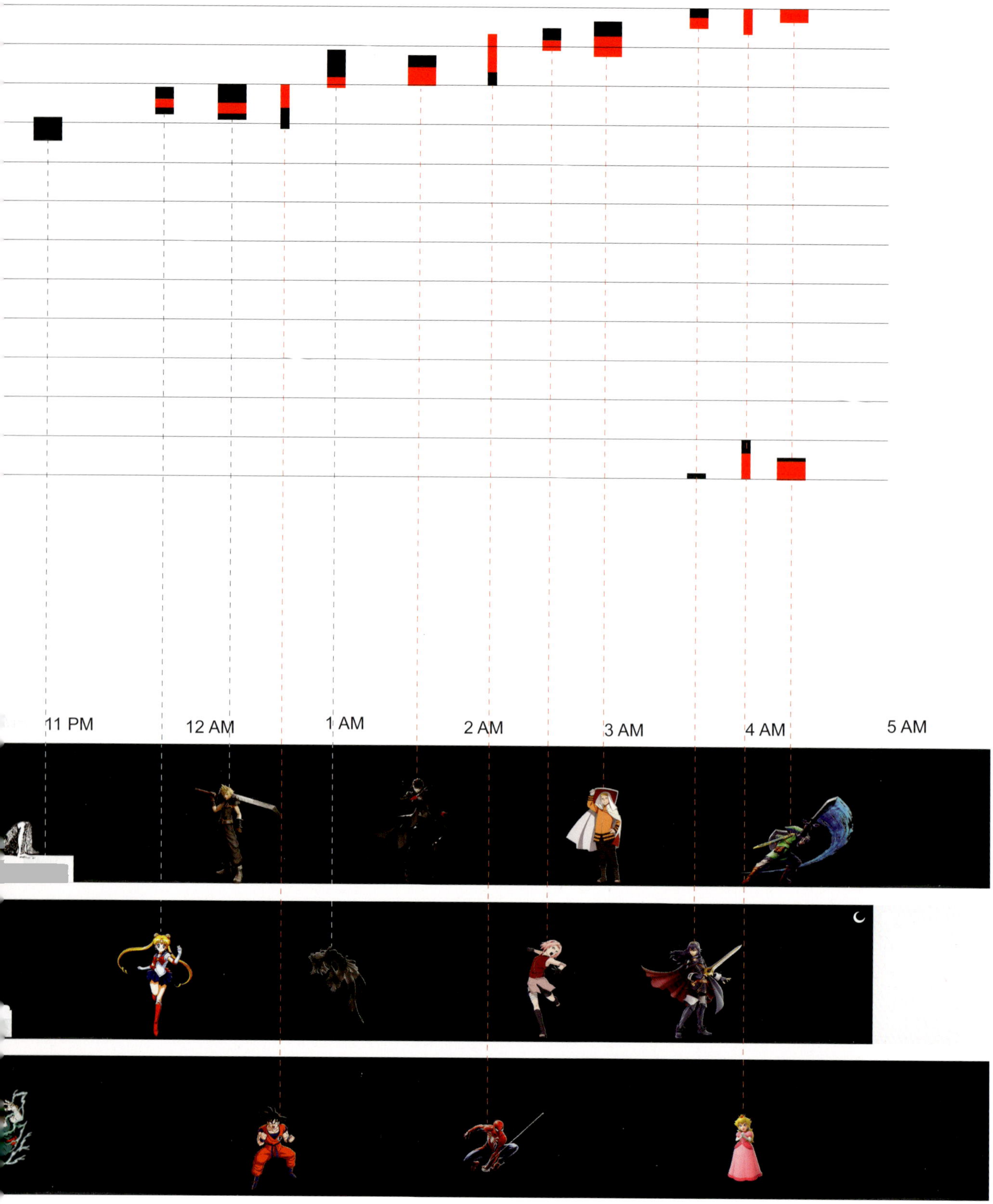
11 PM
12 AM
1 AM
2 AM
3 AM
4 AM
5 AM

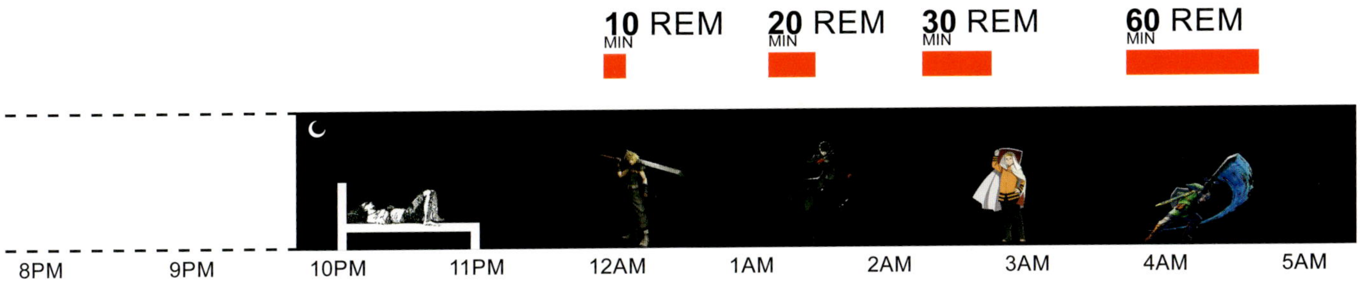

MAN
2 HRS UNREAL
25%

WOMAN
2 HRS UNREAL
25%

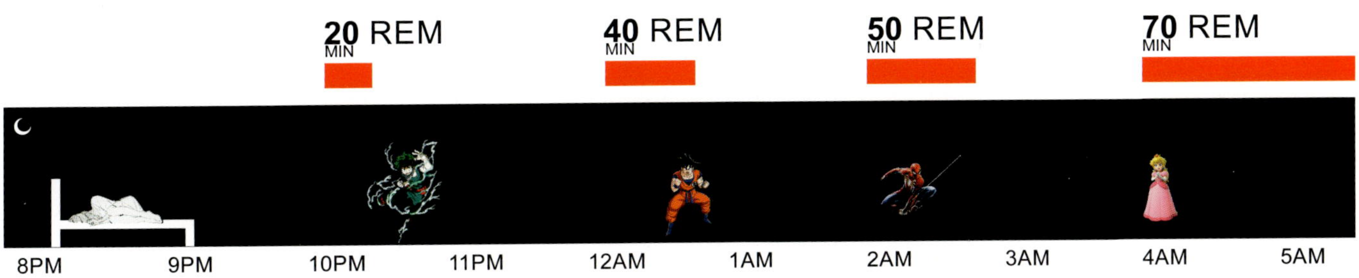

CHILD
3 HRS UNREAL
27%

SLEEP

MAN / WOMAN / CHILD

REAL + VIRTUAL: DAILY ACTIVITY CYCLE

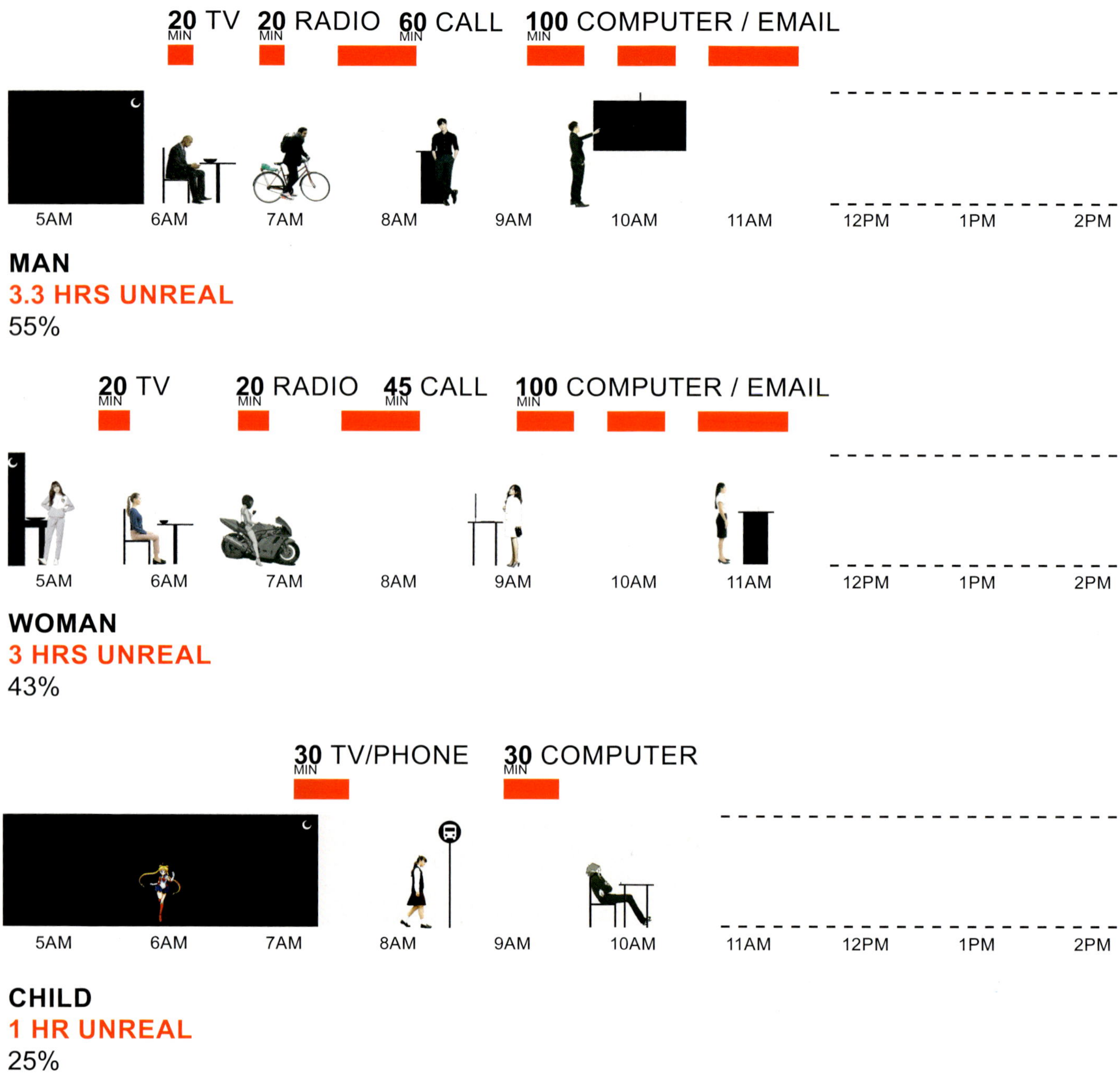

MORNING

MAN / WOMAN / CHILD

By researching the daily cycle of a man, woman, and child in their normal day on a vairiety of devices it is possible to document the relationship between how much time is being spent in the real world vs. that of the virtual world. The possibility of living a balanced life in the future is increasingly being challenged by the introduction of new time-consuming technologies into our daily lives.

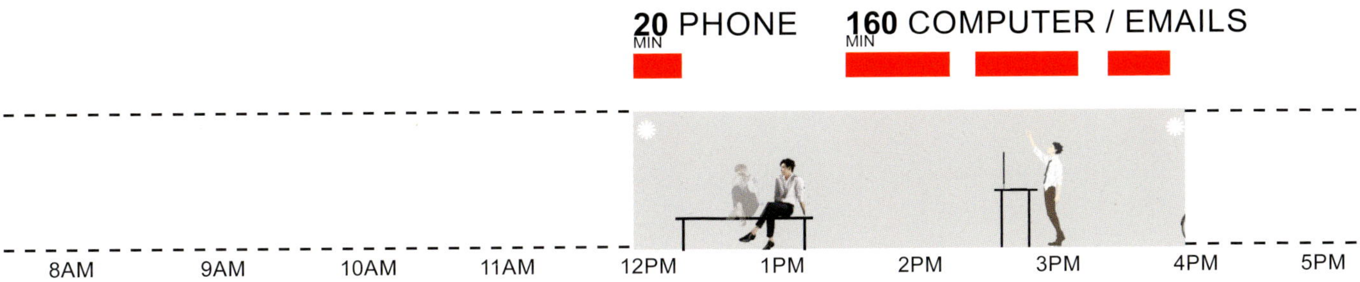

MAN
3 HRS UNREAL
75%

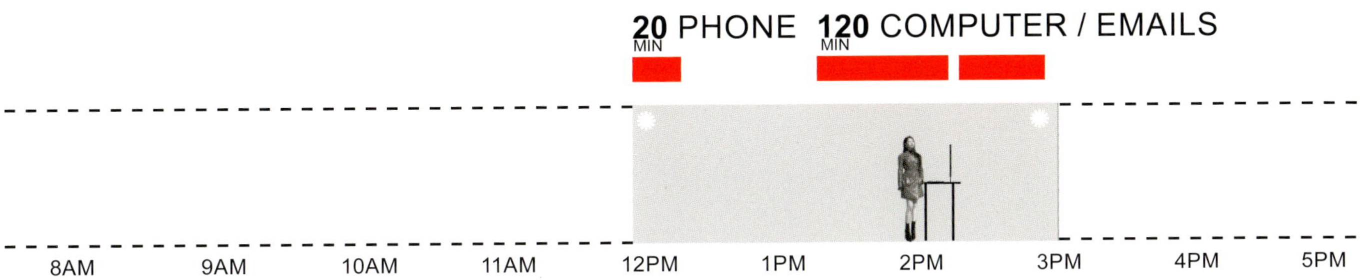

WOMAN
2 HRS UNREAL
77%

CHILD
5 MIN UNREAL
2%

AFTERNOON

MAN / WOMAN / CHILD

REAL + VIRTUAL: DAILY ACTIVITY CYCLE

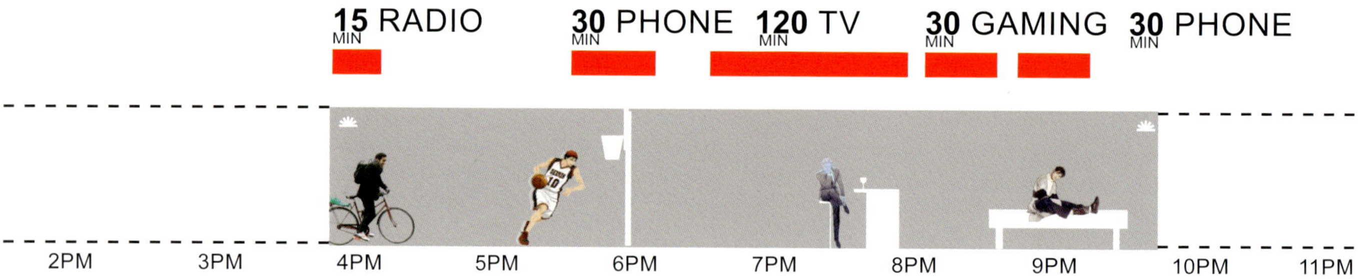

MAN
3.75 HRS UNREAL
62%

WOMAN
3.25 HRS UNREAL
50%

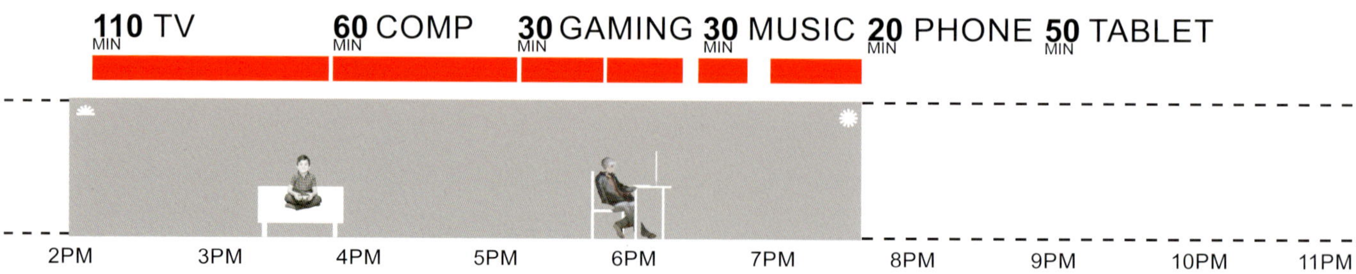

CHILD
5 HRS UNREAL
83%

LATE AFTERNOON

MAN / WOMAN / CHILD

Through Google Trends documentation of a man, woman, and child, the pattern of using technology in the virtual world clearly increases in the late afternoon. By analyzing the behavioral timelines, the increasing use of devices is affecting the day and night patterns of the man, the woman, and the child.

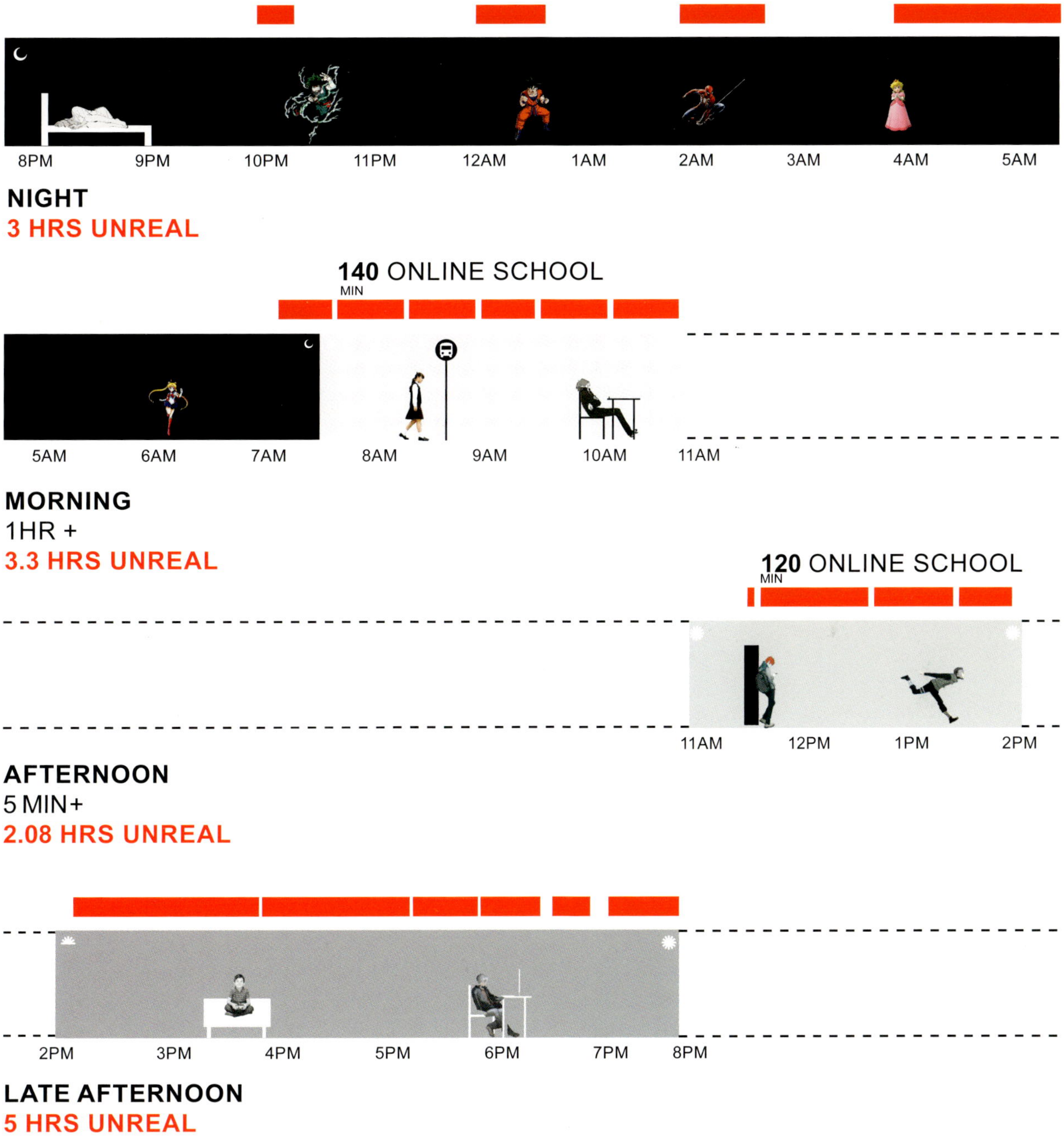

CHILDREN DURING COVID-19[6]

VIRTUAL DURING COVID-19

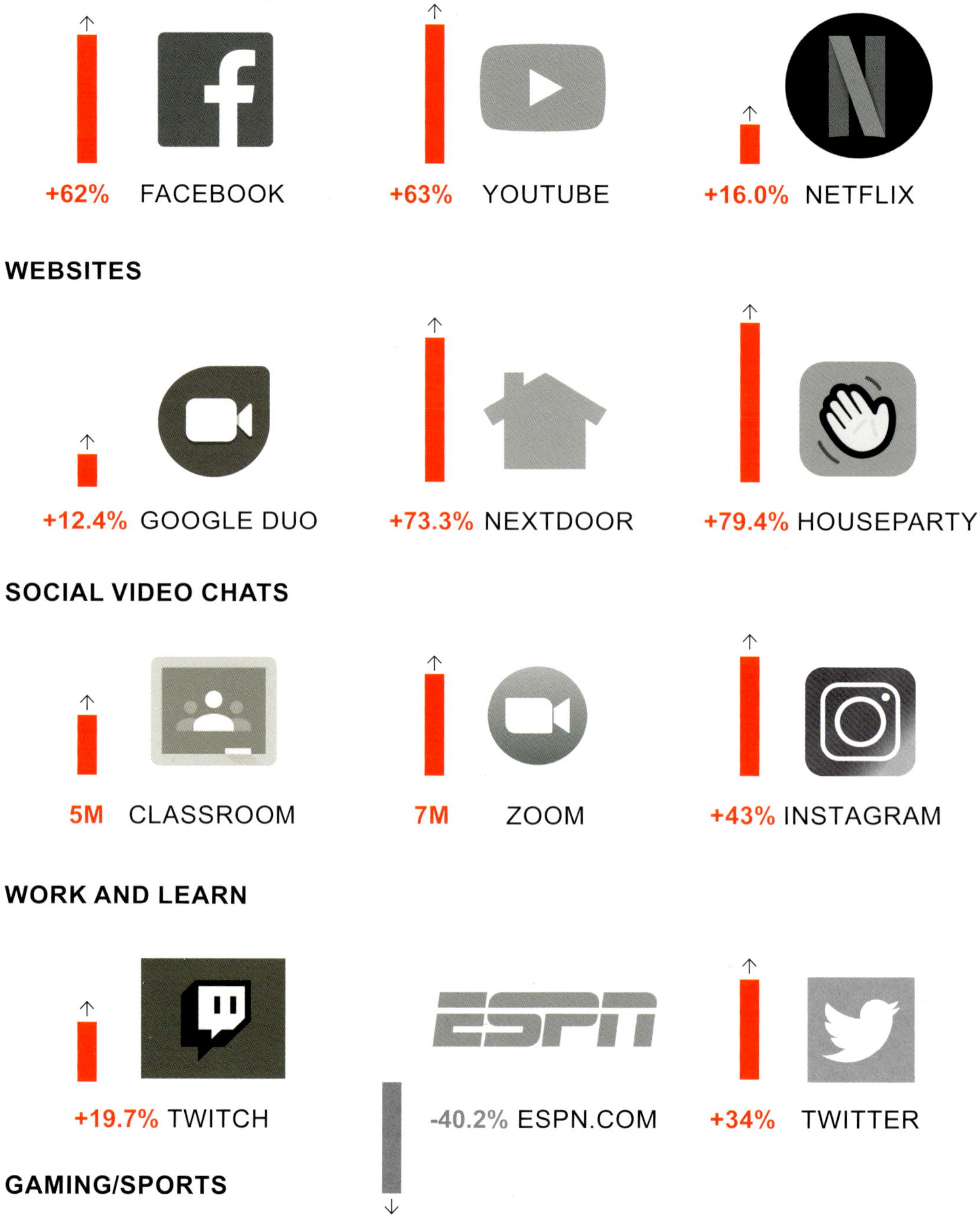

INCREASED USAGE IN VIRTUAL PLATFORMS[7]

The pandemic had a dramatic effect on the patterns of online behavior in children. In addition to remote learning, isolation in the real world also had children turning to the unreal, virtual world for their social connections and entertainment.

BLUE LIGHT CONSUMPTION

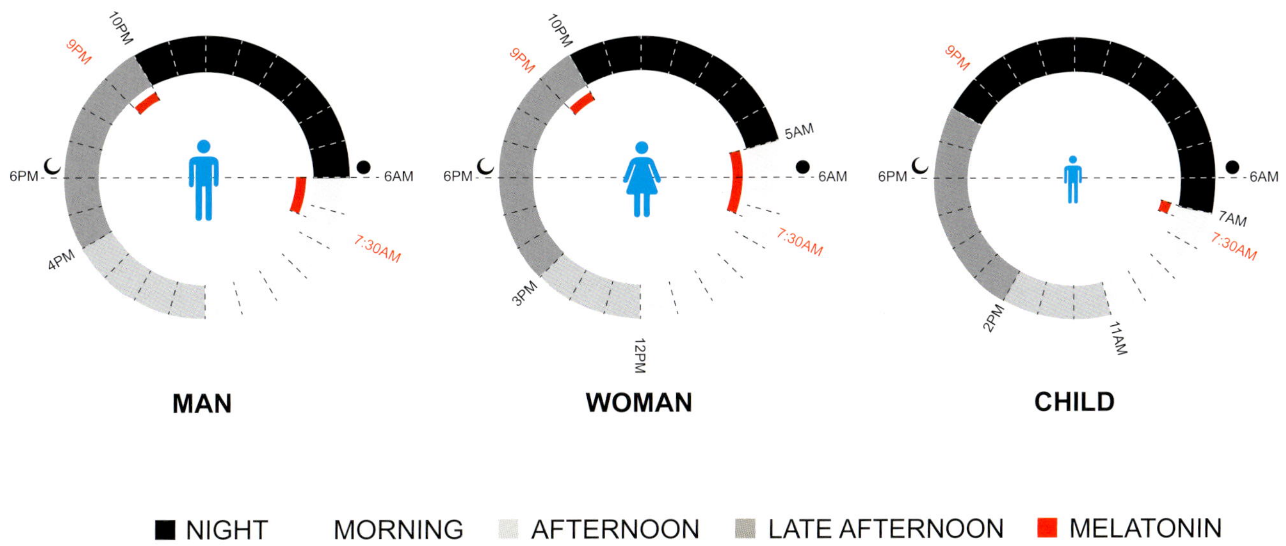

In *The Matrix*, Morpheus says to Neo, "Have you ever had a dream, Neo, that you were so sure was real? What if you were unable to wake from that dream, Neo? How would you know the difference between the dream world and the real world?"

The human perception of reality is simply created by electrical signals in our brain, no mattter the source. We're hardwired to process our five senses in the same way regardless of the environment. That means that as far as our brains are concerned, the fear, joy, or even love that we experience in the virtual world "feels" just as real as it does when your headset is off.[9] As the Metaverse grows in popularity and sophistication, it will likely become even more immersive, further blurring the lines of reality. Even if you've never played a video game, you know what it's like to experience life in another reality. Dream cycles of REM mimic the virtual world. The impulses interpreted in the brain are similar in both the real and virtual worlds, and with the introduction of the Metaverse into a three-dimensional space, it is now possible to create a new reality. We have all had dreams at night and perhaps this represents the first experience we have had of going into a virtual world, a place that is only familiar to each individual. It is often the case that we have recurring dreams where true emotions can be felt. Not unlike the virtual world, a dream can be fully immersive and is only limited by the imagination.

RAPID EYE MOVEMENT (R.E.M): 4 CYCLES

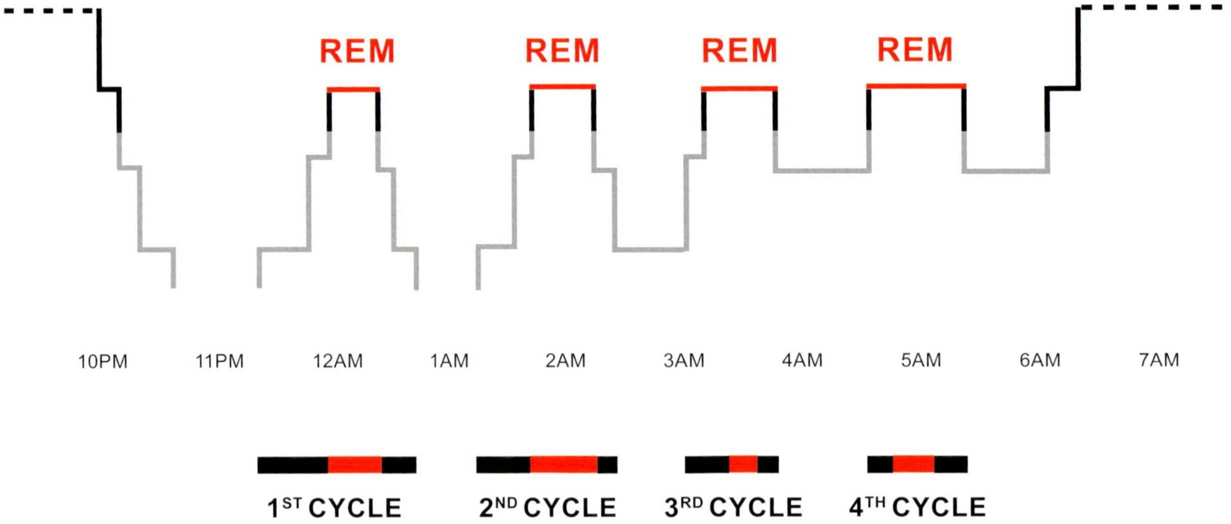

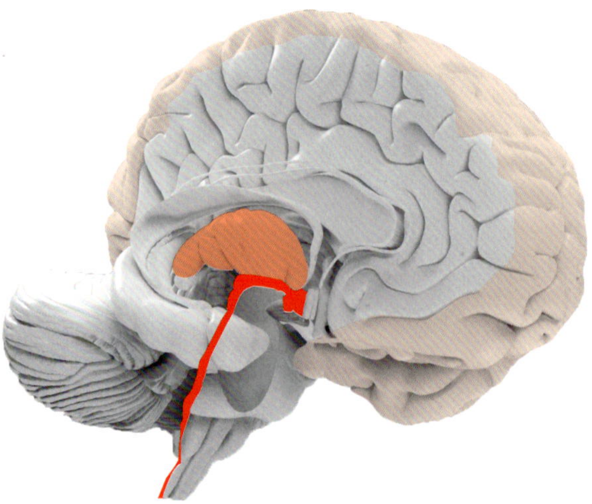

REM & DAYDREAM
NEURAL ACTIVATION
SUBCONSCIOUSNESS

The pattern of sleep with rapid eye movement is a window to the subconscious that can be conceived as being on the spectrum of multiple realities. These dream cycles occur in segments during the night and move between being the real world and in a dream state. This reality can directly inform how we think in the real world as interpreted in the field of psychology.

TIMELINE OF INNOVATIONS

	Innovation	Years	Today
AUTOMOBILES	**FIRST VEHICLE** POWERED BY GASOLINE INVENTED: 1886[10]	**135 YEARS**	**71 MILLIONS** OF VEHICLES SOLD GLOBALLY: 2021[23]
	FIRST STEAM POWERED ENGINE INVENTED: 1698[11]	**323 YEARS**	**150,500 KM:** THE US HAS THE LARGEST GLOBAL RAILROAD NETWORK[24]
	FIRST VEHICLE WITH AUTOMATIC TRANSMISSION: 1948[12]	**73 YEARS**	**96%** OF AMERICANS DRIVE AUTOMATIC TRANSMISSION VEHICLES[25]
	FIRST DRIVERLESS CAR INVENTED: 1939[13]	**82 YEARS**	**1,400 SELF-DRIVING CARS**: THE US, 2020[26]
	FIRST US INTERSTATE HIGHWAY SYSTEM INVENTED: 1956[14]	**65 YEARS**	**4.1 MILLION** MILES OF HIGHWAYS: US, 2020[27]
COMMUNICATION	**FIRST HANDHELD MOBILE PHONES** MASS PRODUCED: 1973[15]	**48 YEARS**	**15 BILLION** MOBILE DEVICES IN THE WORLD: 2021[28]
	FIRST SMARTPHONE INVENTED: 1994[16]	**27 YEARS**	**6.4 BILLION** GLOBAL SMARTPHONE USERS: 2021[29]
	FIRST WORKABLE INTERNET INVENTED: 1960[17]	**61 YEARS**	**4.66 BILLION** GLOBAL ACTIVE INTERNET USERS: 2021[30]
	10 MILLIONS ZOOM USERS IN THE WORLD, 2019[18]	**2 YEARS**	**300 MILLIONS** ZOOM USERS IN THE WORLD: 2020[31]
SPACE	**DISCOVERY: EARTH IS ROUND**, 500 BCE[19]	**2,469 YEARS**	**FIRST HUMAN LANDING ON THE MOON:** 1969[32]
	32,768 BITS OF RAM MEMORY: APOLLO 11 COMPUTER LANDING ON THE MOON: 1969[20]	**52 YEARS**	**3 GIGABYTES** OF RAM MEMORY: IPHONE X: 2021[33]
	FIRST SATELLITE LAUNCHED INTO THE EARTH'S ORBIT: 1958[21]	**63 YEARS**	**6,542 SATELLITES** IN EARTH'S ORBIT[34]
	FIRST ROCKET LANDED BACK ON EARTH FROM SPACE: 2015[22]	**6 YEARS**	**FIRST CIVILIAN SPACE TRAVEL**: JULY 2021[35]

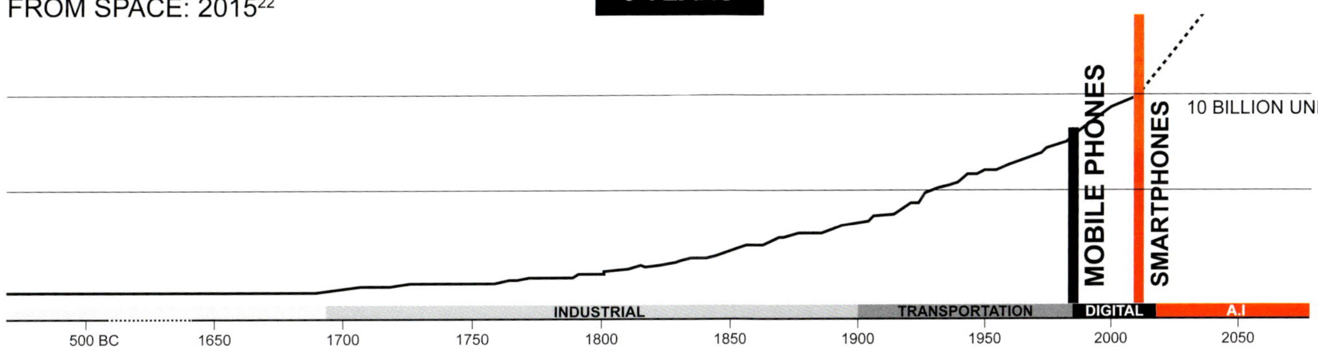

EVOLUTION OF TECHNOLOGY

THE SPEED OF EVOLUTION: TECHNOLOGY

GAMES: MIGRATING TO THE VIRTUAL WORLD

1.6 BILLION GLOBAL GAMERS: 2013[36]	7 YEARS	**2.69 BILLION** GLOBAL GAMERS: 2020	GAMERS
517 MILLION GAMERS: CHINA, 2014[37]	6 YEARS	**665 MILLION** GAMERS: CHINA, 2020[47]	GAMERS
155 MILLION GAMERS: US, 2015[38]	5 YEARS	**244 MILLION** GAMERS: US, 2020[48]	GAMERS
27% (66 MILLION) OF AMERICAN ADULTS ARE GAMERS: 2015	5 YEARS	**68% (163.3 MILLION)** OF AMERICAN ADULTS ARE GAMERS: 2020[49]	GAMERS
26% (19 MILLION) OF AMERICAN KIDS ARE GAMERS: 2015	5 YEARS	**70% (51.1 MILLION)** OF AMERICAN KIDS ARE GAMERS: 2020[50]	GAMERS
FIRST COMPUTING SYSTEM WITH BINARY CODES INVENTED: 1938[39]	78 YEARS	**2 BILLION COMPUTERS** OWNED GLOBALLY: 2019[51]	
FIRST VIDEO GAME: TENNIS FOR TWO: 1958[40]	63 YEARS	**1 MILLION** GAMES RELEASED ON PLATFORMS: 2019[52]	GAMES
FIRST ONLINE MULTIPLAYER GAME: MULTIUSER DUNGEON: 1980[41]	41 YEARS	**$42 BILLION**: REVENUE OF MASSIVELY MULTIPLAYER ONLINE GAMES: 2020[53]	GAMES
FIRST VIRTUAL REALITY GAME: SWORD OF DAMOCLES: 1968[42]	53 YEARS	**$1.1 BILLION:** VIRTUAL REALITY GAMING REVENUES: 2021[54]	GAMES
$70.98 BILLION: GLOBAL GAME REVENUES: 2000[43]	20 YEARS	**$159.3 BILLION:** GLOBAL GAME REVENUES IN 2020[55]	GAMING MARKET
$18.5 BILLION: CHINA'S GAME REVENUES: 2014[44]	6 YEARS	**$40.85 BILLION:** CHINA'S GAME REVENUES: 2020[56]	GAMING MARKET
$22.2 BILLION: US' GAME REVENUES: 2014[45]	6 YEARS	**$37 BILLION:** US' GAME REVENUES: 2020[57]	GAMING MARKET
$1.4 BILLION: GLOBAL MOBILE GAME REVENUES: 2000[46]	20 YEARS	**$80 BILLION:** MOBILE GAME REVENUES[58]	GAMING MARKET

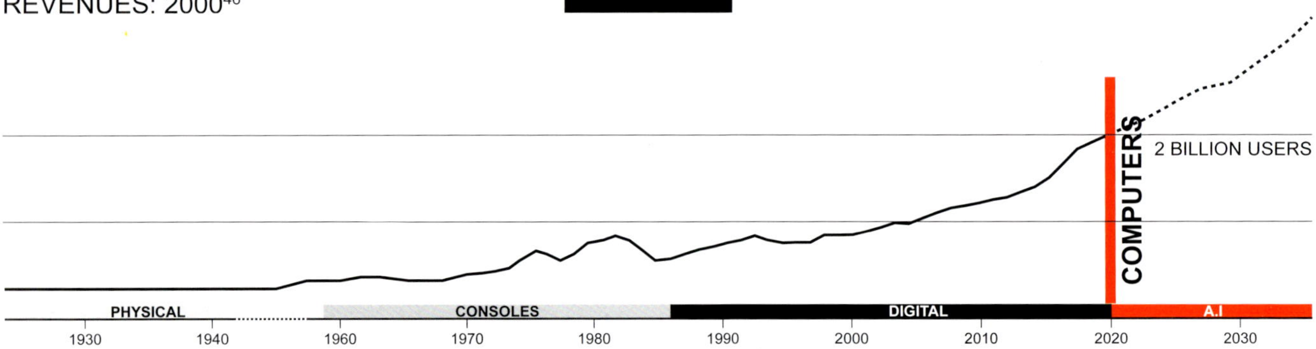

EVOLUTION OF THE VIRTUAL WORLD

YEARS OF EVOLUTION — SIGNIFICANT EVOLUTION

TRUTH

Truth has become merely a matter of perception. In this age of technology, the line between the real and unreal is becoming increasingly blurred. Morpheus said in *The Matrix* in 1999, "The Matrix is everywhere. It is all around us, even now in this very room. You can see it when you look out your window, or when you turn on your television. You can feel it when you go to work, when you go to church, when you pay your taxes. It is the world that has been pulled over your eyes to blind you from the truth." Throughout history, societies have been built on common myths. "There are no gods, no nations, no money and no human rights, except in our collective imagination," says Yuval Noah Harari, historian and author of *Sapiens*[59]. These myths are how humans make sense of the world and build trust. But today, humans have a hard time knowing what or who to believe thanks in part to the proliferation of technology. We've all heard the phrase, seeing is believing, but we know that's no longer true. Photography was designed to capture reality, but Photoshop, CGI, and filters create such a convincing illusion of truth that we must now question our own perceptions. Artificial intelligence and machine learning take this manipulation even one step further with emerging technology such as Deep-fakes, in which a person and their voice can be transformed into someone else's likeness. With the right combination of skill and imagination, anything is possible, as truth has become harder to define. The Information Age gave us access to all of the world's knowledge at our fingertips, yet misinformation runs rampant on the internet. It can be hard to decipher fact from fiction when both are presented as truth. Can faith and science find a way to co-exist so humans can rebuild trust in one another? Architecture has historically been revered as seeking truth to become a representation of its own time. Historians classify architecture into time periods and this becomes the accepted interpretation of a larger truth of past realities that encompasses—the fine arts, architecture, music, etc. Technological revolution and a global climate crisis on a biblical scale have created an opportunity for architecture to navigate between the reality of science and the social construct of faith. With augmented reality the INbetween has the possibility of becoming a dynamic architecture that can explore the polarity of science + faith and how architecture can begin to live and to be experienced INbetween these two different views of reality.

REAL

PLATO'S CAVE

ALLEGORY OF THE CAVE 360 BCE

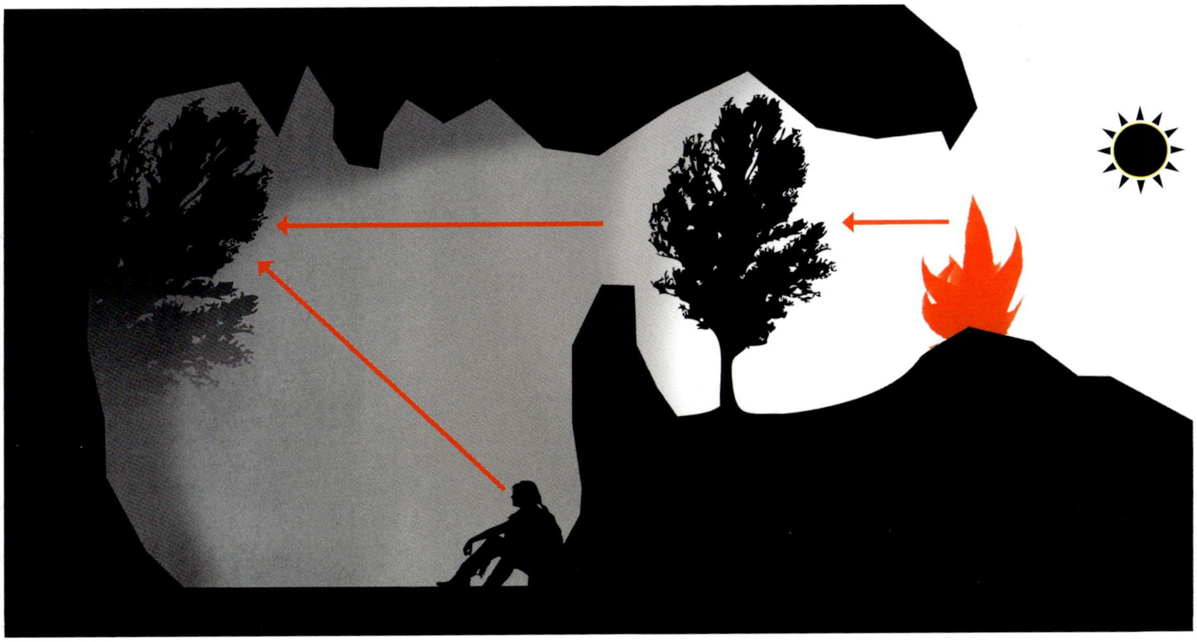

REALITY	PERCEPTION	CONTENT	TRUTH
CAVE	**SHADOW / BRAIN**	**TREE**	**SUN**

Socrates, in 360 BCE, questioned the meaning of beauty, and by extrapolation truth. As a metaphor, he conceived of a cave with the sun shining through the opening, casting the shadow of a tree on the walls of the cave. The observer seeing the shadow on the rock thought this was real and the shadow must be beauty and truth itself. He further questioned that even if the observer saw the actual tree, they still did not see true reality, since the sun was in fact allowing you to see the tree. The sun was truth and therefore was true beauty.[60]

In the contemporary world the truth has become even more obscure. The introduction of technology has created a situation where the tree and its shadow are seen projected on a screen, thus removing the observer even further from nature and the truth. Instead of a solitary experience in a cave, the observer now understands the tree through the eyes of someone photographing the tree. This edited view of truth calls into question what is beauty and how has the media transmitted multiple realities? The original truth has become more obscure

UNREAL

VIRTUAL CAVE

DIGITAL AGE 2020

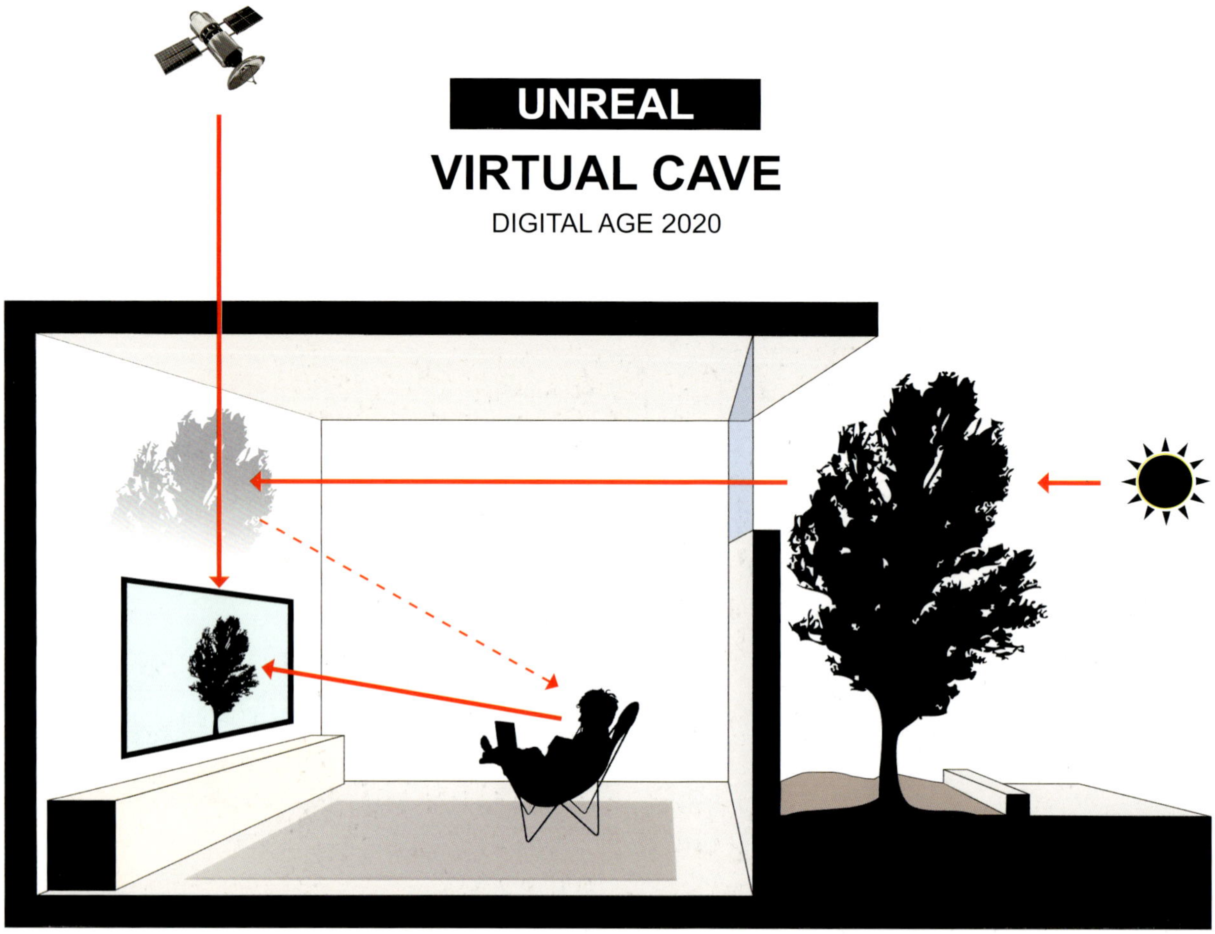

since the tree needs the sun to illuminate the misinformation of the media. Technology today has further removed the need for the sun with the original tree becoming simply a digital image. This opens the door to not knowing what is true in nature, or the representation of a true reality as seen in the media. The human need for understanding science (the rational and empirical) and faith (a belief system with an unknown conclusion) is rapidly dividing the agreement of common truths among the public on a daily basis.

Since the Renaissance, the methodolgy of science's pursuit of truth lies in stark contrast to that of faith. The rise of populism in the twenty-first century marks the battle within humans to marvel at the laws of the universe, but yearn for a belief that humans don't understand. Historically architecture has embodied the principles of the secular and the spiritual. The architect who designs a building that does not yet exist, is by its very nature, an optimist whose truth lies in imagining what a building might be as a dream for creating a better future.

The Allegory of the Cave was a significant historical step in trying to define reality and the need to unpeel our understanding of the Truth. By conceiving of the shadow of a tree in a cave, Plato began the process of unpacking the twenty-first *century's dilemma with dealing with the real and the virtual world, misinformation of reality, and ultimately humanity's loss of understanding what is true.*

REAL: HOMOGENEOUS

TRIBAL
PLEISTOCENE ERA

CITY CONCEPT
6500 BCE

NATION EMPIRE
27 BCE - 476 CE

UNREAL: HETEROGENEOUS

TEXTUAL
1600

TELEGRAPHIC
1830

RADIO
1910

TRUTH | A NEW EXPERIENCE

COMMUNITY: GOING FROM REAL TO UNREAL

CITY
9TH - 15TH CENTURY

TRIBAL NATIONS
9TH - 12TH CENTURY

EMPIRES
17TH CENTURY

GLOBAL SUBCULTURE
1960

WEB / SMARTPHONE
1980

VIRTUAL REALITY
2020

In the arc of history, humanity has been moving from communicating in small tribes of approximately 150 people during the Pleistocene era,[61] to cities of thousands 5,000 years ago, to the evolution of empires of millions. With the advancement of technology, the city has now evolved into 4.6 billion people instantly communicating using the internet[62] over a new concept of being in a tribe.

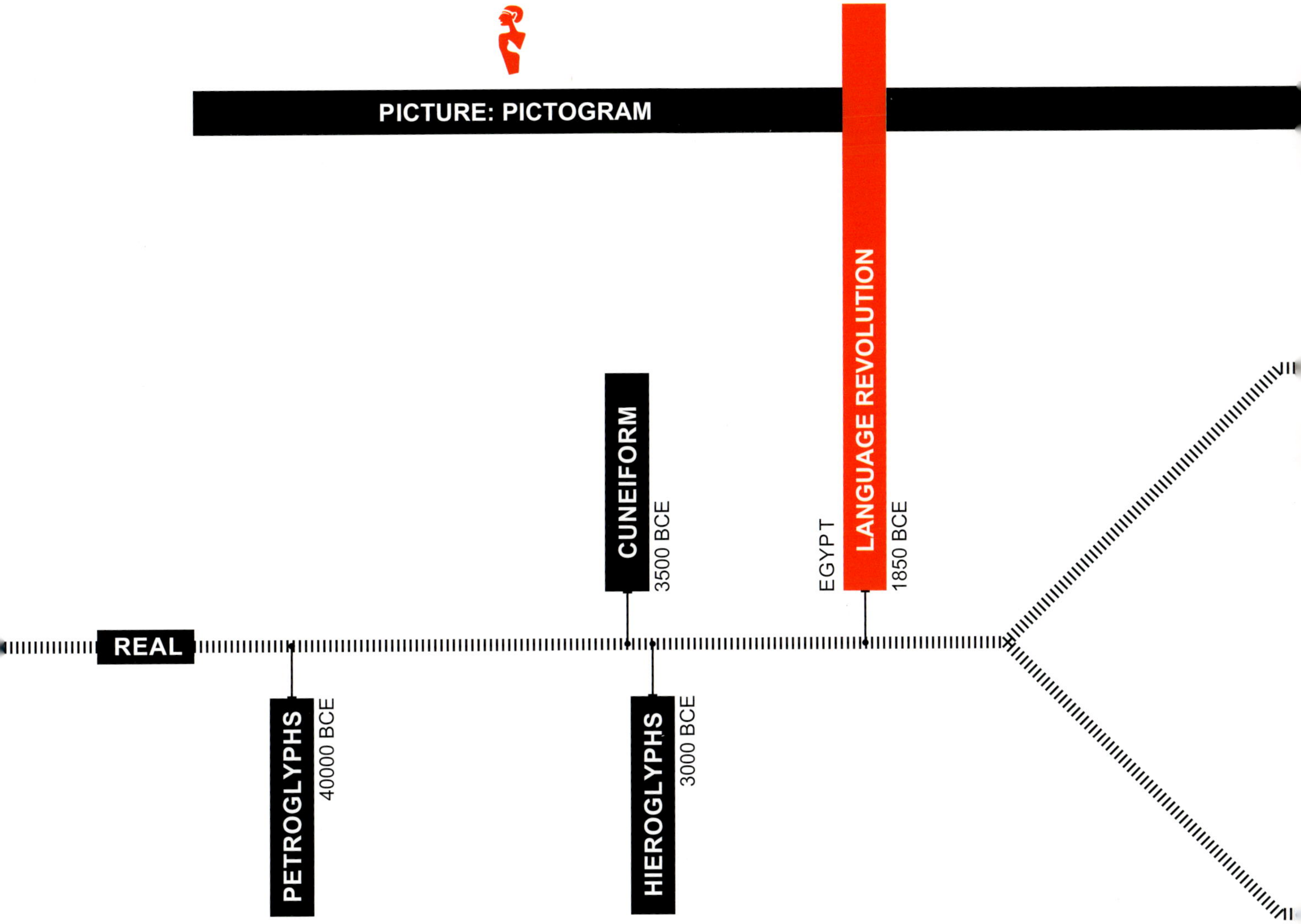

PICTOGRAM

Hieroglyphs evolved as a form of pictograms. Hieratic was the earliest form of cursive script. Vowels used for the first time.

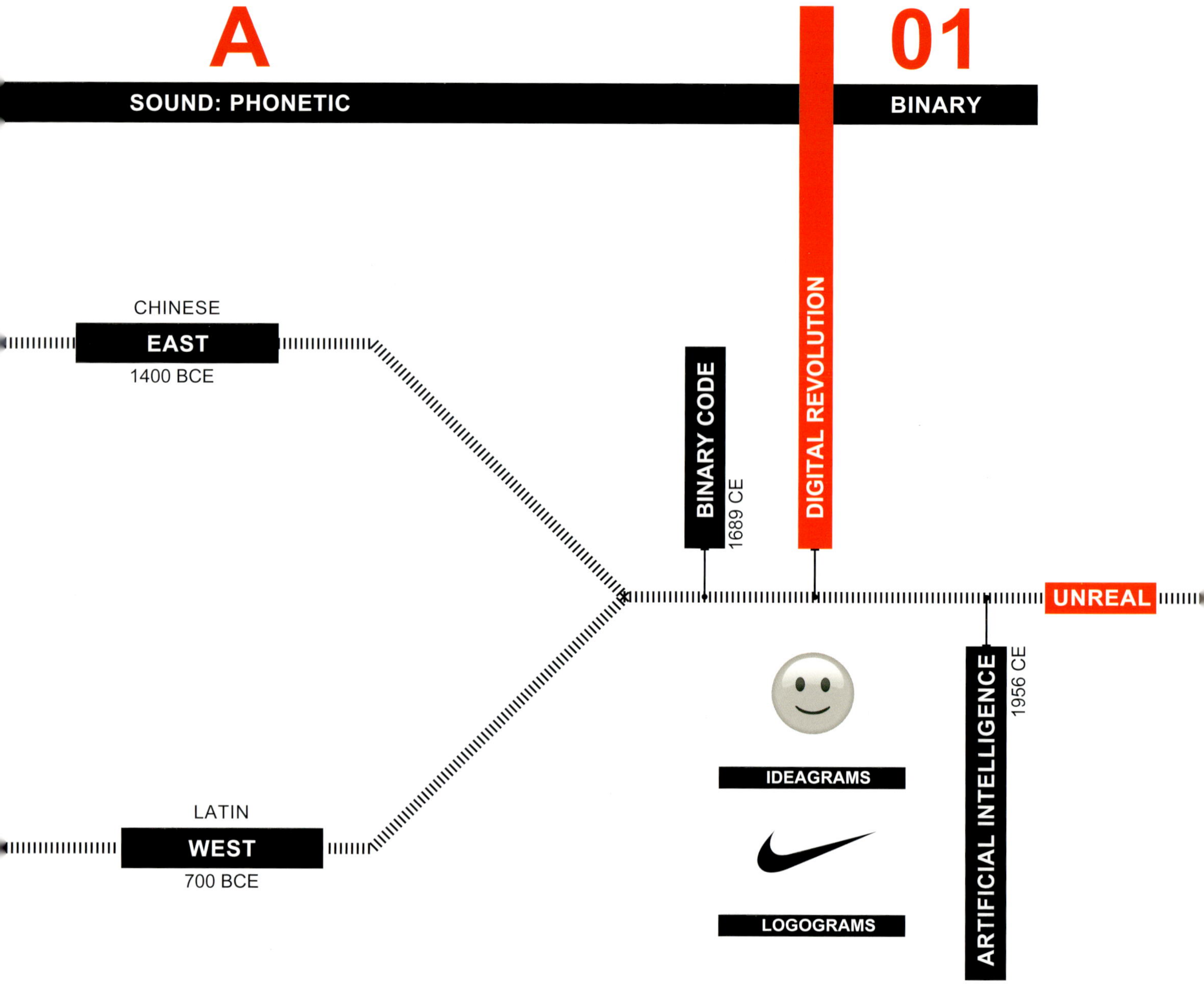

PHONETICS

A writing and speaking system that operates between making symbols and sounds.

BINARY + CODING

The new math is introduced as abstract and provides a method that becomes a precursor to coding.

Written communication between humans began with hieroglyphics using pictures and pictograms. It wasn't until 1800 BCE that the picture transformed into phonetic sounds that were typically derived from its hieroglyphic origins. This became the basis of language as we know it today until the binary computer code became the new languages for humans to talk directly to machines.

Historically architecture has many principles based in natural phenomena in the real world, such as the sun, the wind, and the rain. One such element is color, which represents the way humans actually interpret the world.[65] In reality color is actually an illusion that is interpreted by the brain[66] differently in each individual and is therefore unreal by its very nature. What you see on this page isn't really color, it's reflected light interpreted by our visual cortex. Humans have three types of cone cells in our retinas, allowing us to see about a million colors. But some species have the capacity to see colors that we can only dream of. Birds, for example, can see ultraviolet light thanks to a fourth cone.[67] So, while we see blue and green, birds may look at this page and see something entirely different.

In graphic design, there are at least three ways to define color: RGB, CMYK, and Pantone.[68] RGB uses combinations of red, green, and blue to produce new colors in digital and electronic forms. CMYK uses cyan, magenta, yellow, and key black combinations for print media. But, as you are probably aware, the same color can look very different from screen to screen and in various forms of print—that's where Pantone comes in. This company has created a standardized system based on physical swatches to define which color is the same regardless of the media used.

So, if color in and of itself is unreal, the idea of green screens takes us even further away from reality into the virtual world. Created in 1940, Chroma-Key uses technology to replace the color green with another source.[69] This technology is widely used in weather forecasts, but it's also used in other forms of entertainment like TV and film. Today, production companies often use this technology not just to alter backgrounds, but also the to change the characters themselves.

The human psyche can also be manipulated with colors.[70] Commercial brands and architecture carefully choose hues based on the emotion of the message they want to create. Blue is thought to convey feelings of loyalty, stability, and calmness, whereas red might be connected to feelings of love and desire, or danger depending on the context. But these feelings aren't inherent to the colors, they're cultural constructs. For example, different cultures associate luck with different colors—red is used in

PANTONE: 354C **CMYK**: 85 0 98 0 **RGB**: 0 174 79

CHROME KEY GREEN[64]

China and green in Ireland. Pink and blue weren't associated with the current gender norms until the 1940s in the US Prior to then, the colors for gender were reversed.

Artists have always used color in unique ways. In the late 1800s, divisionism and pointillism began using dots, or patches of color which are mixed optically, instead of mixing pigments. Decades later, Josef Albers explored the magic of color even further by illustrating the relativity of color, the afterimage effect, and other optical illusions. In his book, Interaction of Color, he said, "In visual perception a color is almost never seen as it really is—as it physically is. This fact makes color the most relative medium in art."[71] Color creates visual effects in architecture too. Color alters depth perception, bringing design elements to the front or back, depending on the shade. Darker shades can also create the illusion of a shorter space, whereas lighter colors can elongate a space. These techniques are also utilized in the unreal world of VR, where entire architectural worlds are built pixel by pixel.

The field of phenomenology has expanded in the twenty-first century where color in architecture has become interpreted subjectively, rather than universally. A classic color experiment was posted on the internet where some people saw a black / blue dress, while others saw a white / gold dress. The real dress was black / blue, but 59% saw the color dress as white / gold.[72] This notion of color becoming an individual perception that varies has broadened the traditional notion of the color wheel as being a gauge of the ultimate truth of color. Contemporary artists such as James Turrell, Carlos Cruz Diez, and Robert Irwin have all worked in two and three dimensions to explore the way color and light is changed by the individual's perception in a particular space. In addition, the desktop computer's ability to make 16 million color variations[73] has also been a game changer in the designer's capability to manipulate colors easily in a way unknown in history.

As color moves toward the potential use of biomimicry, the unlimited virtual choices of color selection, and developing an understanding of how color is interpreted in the brain, allowing the twenty-first century's toolbox in architecture to increase exponentially.

PANTONE: 278C **CMYK**: 99 76 0 0 **RGB**: 0 83 165

CHROME KEY BLUE

Color is one of the tools of traditional architecture that has helped people understand the real world of materials and decoration. Today, through the field of phenomenology the green and blue screen have transformed the virtual world and our perception of reality by allowing images to trick the eye into seeing different environments or spaces over time and location.

POKÉMON GO

A PORTAL INBETWEEN THE REAL AND THE VIRTUAL

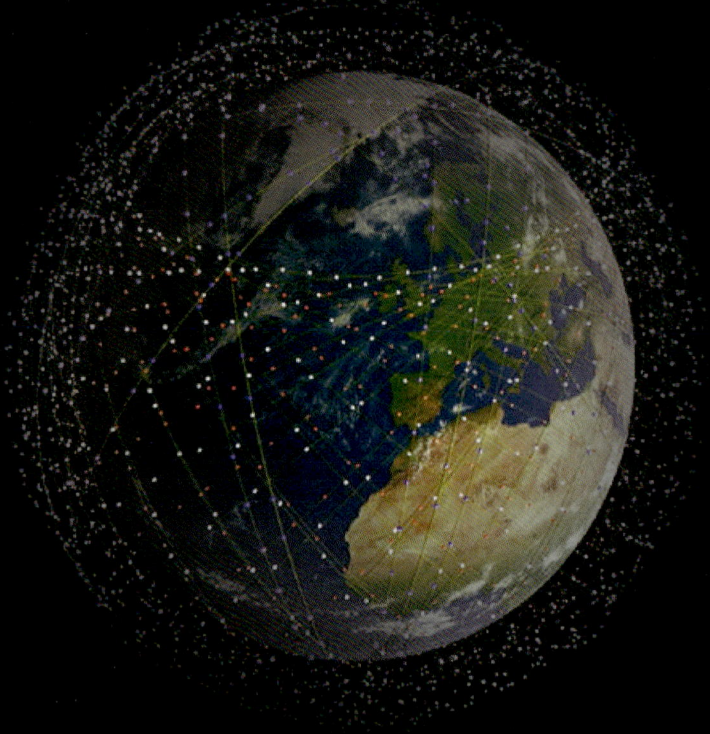

CONSTELLATION STARLINK
4,408 SATELLITES CONNECT THE WORLD ONLINE[74]

"WE'VE LAUNCHED AND NOW HAVE ACTIVE SATELLITES"
ELON MUSK, INTERVIEW MWC 2021

Augmented Reality creates a computer image that can be inserted into the physical world and can be viewed on multiple devices. This new technology will enhance the normally static environments of architecture creating a world INbetween as demonstrated by the popular Pokémon GO, and will be instantly available to everyone with the installation of Elon Musk's Constellation Starlink.

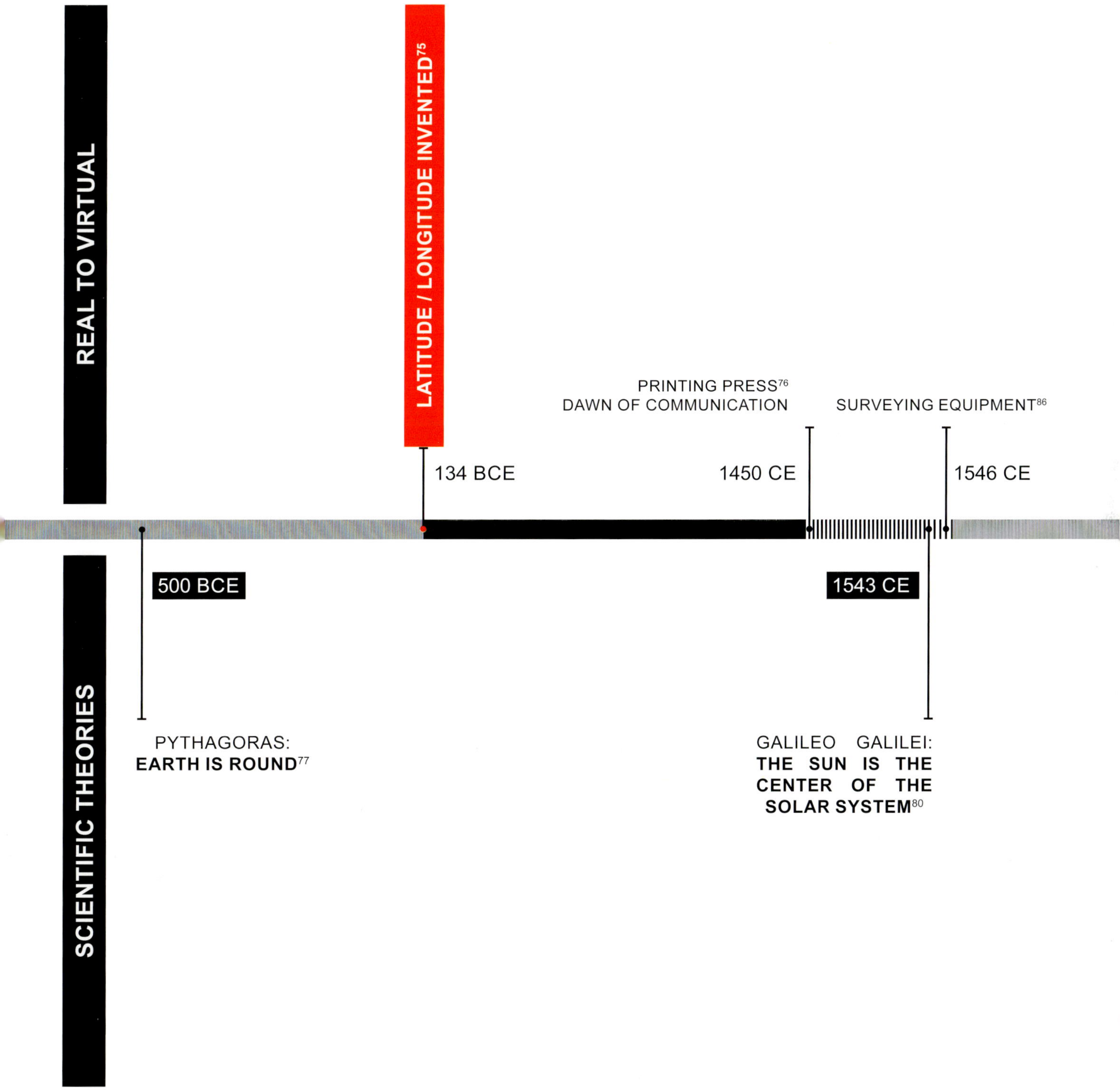

TIMELINE: HISTORICAL SCIENTIFIC THEORIES

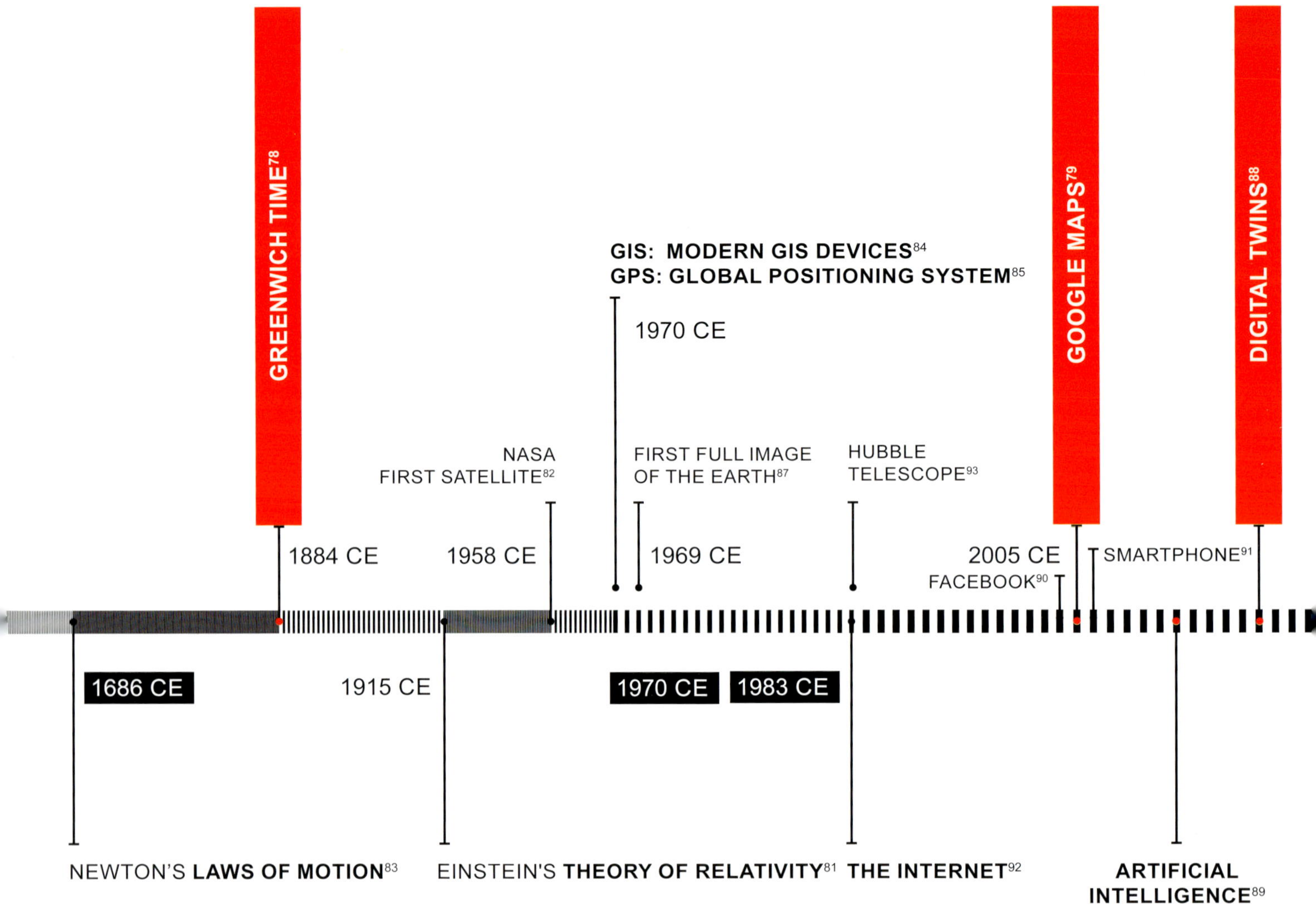

Human's desire to understand the world through mapping has evolved from the real to the virtual. From the earliest need to understand the real location of stars for navigating ships, to the discovery in 134 BCE of latitude and longitude as a method to organize travel, until the complete mapping of the earth as a digital twin through real time enhancement using artificial intelligence and satellites.

NEWTON'S LAWS OF MOTION

INERTIA

AN OBJECT WILL NOT CHANGE ITS MOTION UNLESS A FORCE ACTS ON IT.

MOVEMENT

A FORCE APPLIED TO AN OBJECT.

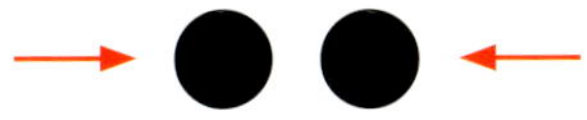

REACTION

FOR EVERY ACTION THERE IS AN EQUAL AND OPPOSITE REACTION.

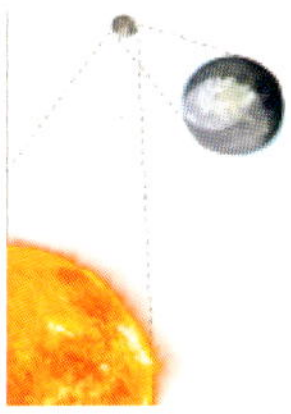

GENERAL RELATIVITY

ALBERT EINSTEIN IN 1916 DEVELOPED THE THEORY OF RELATIVITY THAT ESTABLISHES THE RELATIONSHIP OF SPACE AND TIME TO GRAVITY. THE EQUATIONS CONFIRM THE NOTION OF ENERGY BEING CONTAINED IN SPACETIME.

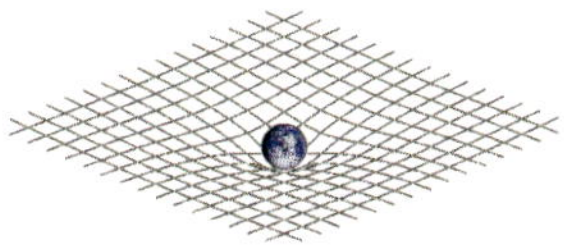

SPACE TIME

EINSTEIN POSTULATED THAT OBJECTS DON'T SIMPLY ATTRACT EACH OTHER BY INVISIBLE FORCES, BUT RATHER THAT GRAVITY CONTINUALLY CURVES, OR WARPS SPACE TO ACCOMMODATE OBJECTS. THE GREATER THE SIZE OF THE OBJECT THE MORE IT WARPS SPACE AROUND IT.

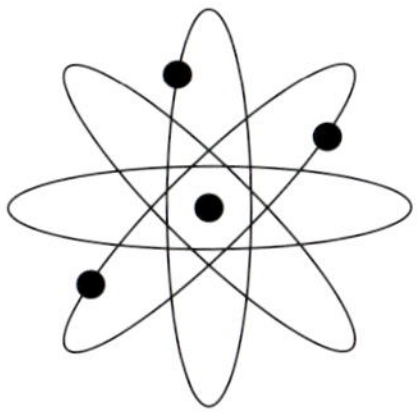

QUANTUM THEORY[95]

QUANTUM PHYSICS DESCRIBES THE PROPERTIES OF NATURE AT THE SCALE OF ATOMS AND SUBATOMIC PARTICLES. IT IS THE FOUNDATION OF MODERN PHYSICS. THE LAWS OF PARTICLES DEFY THE NOTIONS OF EINSTEIN'S THEORY OF RELATIVITY.

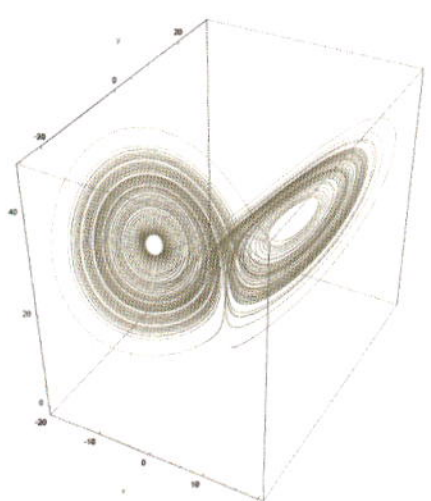

CHAOS THEORY[94]

CHAOS THEORY IS A PARADOX BECAUSE AS A FIELD OF MATHEMATICS IT STATES THAT TWO OPPOSING CONDITIONS CAN EXIST AT THE SAME TIME. THE DETERMINISTIC AND THE UNPREDICTABLE CAN CO-EXIST AT THE SAME TIME IN ONE REALITY.

"THE ONLY THING THAT'S REAL IS REALITY"

JAMES HALLIDAY
CREATOR OF OASIS

READY PLAYER ONE
ERNEST CLINE 2011[96]

The digital age began with the development of science during the Renaissance. The evolution of technology has been exponential and now connects billions of people instantly. The Matrix is one of the first visualizations of the Metaverse as not just viewing a computer from the outside, but being able to actually step into the virtual world from the inside with a new understanding of space.

NO BORDERS

In the book, *The World of Yesterday* by Stefan Zweig in 1942, he wrote, "Before 1914 the earth belonged to the human race. Everyone could go where they wanted and stay there as long as they liked: no permits, no visas, no passports. The borders today are a tangled fence of red tape, but then they were nothing but symbolic lines on the map."[97] The very definition of architecture is being reimagined in this era of three competing realities: the real, the augmented, and the virtual. The virtual world has allowed architecture to ignore the physical constraints of the real world such as sun, wind, rain, and even gravity. In a virtual world, the only limit is an architect's imagination. But, in the augmented world INbetween, there is an opportunity for architects to use creativity to harness the best of both worlds. Augmented reality in architecture can use color, light, and the horizon to alter people's perceptions as we've seen in the work by contemporary artists, or for controlling cars, visualizing ecological climate changes, healthcare, manufacturing, or genetically engineering renewable materials. With the advent of technology, physical borders are no longer a barrier. Now the only borders that remain are social constructs: politics, money, and religion, created to help people make sense of the world around them. Now that our worlds are expanding and we're communicating on a global scale, these same social constructs meant to bring people together are what's actually keeping people apart. The future poses a tug of war between the divided reality we know and an attempt at building a new reality in the virtual world, free from borders. In the past, architecture has been framed by the distinction between public and private life, but that line is being blurred as the world spends more and more time in virtual spaces. So we must ask: who will now own this new public life? Will architects' previous focus on plazas, courtyards, and boulevards make way for public spaces designed to be augmented by apps? Who will design the architecture of the virtual world—a decentralized Metaverse with unlimited potential INbetween becomes a new design sensibility that will influence how we see the reality of a world of *no borders* in multiple realities.

WAVES: EVOLUTION OF WAVELENGTHS[98]

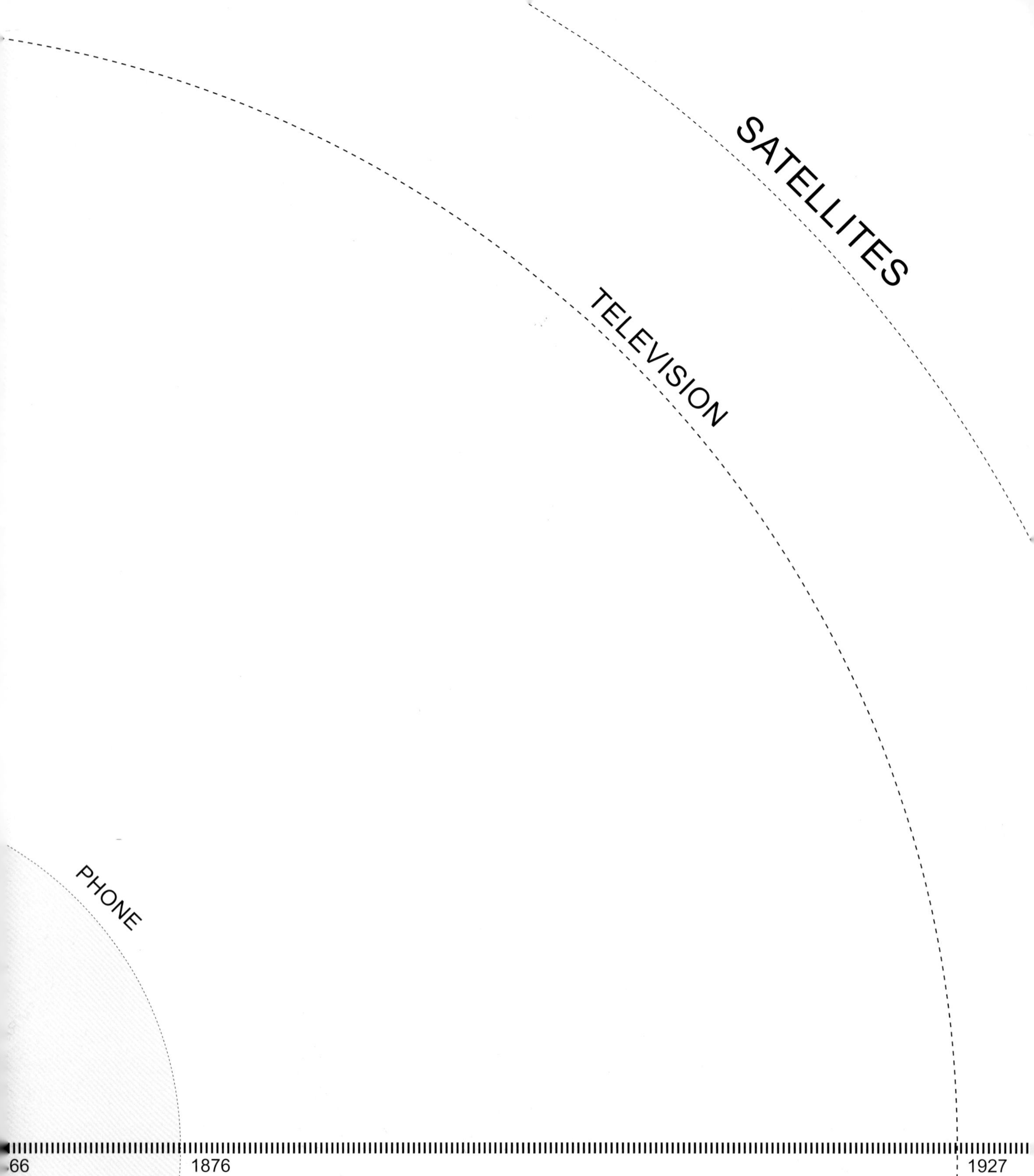

The evolution of communication technology has been accelerating exponentially since the telegraph was invented in the 1830s by Samuel Morse. From the word of mouth and the written word, we now have the capability to converse globally on an interplanetary scale. The expansion of distance, speed, and visual communications have dramatically increased in the last two centuries.

1969

LANDING ON THE MOON

It was considered a turning point in the history of mankind: landing on the moon in 1969. For the first time in history there was a photograph of the earth's image from space as a complete sphere. The fragility and majestic presence of a planet supporting life as viewed without national borders or highways circling the earth. This view of the earth that was written by poets and debated by science as being flat or round had been a source of speculation that now had visual proof of its authenticity. Humans' desire to understand the earth has led the species to try to harness its resources and ultimately attempt to conquer nature. The idea of setting foot on the moon was a concept that appeared to be science fiction prior to 1969. The act of landing on the moon has forever changed humans' view of themselves in the universe.

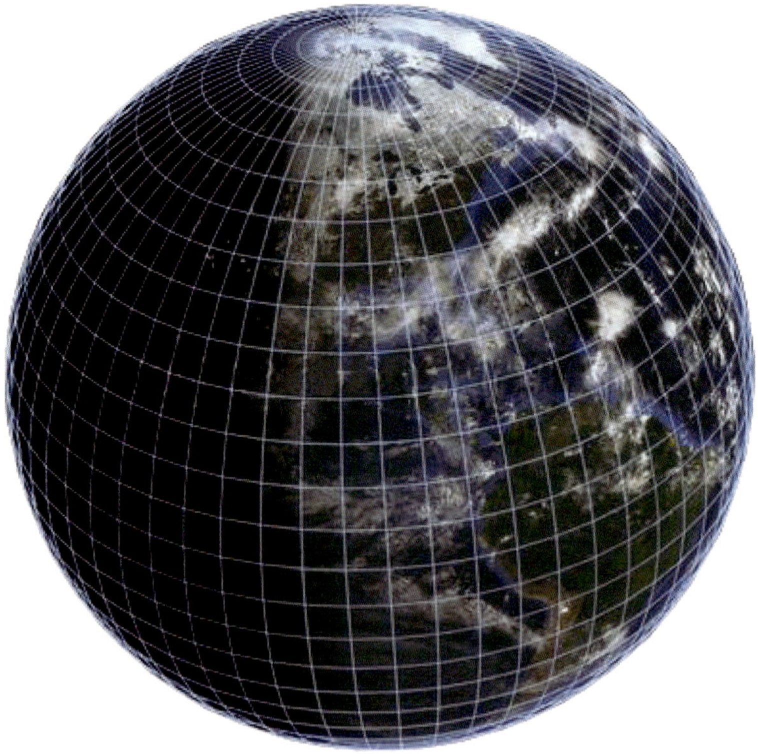

2030

DIGITAL TWIN OF THE EARTH

A digital twin is a virtual representation that recreates the original, physical source that can communicate with machines as a digital image. NASA began using this term in 2010 to improve the simulation of space crafts. Currently, Google Maps is in the process of creating a digital twin of the earth that uses geospatial data in real time with simulation and machine learning to make a digital twin of the earth. This process fulfills a yearning of humans to understand and catalog their environment. Explorers, scientists, and adventurers have taken risks throughout history to go a step further in mapping. The Voyager 2 probe, which left earth in 1977, has reached the end of our Solar System on November 2018 when it went into interstellar space. This monumental step was one more exploration in extending the reach of our digital twin.

The iPhone 5S *had 1,300 times more processing power in 2013 than the* Apollo 11 *computer had in going to the moon in 1969.*[99] *The vision for a complete digital twin of the earth will essentially recreate all known life forms and geographies in real time of the earth. Google Maps has begun this process.*

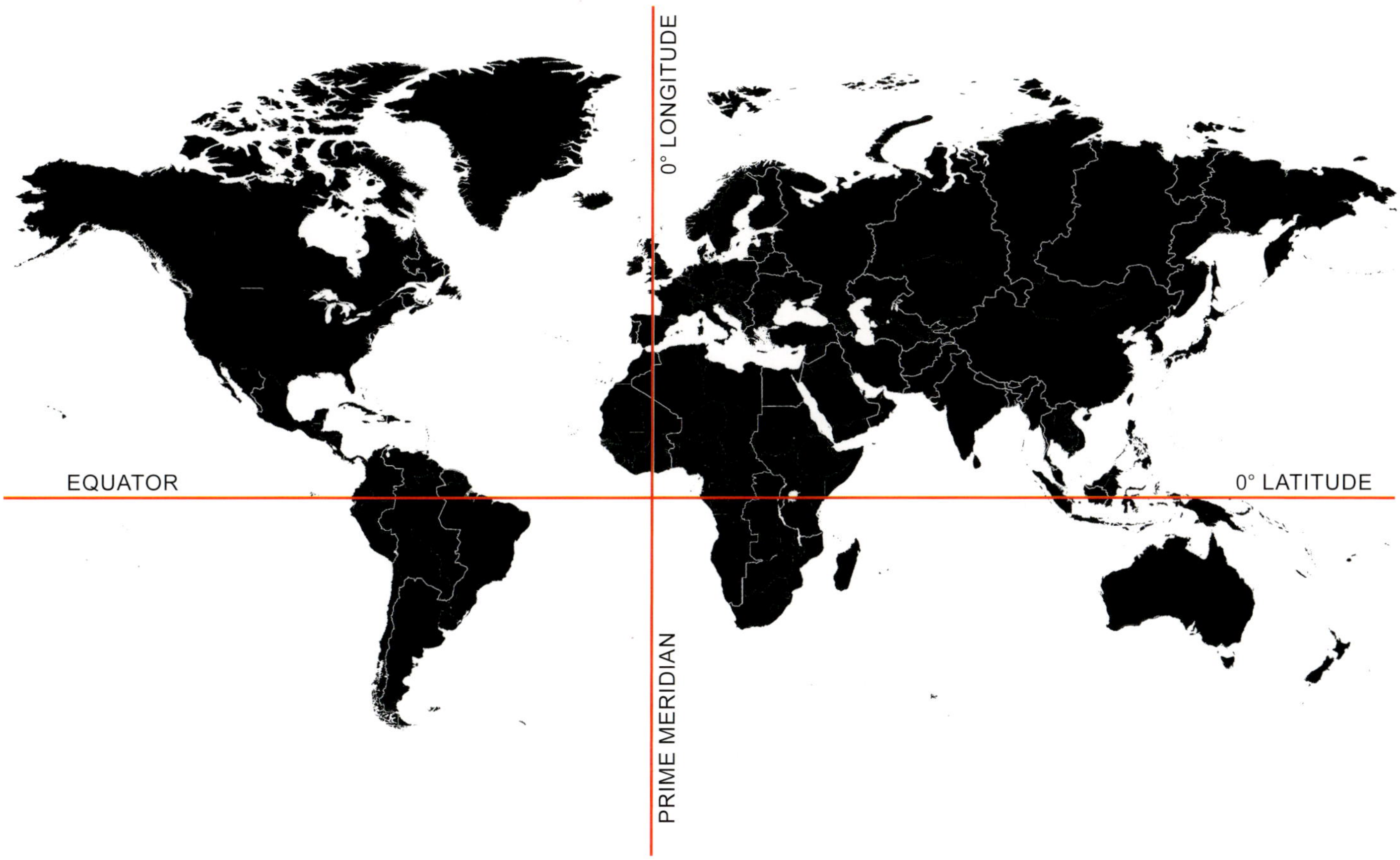

GEOGRAPHICAL COORDINATE SYSTEM

VIEWING THE WORLD IN X + Y PLANES

3RD CENTURY BCE

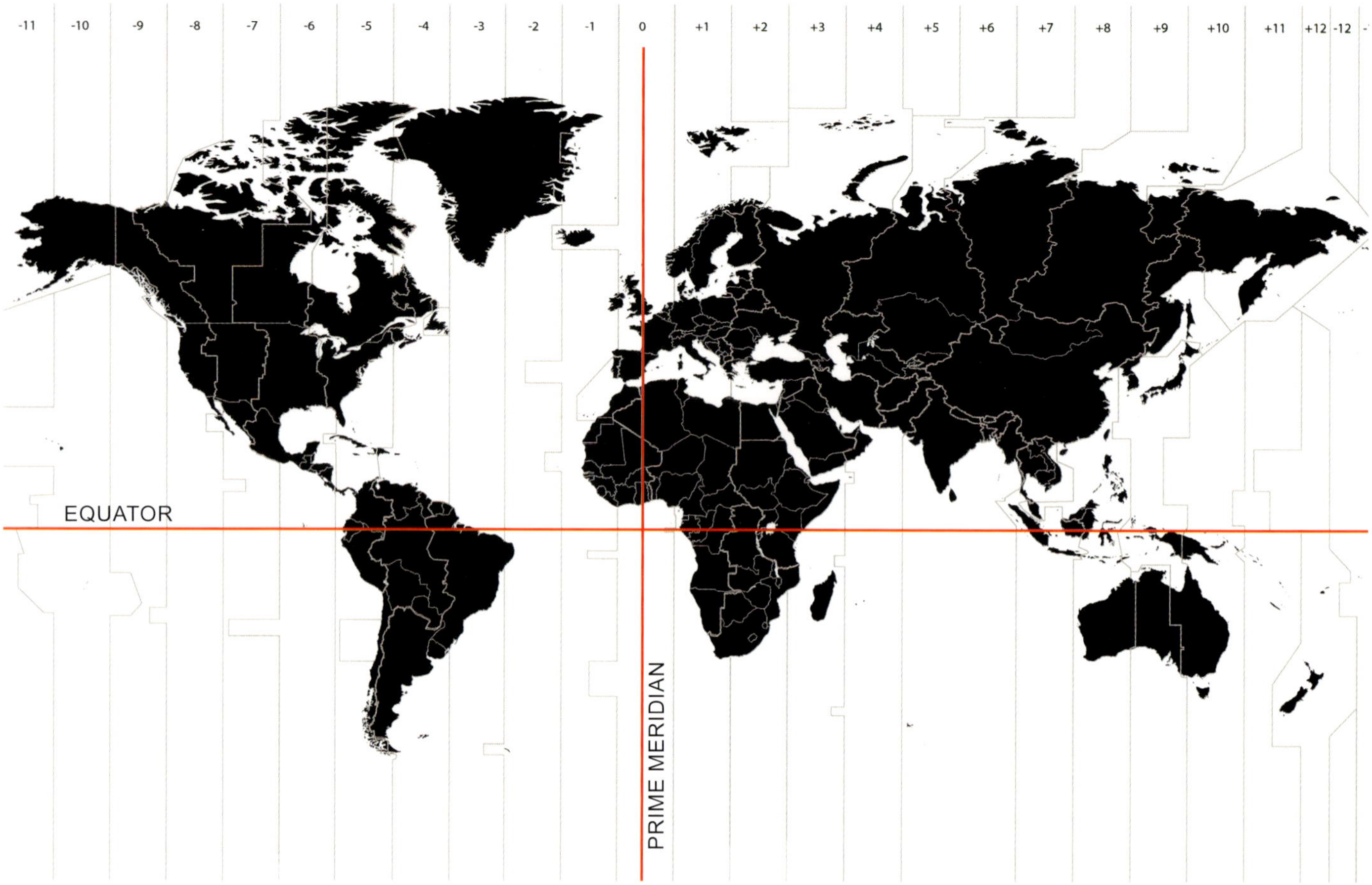

STANDARDIZED TIME ZONES

GREENWICH, ENGLAND

1884 CE

History has continually tried to understand the earth in a variety of evolving mapping methods. Human created latitude and longitude to locate a direction and a place. With the invention of the railroads a unified time sequence was needed, so we created standardized time zones in 1884 CE.[100] *This mapping of the earth created a space / time relationship that is still being used today.*

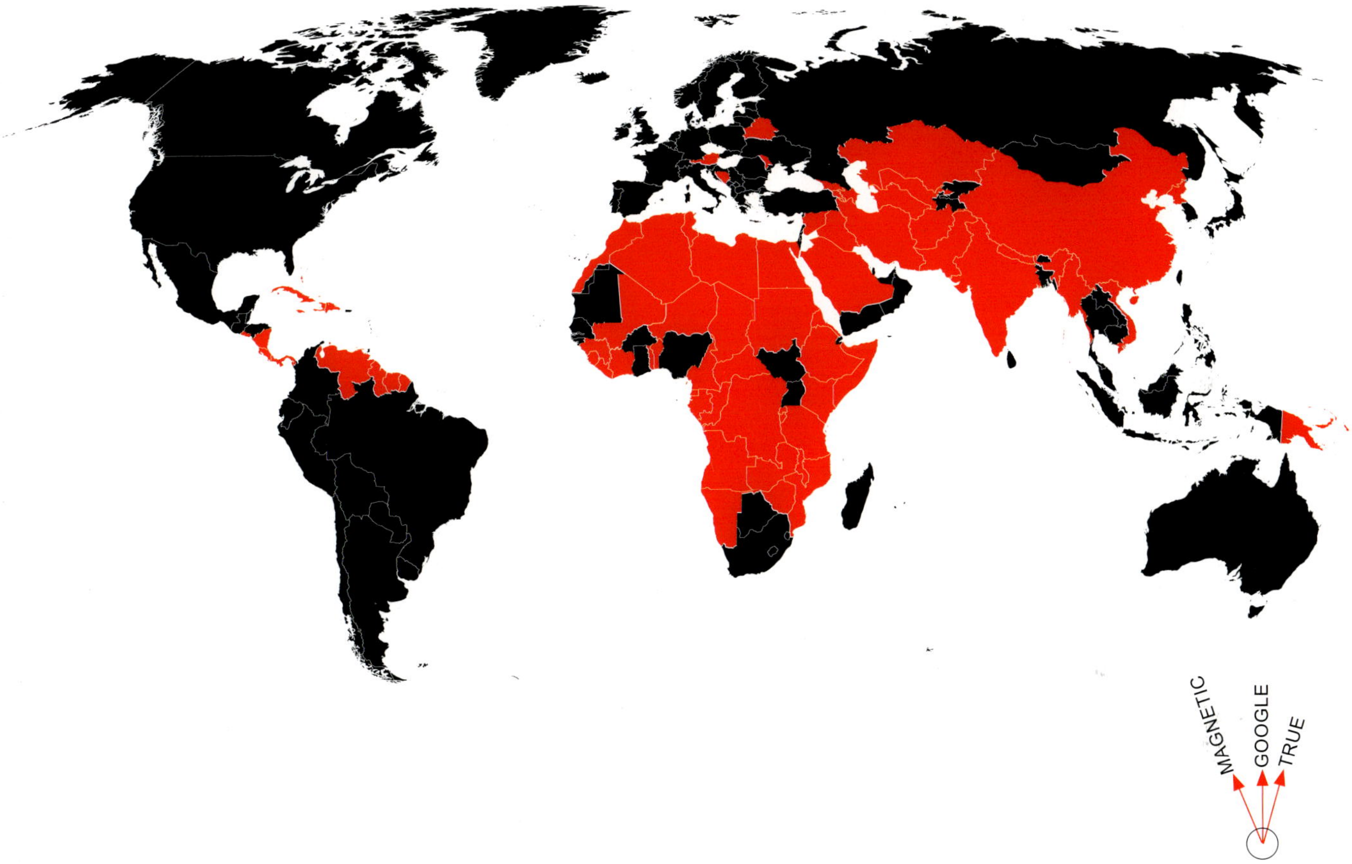

GOOGLE MAP: 50% OF THE WORLD 2021[101]

■ GOOGLE MAPS ARE AVAILABLE ■ GOOGLE MAPS ARE UNAVAILABLE

NATURE
19TH CENTURY

AUGMENTED NATURE
20TH CENTURY

AUGMENTED REALITY
21ST CENTURY

DIGITAL TWIN: GOOGLE
21ST CENTURY

In the nineteenth century cartography was the only way to represent nature. In the twentieth century the landscape was augmented with highways, elevators, and air conditioning, while lighting changed the day-night cycle of cities. As Google creates a digital twin and maps the earth, a new, limitless world of digital real estate and new places to meet is being created in the Metaverse.

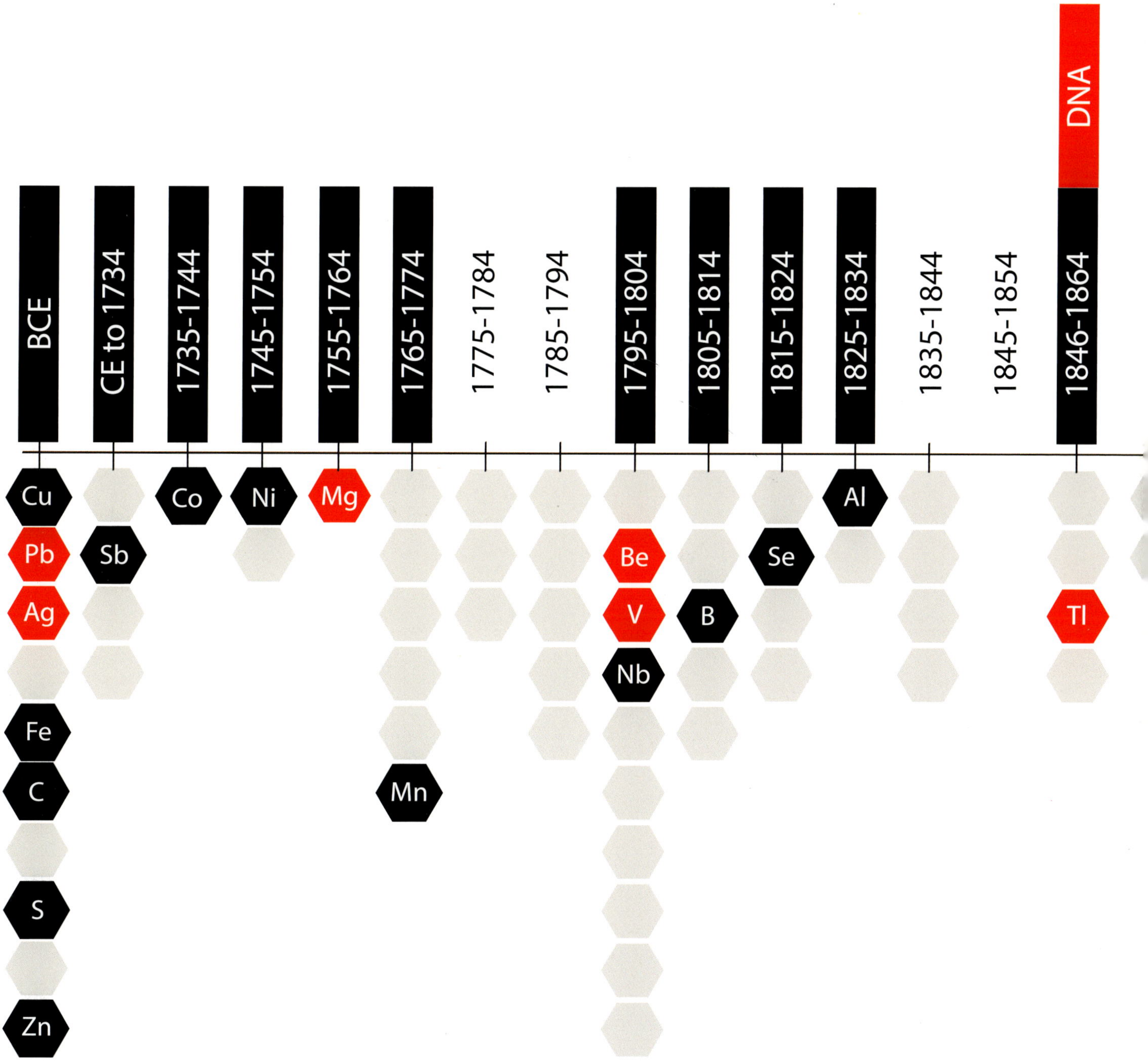

NO BORDERS | DIGITAL TWINS

TIMELINE: DISCOVERY OF NEW MATERIALS[102]

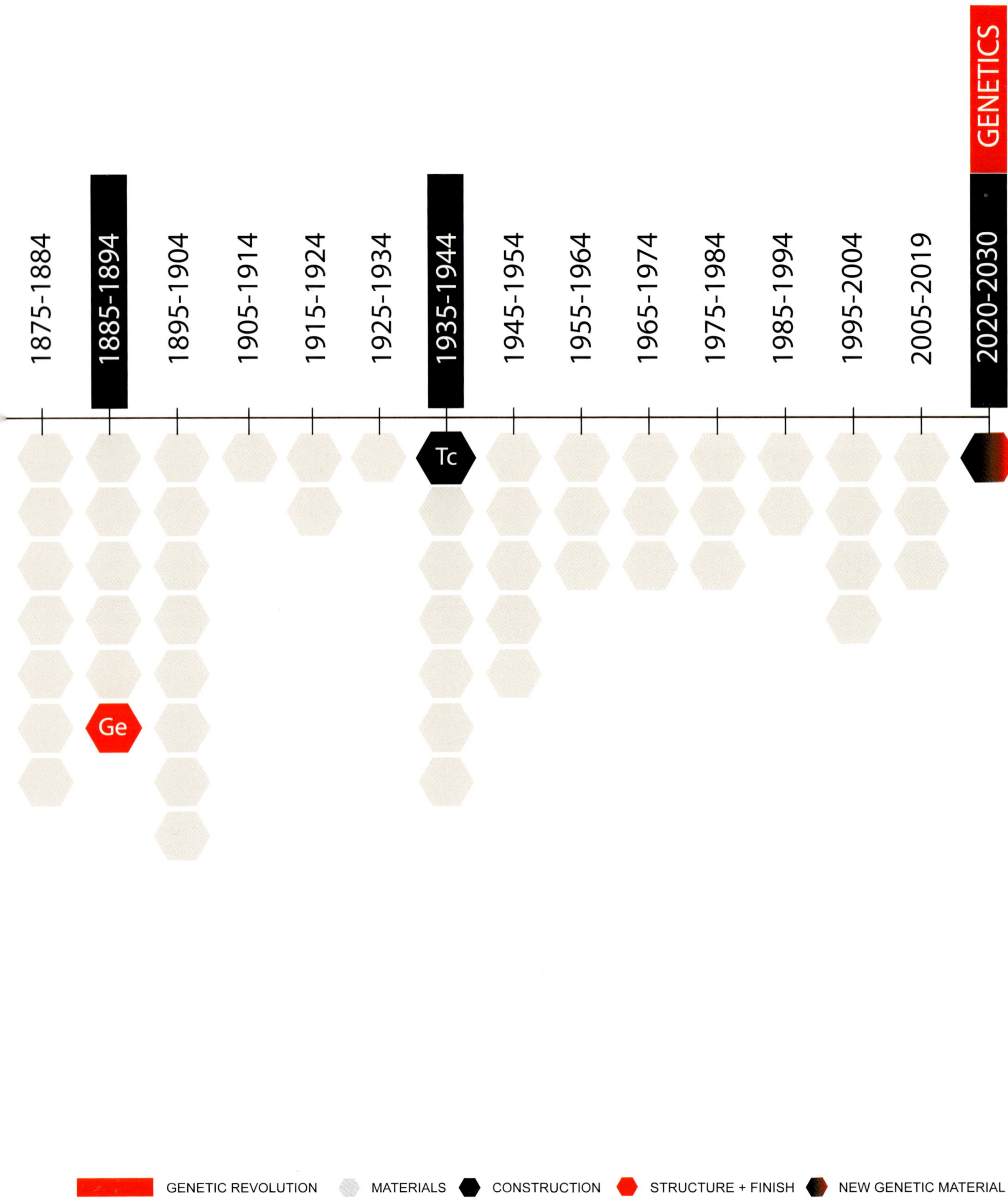
GENETICS
1875-1884
1885-1894
1895-1904
1905-1914
1915-1924
1925-1934
1935-1944
1945-1954
1955-1964
1965-1974
1975-1984
1985-1994
1995-2004
2005-2019
2020-2030
Tc
Ge
GENETIC REVOLUTION
MATERIALS
CONSTRUCTION
STRUCTURE + FINISH
NEW GENETIC MATERIAL

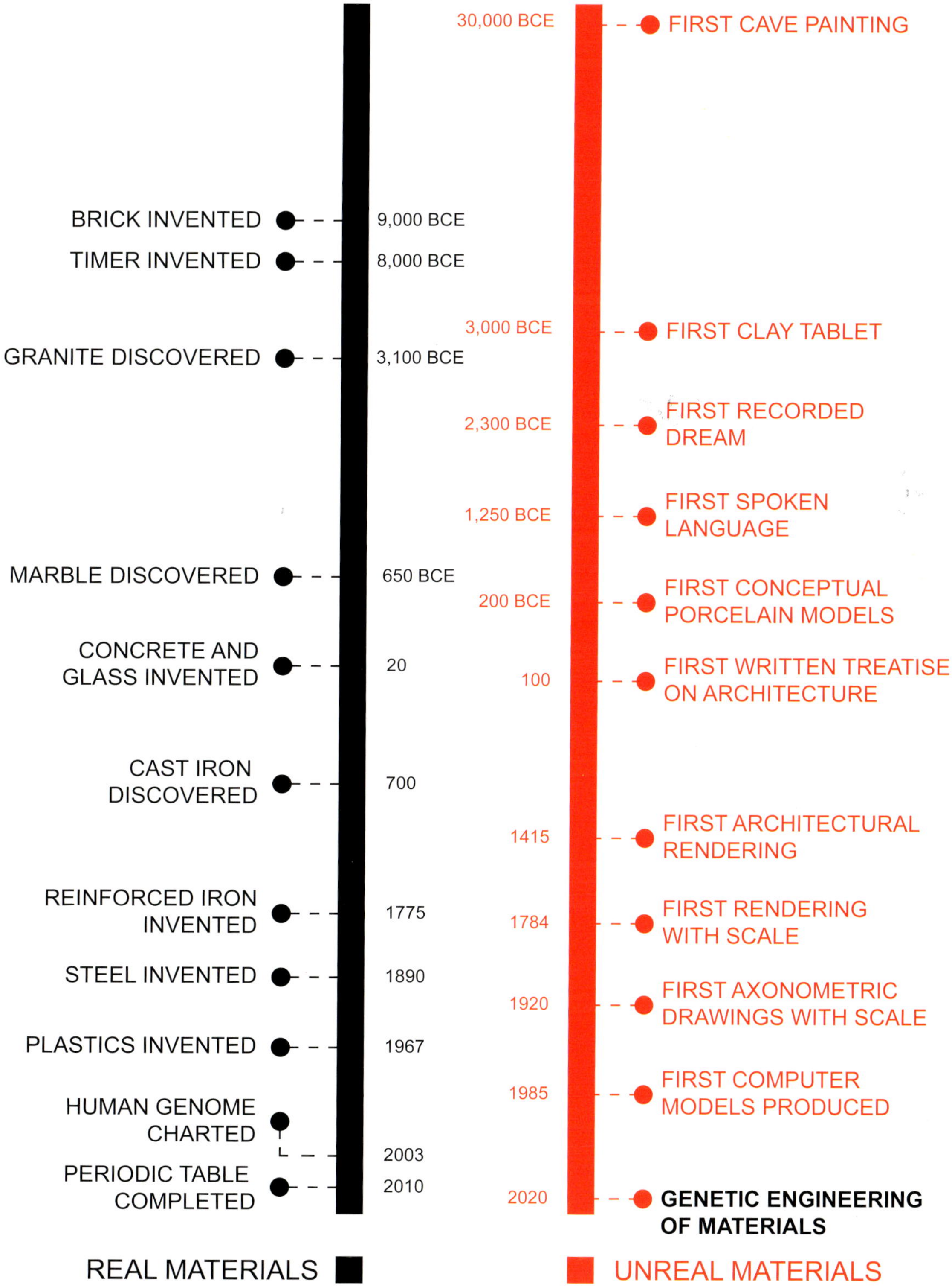

THE EVOLUTION OF MATERIALS

NO BORDERS | DIGITAL TWINS

DISCOVERY: REAL + UNREAL MATERIALS

THE GENETIC ENGINEERING OF NEW MATERIALS

The history of real materials on the periodic tables can be contrasted to the development of genetic engineering of DNA that can create new materials not known in nature. Developed at the MIT Lab with Neri Oxman, this new way of making materials can have the potential of inventing a new field that is revolutionary by evolving INbetween the real and the virtual.

GAME ON

Sherry Turkle, in a book entitled *Alone Together*, wrote in 2011, "The phone call was prompted by a new book about robots by David Levy, a British-born entrepreneur and computer scientist. In 1968 Levy, an international chess master, famously wagered four artificial intelligence (AI) experts that no computer program could defeat him at the game. In the subsequent decade Levy won his bet. The sum was modest, 1,250 British pounds, but the AI community was chastened. They had overreached their predictions for their young science. It would be another decade before Levy was bested in chess by a computer program Deep Thought, an early version of the program that beat Gary Kasparov, the reigning chess champion in the 1990s."[103] The twentieth century, with the invention of the automobile and the allure of the 1950s highway system by President Eisenhower in the US, foreshadowed the flight of the urban to the suburban. Increased mobility, along with the appeal of additional space, sent rural and urban dwellers packing for the suburbs.[104] Bucking the trend, by the early twenty-first century, the world's urban population had surpassed 50% by now living in cities rather than the rural communities.[105] In the same way that Eisenhower planned ahead for America's future and security, architects must also plan for a future that is pure speculation and design for what they believe the future will be. This design space exists INbetween the reality of the moment and the possibilities of the unknown, much like the worlds created in modern gaming. In 2020, 67% of the adults in the US (227 million) are gamers.[106] This adoption of gaming into the culture is an indication of the mass appeal of the unreal for entertainment and business that is now shifting into the virtual world beyond gaming. The twenty-first century represents a new migration that is entirely different than the physical movement of the population in the twentieth century. The notion of Game On acknowleges this new trend of people's increasing time being spent on devices and moving away from the reality of the here and now into the virtual world.

1876
1ST PHONE
INVENTED BY ALEXANDER GRAHAM BELL

1970
1ST DIGITAL PHONE
AT&T PICTURE PHONE

1980
1ST MOBILE PHONE IN JAPAN
AEG TELECAR

1985
TRANSPORTABLE PHONE
VODAFONE

2000
1ST MOBILE PHONE WITH CAMERA
SHARD J-SHO4

1994
1ST TOUCH SCREEN PHONE
IBM SIMON

2008
1ST ANDROID
HTC DREAM

2011
1ST SMARTPHONE A.I.
SIRI: VIRTUAL ASSISTANT

2015
6TH SENSE TECHNOLOGY
CAMERA: PHYSICAL TO VIRTUAL WORLD

COMMUNICATION MEDIUMS

1936
1ST COMPUTER
INVENTED BY ZONRAD ZUSE

1942
1ST DIGITAL COMPUTER
THE ABC

1976
1ST APPLE COMPUTER
STEVE JOBS AND STEVE WOZNIAK

1983
1ST PORTABLE COMPUTER
IBM COMPAQ PORTABLE

1984
1ST GRAPHIC USER INTERFACE
APPLE MACINTOSH

1991
WORLD WIDE WEB
WWW.COM

1998
GOOGLE
MOST USED INTERNET SEARCH ENGINE

2004
FACEBOOK
MOST USED SOCIAL MEDIA PLATFORM

2015
YOUTUBE
MOST USED VIDEO SHARING PLATFORM

MOVING TO A VIRTUAL WORLD

Before technology, humans could only communicate and keep up with the lives of about 150 people.[108] *Since the invention of the telephone in 1876 the size of our communities have expanded exponentially because distance has become less of an obstacle. Social media now allows billions of people to communicate instantly and to be able to keep up with each other's daily lives.*

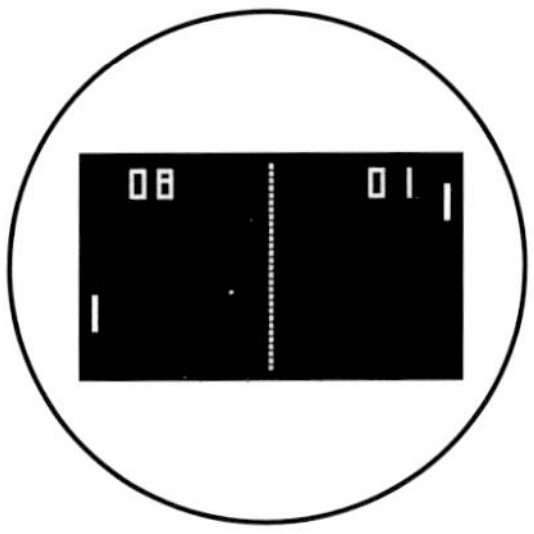

1972
1ST VIDEO GAME
PONG

1982
COLECOVISION
OFFERED ARCADE QUALITY GRAPHICS

1984
1ST CONSOLE WITH BACKWARD COMPATABILITY
ATARI 7800

1995
PLAYSTATION
1ST PLAYSTATION SERIES OF CONSOLE + HANDHELD GAME DEVICES

2000
PLAYSTATION 2
BEST SELLING CONSOLE TO DATE

2001
GAMECUBE
1ST NINTENDO SERIES TO USE OPTICAL DISKS

2005
XBOX 360
FEATURES ONLINE COMPETITION, GAME DEMOS, AND 3RD PARTY ACCESS

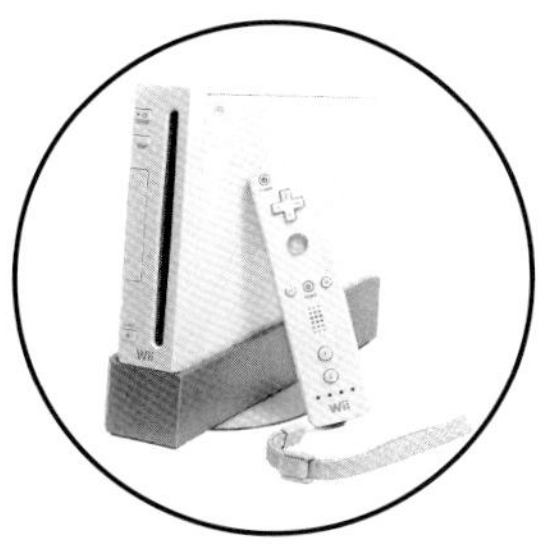

2006
1ST CONSOLE WITH WIRELESS CONTROLLERS
NINTENDO WII

2010
XBOX KINECT
IMPLEMENTED A CAMERA IMAGING SYSTEM PUTTING YOU INTO THE VIRTUAL WORLD

GAME CONSOLES

1957
SENSORAMA SIMULATOR
A MACHINE WITH AN IMMERSIVE SENSORY SYSTEM

1968
1ST VR GAME
SWORD OF DAMOCLES

1985
DATAGLOVE
A DEVICE WORN THAT MONITORS THE PERSON'S MOVEMENTS

2013
HAPTICS MOTION CONTROLLER
ALLOWED TOUCH FEEDBACK TO BRING REALITY INTO VR

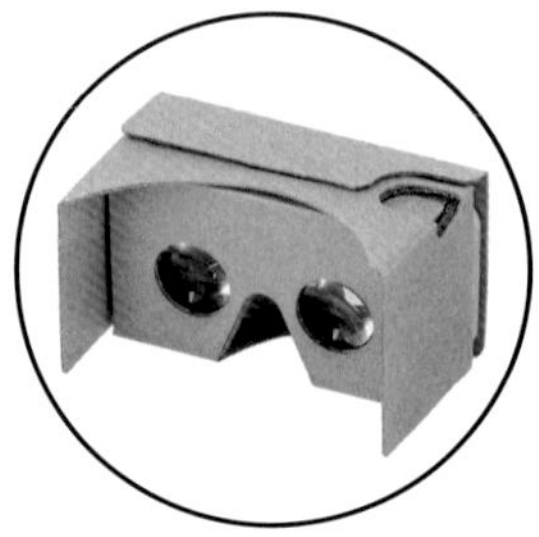

2014
GOOGLE CARDBOARD
A DO-IT-YOURSELF CARDBOARD VR SET

2016
OCULUS RIFT
ONE OF THE MOST TECHNOLOGICALLY ADVANCED VR SETS

2021
FLYING SIMULATION
ADVANCED VR PILOT SYSTEM

2021
HORIZON WORKROOM
FACEBOOK VR WORK ENVIRONMENT

2025
RETINAL PROJECTION GLASSES
RETINAL PROJECTION IN YOUR RETINA

VIRTUAL REALITY

Gaming is getting more immersive as we shift from the real to the unreal. Static games became more physically interactive with the invention of video game consoles and motion controls, which created a more realistic way to experience the evolving virtual worlds. Today people can jump into a virtual world with just a small headset, but soon even the headset may no longer be necessary.

FULL-BODY TRACKING

MONOPOLY GAME

OMNIDIRECTIONAL TREADMILL

BOARD GAMES

STATIONARY EXTERNAL DEVICES

REAL **I N B E**

PHYSICAL **CONSOLES**

1930 1940 1960 1970 1980

In the 1950s the newly created Braniac computers took up the space of an entire room and were only capable of doing simple mathematical problems. It was necessary to stand at the computer with holes punched in a card like a player piano music sheet that were used to communicate with the machine: These stationary external devices were getting smaller every two years according to Moore's Principle.[110] Smaller computer chips created new, more advanced stationary devices, such as full body tracking and omni-directional treadmills, as well as new portable external devices, such as mobile phones, laptops, and tablets. The popularity of these mobile devices has exploded with 15 billion mobile devices in the world in 2021.[111] As society continues its migration from the real world to the virtual, wearable devices such as smart watches and the projected AR / VR glasses will further integrate computers into

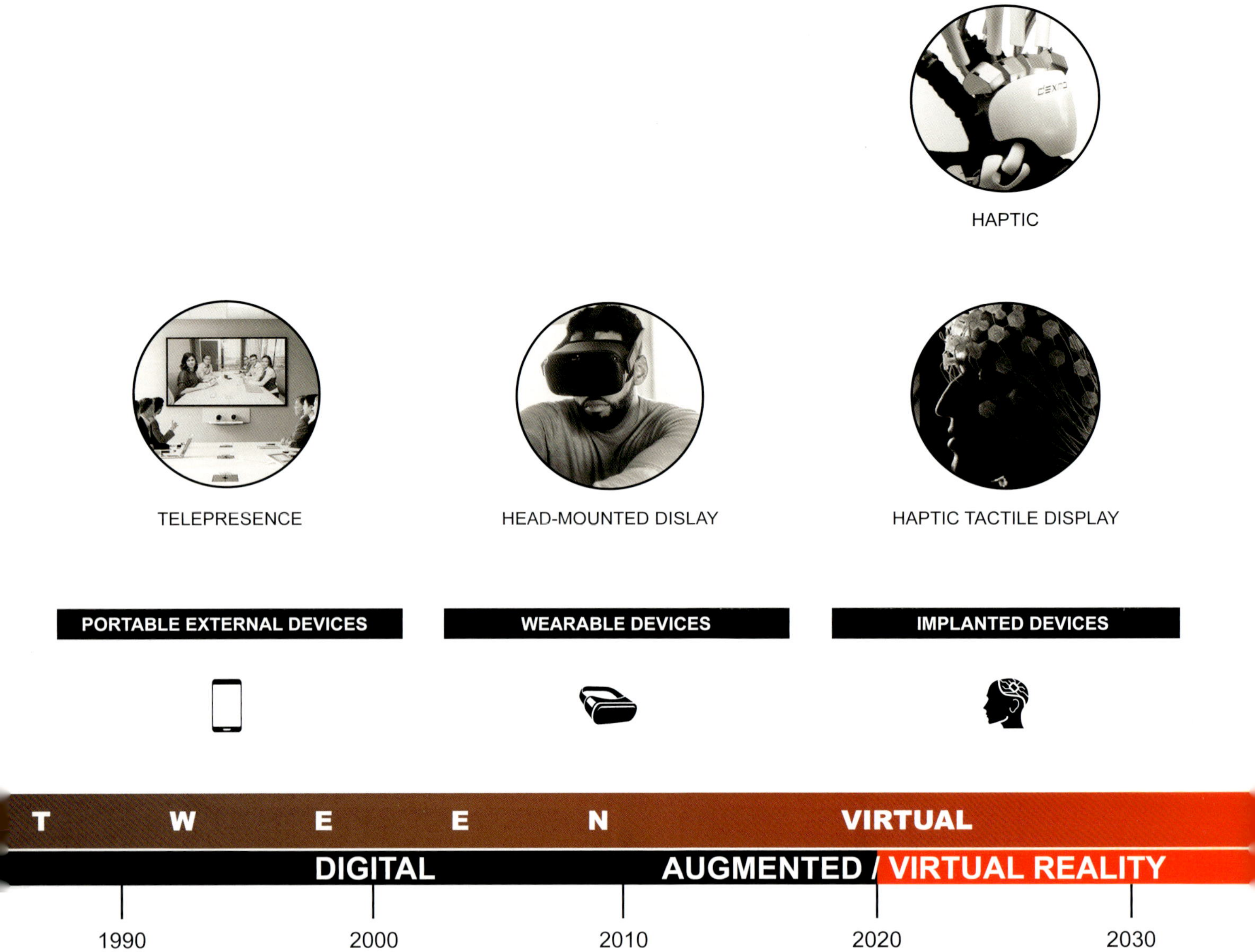

our daily lives, while VR headsets and glasses are allowing people to escape from the real world altogether. It is inevitable that information driven, implanted devices as pioneered by pacemakers and transplants will become commonplace in the near future. The increasing use of technology has created a symbiotic relationship with humans. The evolution of communicating with machines is now transforming how humans live and understand reality. How we will navigate this future with the introduction of AI into the current "smart home" is in its infant stage in terms of allowing sensors, the Metaverse, and Alexa into our homes. As we think about the future of technology it has become clear that we are in the dawn of a data driven, algorithm, robotic world that will only accelerate the use of devices and technology in our daily lives. How well humans will adapt to technology in society remains to be seen in the twenty-first century.

The increasing interdependance of humans, machines, and technology has begun a new evolutionary step for society. The introduction of the Metaverse, artificial intelligence, and machine learning into our daily living has ushered in a new age of technology that will surpass the most augmented realities of humans. This increasing move to the virtual world is now inevitable.

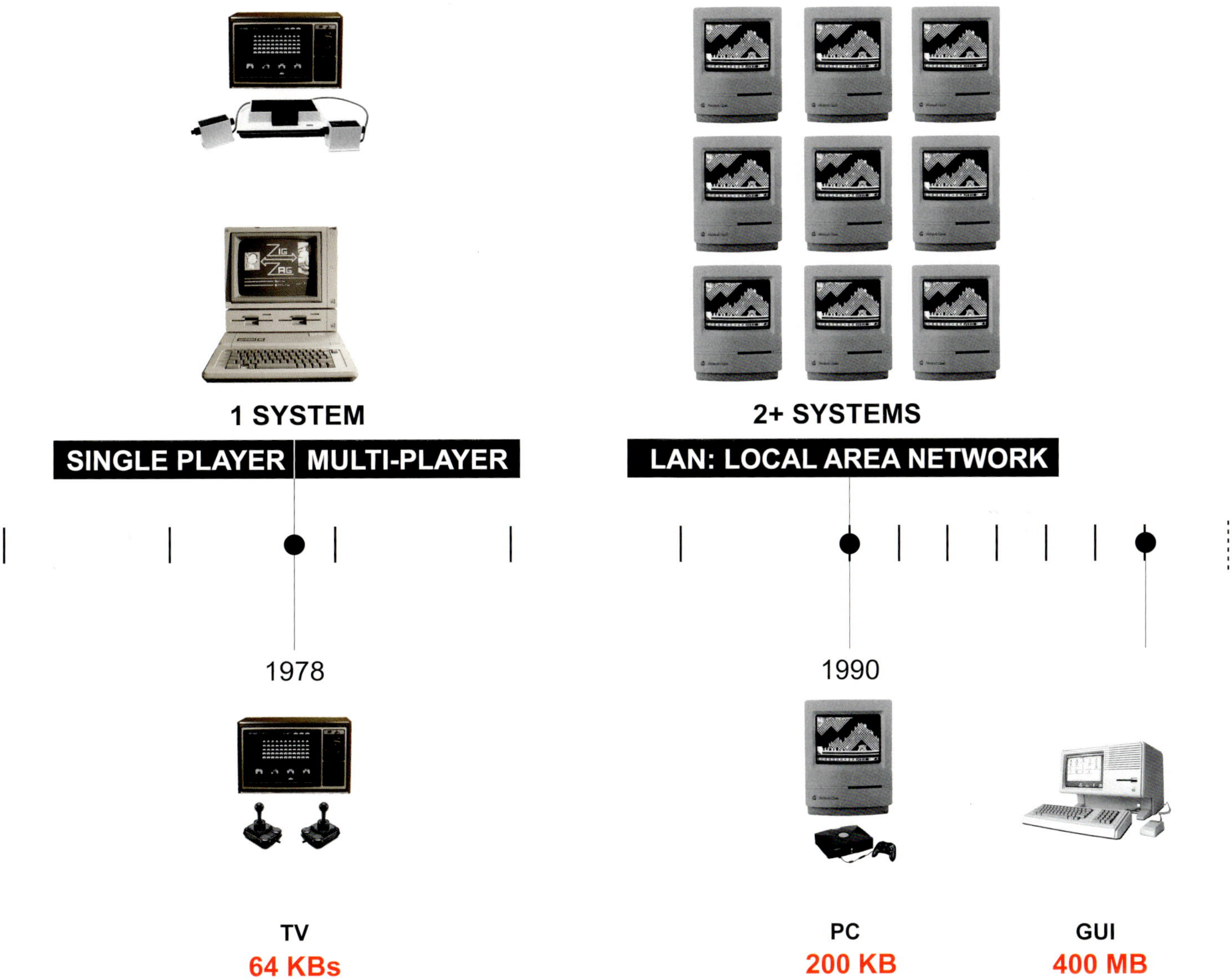

GAME ON | GAMING REALM

TIMELINE: EVOLUTION OF DIGITAL GAMING

WORLD OF WARCRAFT: 116 MILLION DAILY PLAYERS

MMO: MASSIVELY MULTIPLAYER ONLINE

1994

2007

2022

MOBILE
1 GB

THE INTERNET
10 GB

iPHONE
256 GB

CLOUD STORAGE
295 EB

KB: KILOBYTES **MB:** MEGABYTES **GB:** GIGABYTES **EB:** EXABYTES

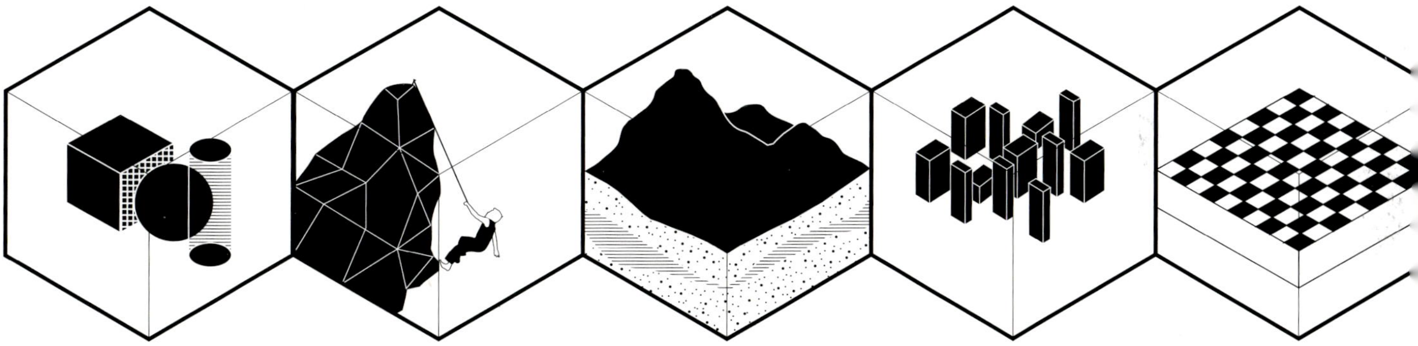

PUZZLERS	ACTION-ADVENTURE	SANDBOX	SIMULATION	REAL-TIME STRATE
SPATIAL PROPERTIES	JOURNEY	OPEN ENVIRONMENT	MIMICRY	PLANNING
LIMBO	*ASSASSIN'S CREED*	*MINECRAFT*	*MINECRAFT*	*WARCRAFT*
FEZ	*STAR WARS JEDI*	*GRAND THEFT AUTO V*	*THE SIMS*	*DOTA 2*
MONUMENTAL VALLEY	*THE WITCHER*	*TERRARIA*	*SIMSCITY*	*LEAGUE OF LEGEND*
PORTAL	*TOMB RAIDER*	*THE SIMS*	*MADDEN NFL*	*AMONG US*
ECHOCHROME	*RESIDENT EVIL 4*	*EVE ONLINE*	*CIVILIZATION*	*STELLARIS*
TETRIS	*LEGEND OF ZELDA*	*GARRY'S MOD*	*ANIMAL CROSSING*	*AGE OF EMPIRES*

GAMING REALM: TYPOLOGY[112]

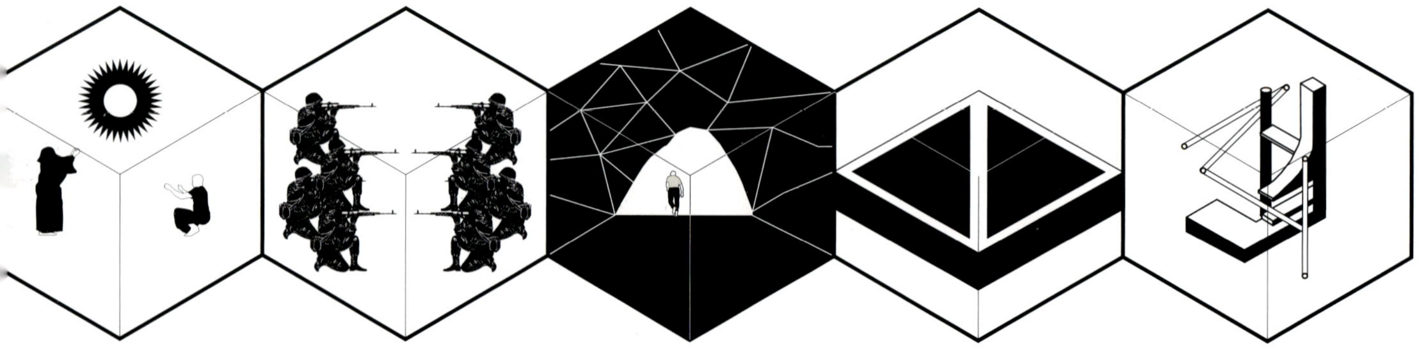

ROLEPLAYING NARRATIVE	SHOOTERS FIRST PERSON SHOOTER	SURVIVAL SUSPENSE	MOBA BATTLE ARENA	PLATFORMER GRAVITY CONQUEST
WORLD OF WARCRAFT	*HALO*	*RESIDENT EVIL*	*LEAGUE OF LEGENDS*	*SUPER MARIO BROS*
FINAL FANTASY XIV	*DOOM*	*THE LONG DARK*	*DOTA 2*	*DONKEY KONG*
UNDERTALE	*GEARS OF WAR*	*DON'T STARVE*	*VALORANT*	*SONIC THE HEDGEDOG*
TO THE MOON	*OVERWATCH*	*THE EVIL WITHIN*	*SMITE*	*CUPHEAD*
STANLEY PARABLE	*CALL OF DUTY*	*ALIEN: ISOLATION*	*HEROES OF THE STORM*	*CRASH BANDICOOT*
THE LAST NIGHT	*COUNTER-STRIKE*	*OUTLAST*	*ARENA OF VALOR*	*ORI & THE BLIND FOREST*

The gaming community has many sub-genres that challenge users in different ways and cater to a variety of interests. The evolution of these typologies on multiple devices has also developed a new mode of interacting with the creation of the Metaverse. This platform does not conform to any rules of the real world and offers an uncharted territory for visualization in this new environment.

THE LAST NIGHT

CAMERA ANGLES TO SHOW THE STATE OF THE WORLD

DARK TO BRIGHT COLOR HUES AS DANGER GROWS FOR THE MAIN CHARACTER

LIMBO

COMPOSITION: RULE OF THIRDS

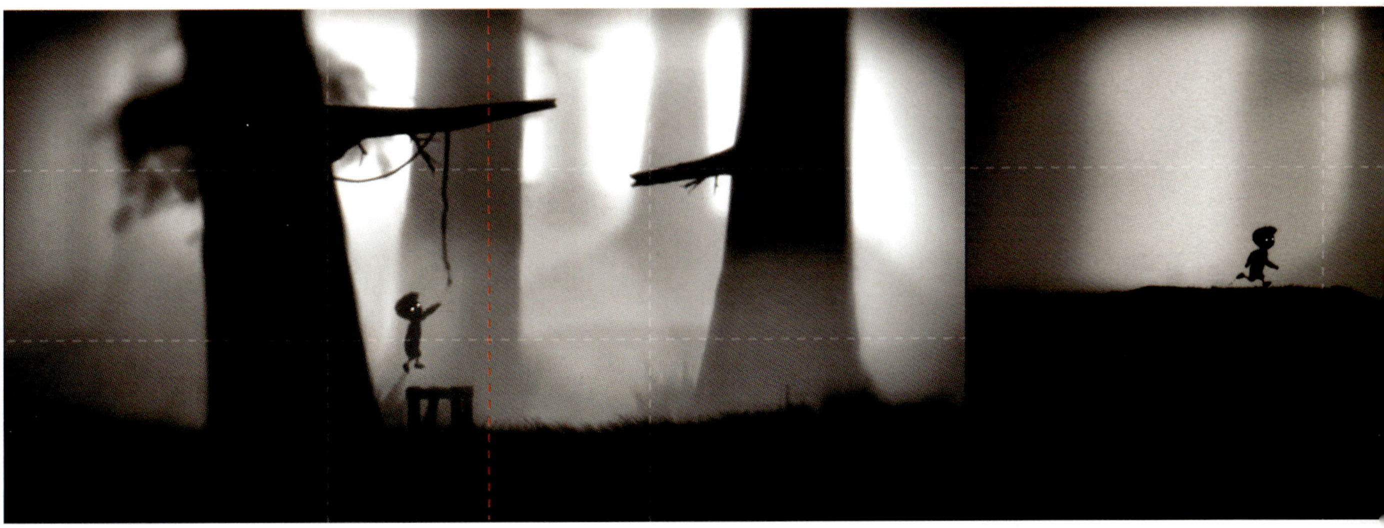

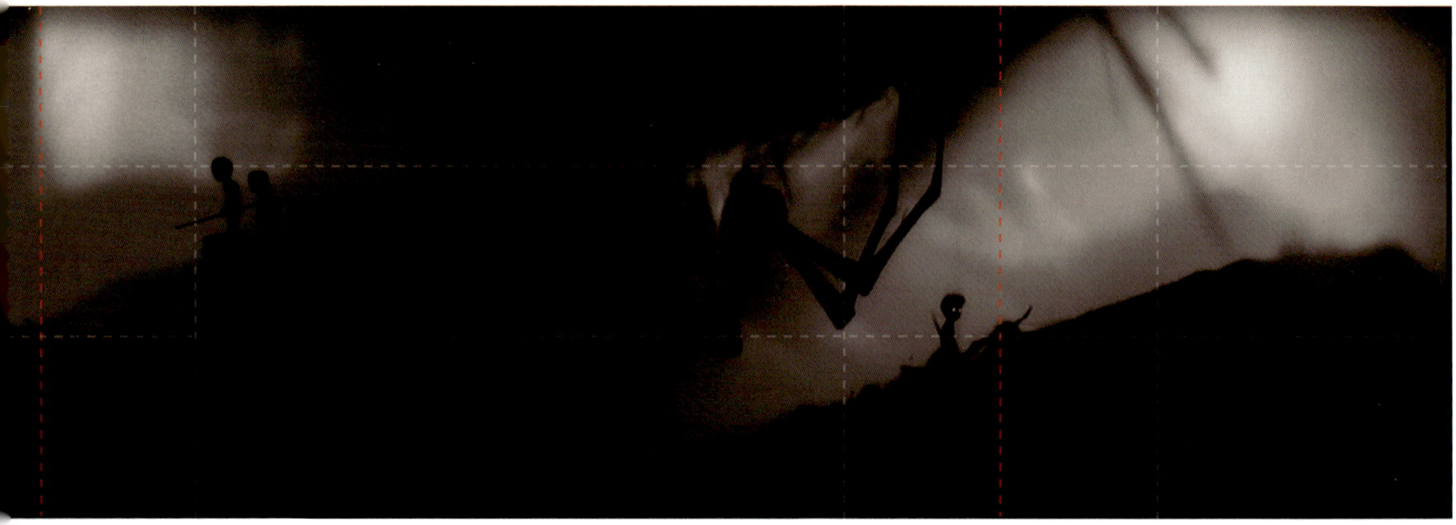

The popular game of LIMBO and the anticipation of THE LAST NIGHT are two games that illustrate some of the current architectural and graphic techniques being used for games. The games use real time movement through space and alter the horizon line to create changing atmospheres and have colors respond to the narrative of the storyline as the player moves through the game.

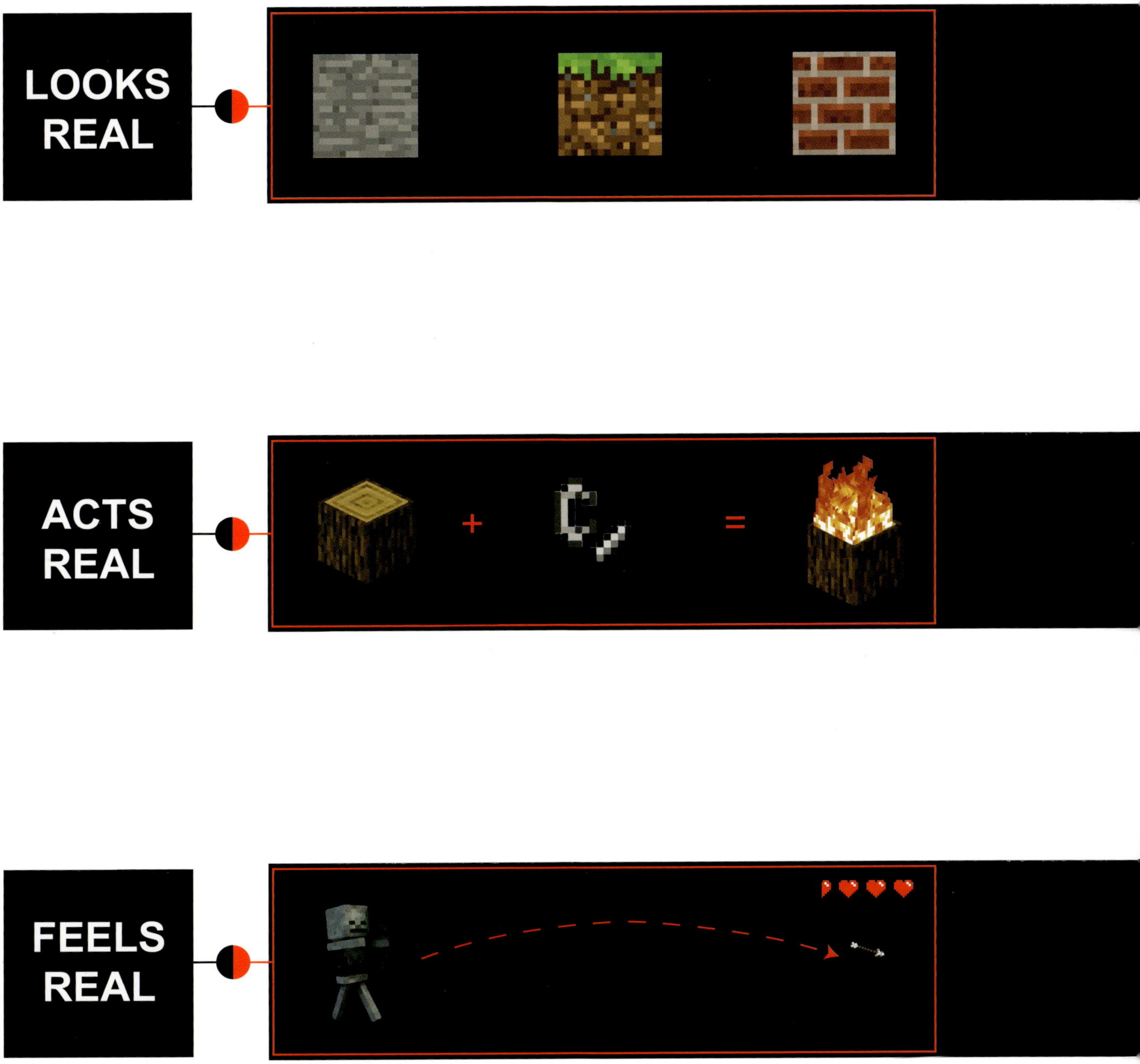
LOOKS REAL
ACTS REAL
FEELS REAL

The material world has established an entirely new language in the virtual world. The materiality of what we see can be represented as an abstraction of what we know. Plato posed the question, "What is the beauty of an apple, is it the perfect shape in our mind when we talk about an apple, or is it the reality of the misshapen apple that exists in reality?" The virtual world creates a perfect digital image of materials from reality, yet we call them by their name from the imperfect real world.

The idea of creating a virtual world that begins to simulate reality by acting real is a coding system that exists in humans to try to capture time and movement in a narrative. As Minecraft has symbols to use flint and steel to create fire, so we suspend our belief in this world-building to imagine that a fire can exist in the virtual world by the stroke of a mouse. Acting real is projected from a digital image to our brain and we allow this deception to become real in our minds.

Feeling emotions in a virtual world is perhaps one of the new, elusive traits in the development of technology. The fact that social media and dating services have evolved as the primary source of meeting on the internet[114] is a testimony to the notion that the virtual world feels real. The Metaverse is also opening the possibility that the open, public forum on the internet will become the new public space, the central plaza, the hangout for meeting others in a social setting.

The very nature of gaming promotes interaction and immersive experiences—first with hand controllers and then with motion control—but soon the Metaverse will take this immersion to a whole new level with three-dimensional games, business meetings, and social interactions.

REAL SOCIETY

VIRTUAL SOCIETY

Sherry Turkle wrote in *Alone Together*, "The advertising for Second Life, a virtual world where you get to build an avatar, a house, a family, and a social life. Finally, a place to love your body, love your friends, and love your life. On Second Life people are represented by their avatars, richer than they are in their first life and a lot younger, thinner, and better dressed." We now have a choice between living our real life, or creating a new, other life in the virtual world.

PRIVATE EXPERIENCE

The illusion of privacy is causing people to shed their inhibitions online.[115] As VR develops and people interact as avatars, they have the ability to take on a new persona amplifying this feeling of anonymity as well as its consequences. This false sense of anonymity can help people who are shy in gaining the courage required to meet people, but it can also make others more aggressive. Meta has even created warnings before its social games on the Oculus

HOME AND FAMILY

to remind you that you're interacting with real people. Family life has also been impacted by digital devices, with the dinner table, once lively with conversation, still brings families together physically, but often family members are often lost in another world on their phones or TV. Now that people rarely talk on phones, they rely on GIFs and memes to communicate their emotions. The desire for instant gratification is becoming apparent with Tik Tok, a company

COMMUNITY

that has reported 167 million videos streaming every minute of every day and is now number one in the number of hits in the world. The past social constructs in society have easily moved into tribal communities in the virtual world with new internet followers grouping themselves around the evolving political, social, and cultural worlds. Often based upon varying levels of what is real, mega-communities are developing virtually that simulate real life events.

As technology evolves, the real and unreal are competing for attention, and the virtual world is winning much more often than it used to. Machines are becoming ingrained in our society and they're redefining our private, family, and community experiences. The exchange between the real and the virtual is beginning to blur the boundaries between our shopping, business, and social life.

AVATAR

"You might have begun your online life in a spirit of cooperation. If you were lonely and isolated, it seemed better than nothing. But online, you're slim, rich, and buffed up, and you feel you have more opportunities than in the real world. So, here, too, better than nothing can become better than something—or better than anything." You may not have realized it, but when you opened a social media account, you were opening Pandora's box—the first step toward creating your very own digital twin. You likely considered what you wanted your brand to be and chose a carefully curated image or photograph to match. But that was just the beginning. Using artificial intelligence and algorithms, social media websites began collecting information about you—your likes, your purchases, and your social connections—and they use that information to build their own profile of you that can be used in marketing. When you create a digital representation of yourself, that avatar has the potential to live on long after you are gone, so a part of you lives forever in the virtual world.[116] The communication technology of the last hundred years, has evolved from the land line telephone to the smart phone with nearly 2 billion active users now on Facebook,[117] which has created one of the largest social experiments in one generation ever undertaken by humans. This migration from the real to the virtual has infiltrated our entire lives. Architects have historically acknowledged the service to and the collaboration of clients. In the twenty-first century the architect's practice of working with the client's images, dreams, visualization, email correspondence, and perhaps even the currency paid for the service through cryptocurrency will be reflected both in the real world and online in NFTs or in digital real estate. Once only envisioned in sci-fi or movies like *The Matrix* or *Ready Player One*, the new possibilities of living in the virtual world have now been realized with the Metaverse redefining the architect's historical reference to place, once revered for defining a great space in the city. The public places and parks are memorable, however, in the virtual world this new sense of place will no longer be a real space, but a virtual environment for our avatars to meet. In the future this virtual place may be more real and public than the reality of the historic civic plaza.

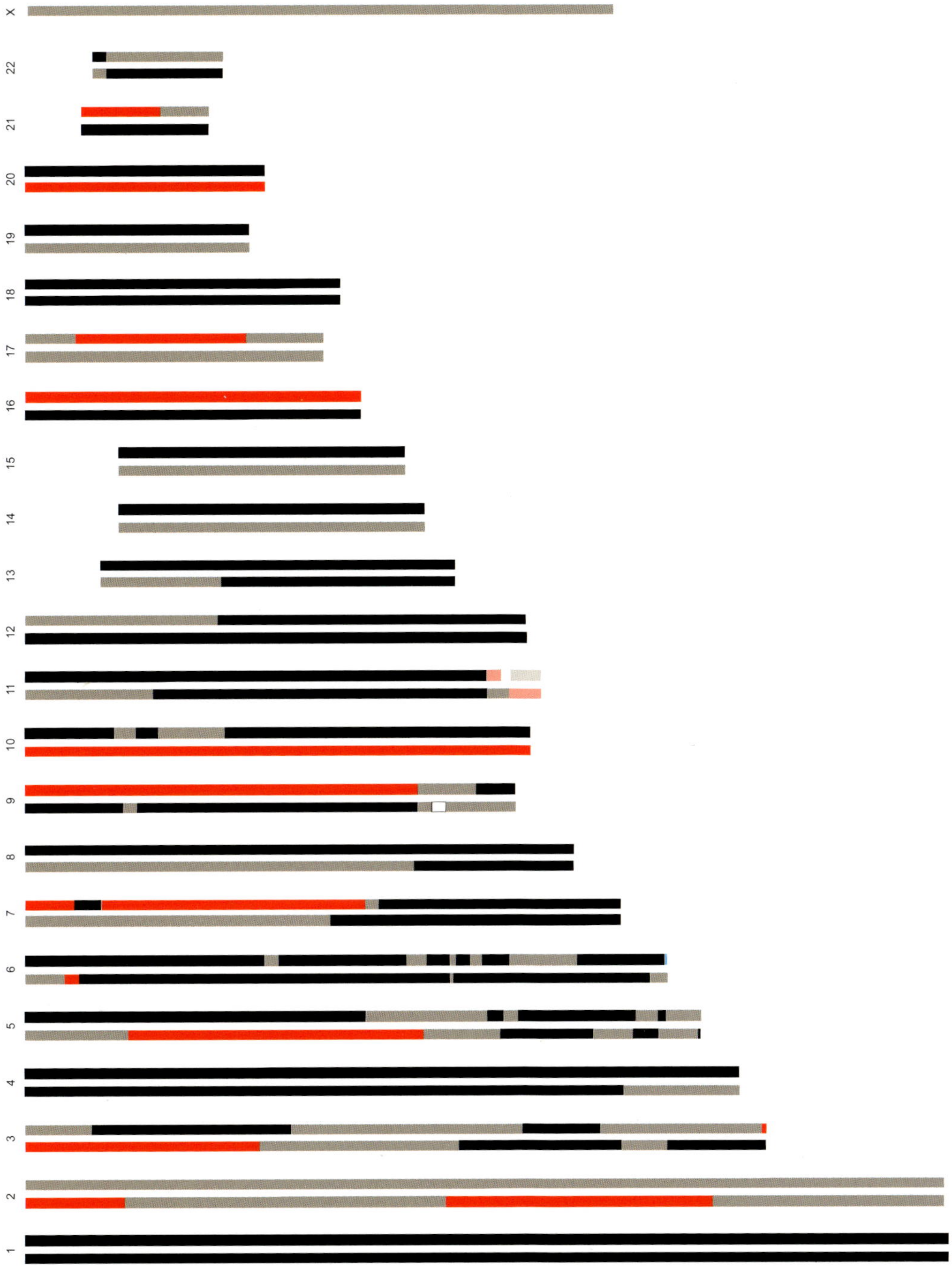

HUMAN ANCESTRY: GENETIC MAPPING[118]

- NORTHERN EUROPEAN
- SOUTHERN EUROPEAN
- WEST AFRICAN
- UNASSIGNED
- EASTERN EUROPEAN
- ASHKENAZI

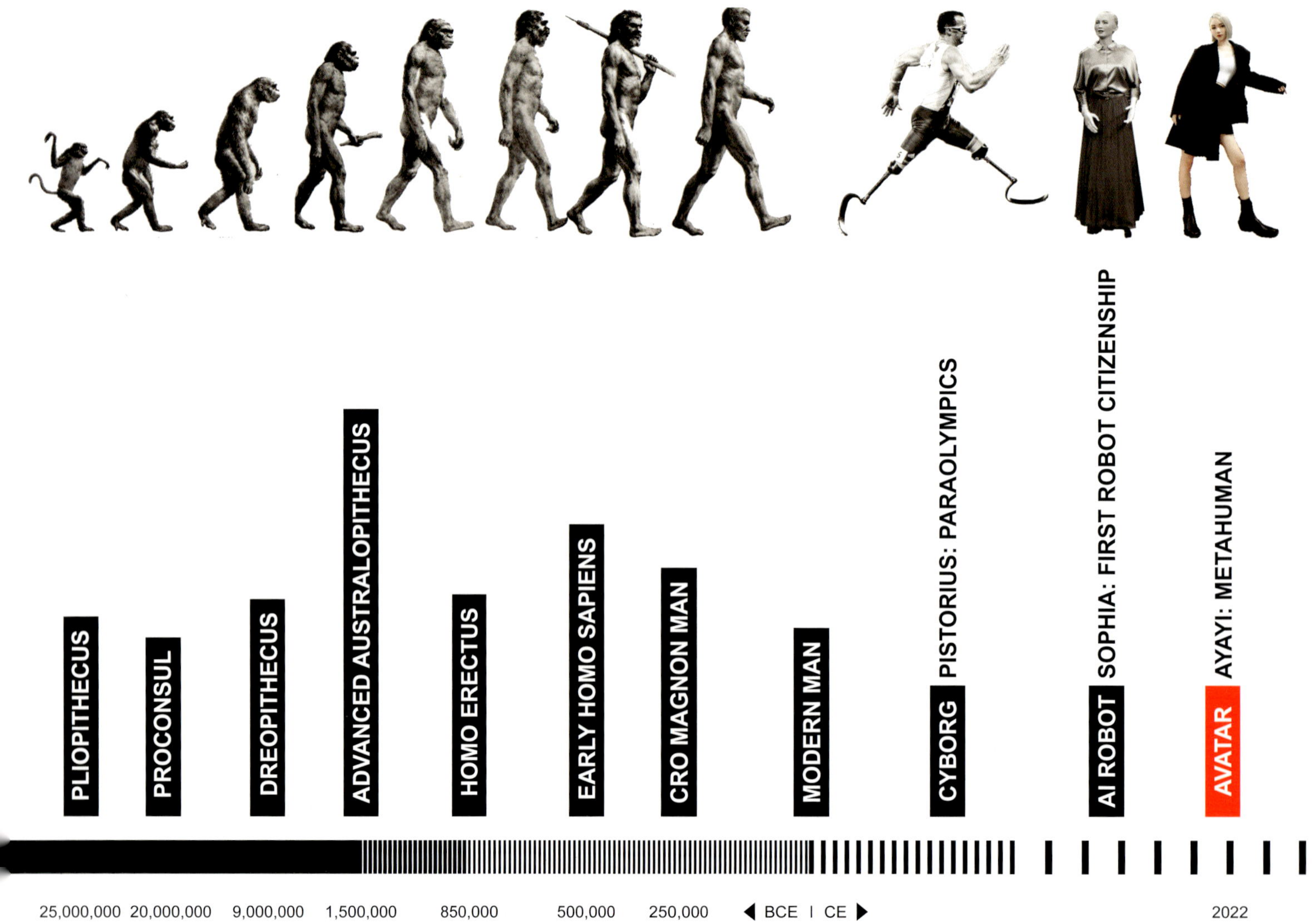

"WE'RE ALREADY A CYBORG."
ELON MUSK 2011[119]

The genetic mapping of the human genomes is one of the most extraordinary advances in the twenty-first century. As humans evolve to become cyborgs (with pacemakers, prosthetics, stents, etc.), and we race to the virtual with avatars, the metahuman using AI and photorealistic images of the body will further blur the lines between what is real and what it is to be human.

PERCEPTION TIME

+ EXPERIENCE OR AWARENESS OF THE PASSAGE OF TIME

1500BCE	1656CE	1884CE	2007CE
SUNDIAL	PENDULUM	GREENWICH MEAN TIME	SMARTPHONE

MECHANICAL TIME

+ DEVICE FOR MEASURING AND INDICATING ELAPSED TIME

1364CE	1571CE	1955CE	2015CE
MECHANICAL CLOCK	WRISTWATCH	ATOMIC CLOCK	APPLE WATCH

TECHNOLOGICAL TIME

+ THE PERIODS & DISCOVERIES OF TECHNOLOGY

1760CE	1947CE	1953CE	1969CE
INDUSTRIAL REVOLUTION	ELECTRONIC REVOLUTION	BIOLOGICAL REVOLUTION	INTERNET REVOLUTION

TIME: PERCEPTION, MECHANICAL, AND TECHNOLOGICAL

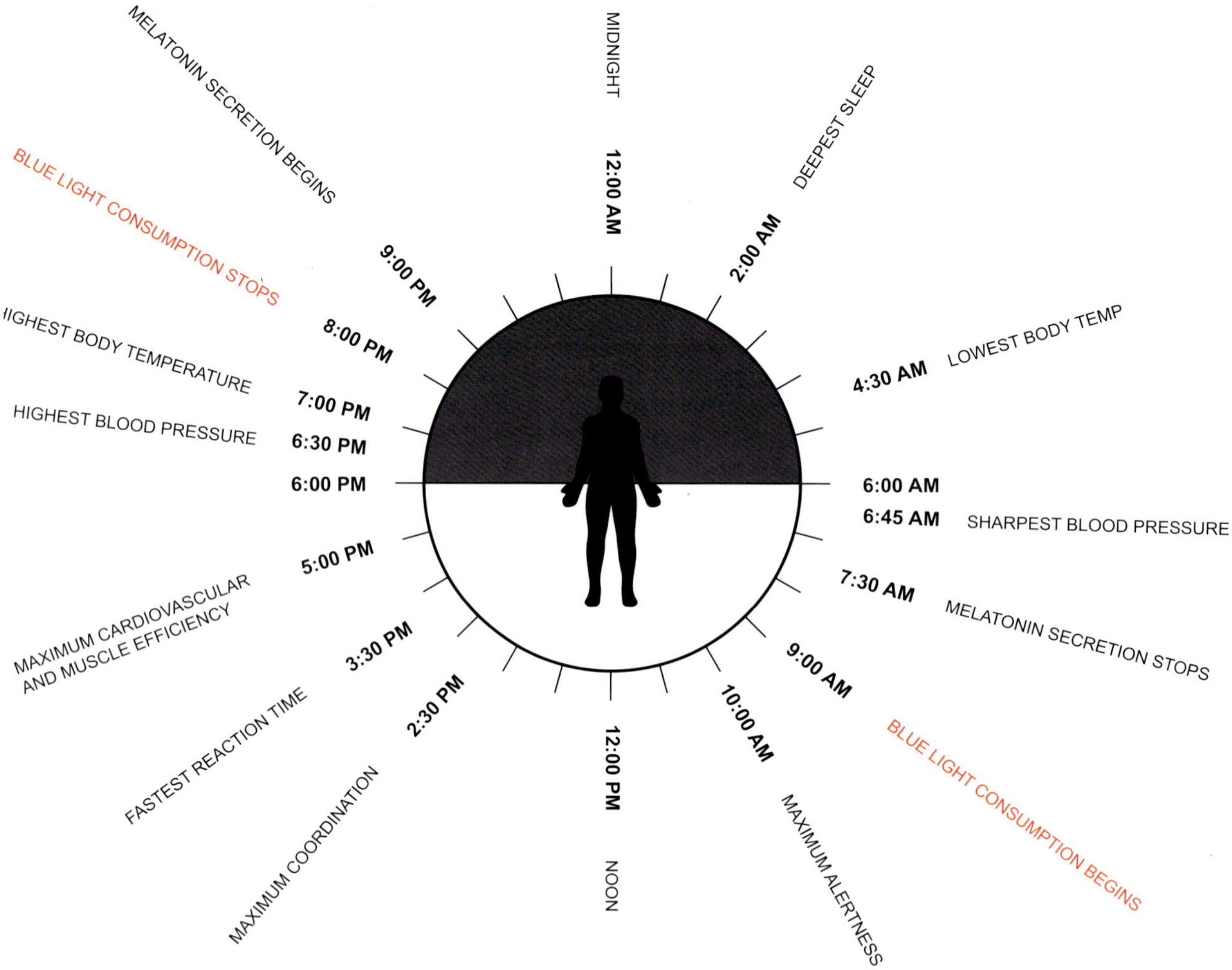

CIRCADIAN RHYTHM[120]

The evolution of technology has produced a new way for humans to relate to reality. The mechanical clock invented in 1360 CE by Henry de Vick created a way to mark time that did not depend upon the movement of the sun. It is the first 12 light bulbs installed in Cleveland, Ohio in 1879 that changed the circadian rhythms of humans forever by allowing cities to extend the daylight into night.

VIRTRUVIAN MAN
Da Vinci (1490)[121]

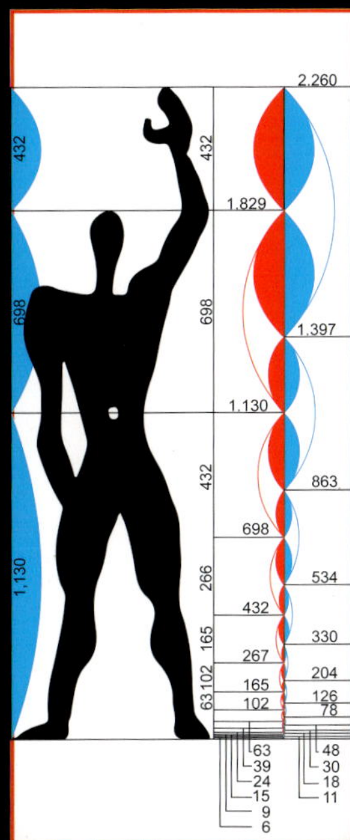

THE MODULOR
Corbusier (1948)[122]

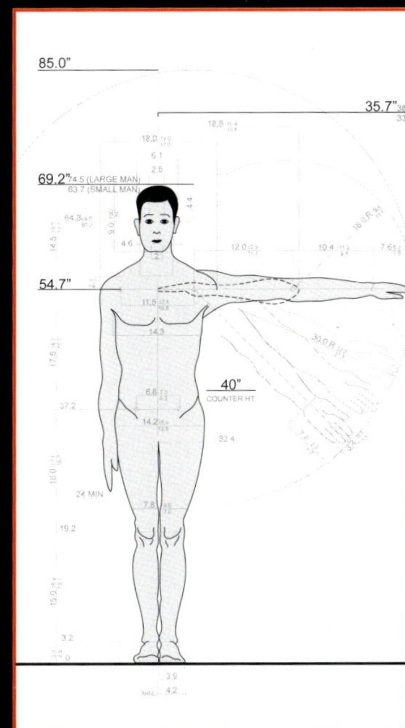

THE MEASURE OF MAN
Henry Dreyfuss (1966)[123]

Society's perception of the body has evolved over time regarding how each culture has portrayed and viewed the body. In 1490 Leonardo Da Vinci drafted the words taken from writings by the architect Marcus Vitruvius Pollio in De Architectura in the first century BCE for a theory based upon an arm span that was equal to the height of a man with a perfectly proportioned body. Leonardo's self-portrait of a man in a circle (the center of a navel, and the genitals being the center of a square) has become an iconic image since the Renaissance. Vituvius's theory of the divine proportion of the human form can be applied to the principles of temple architecture with symmetry, proportion, and weight. In 1948 Le Corbusier also conceived of a man that was based upon a modular system that defined his proportions for sitting, writing, standing, etc. It wasn't until Henry Dreyfuss in 1966 that the notion of culture, individuality, and

PERCEPTION OF THE BODY

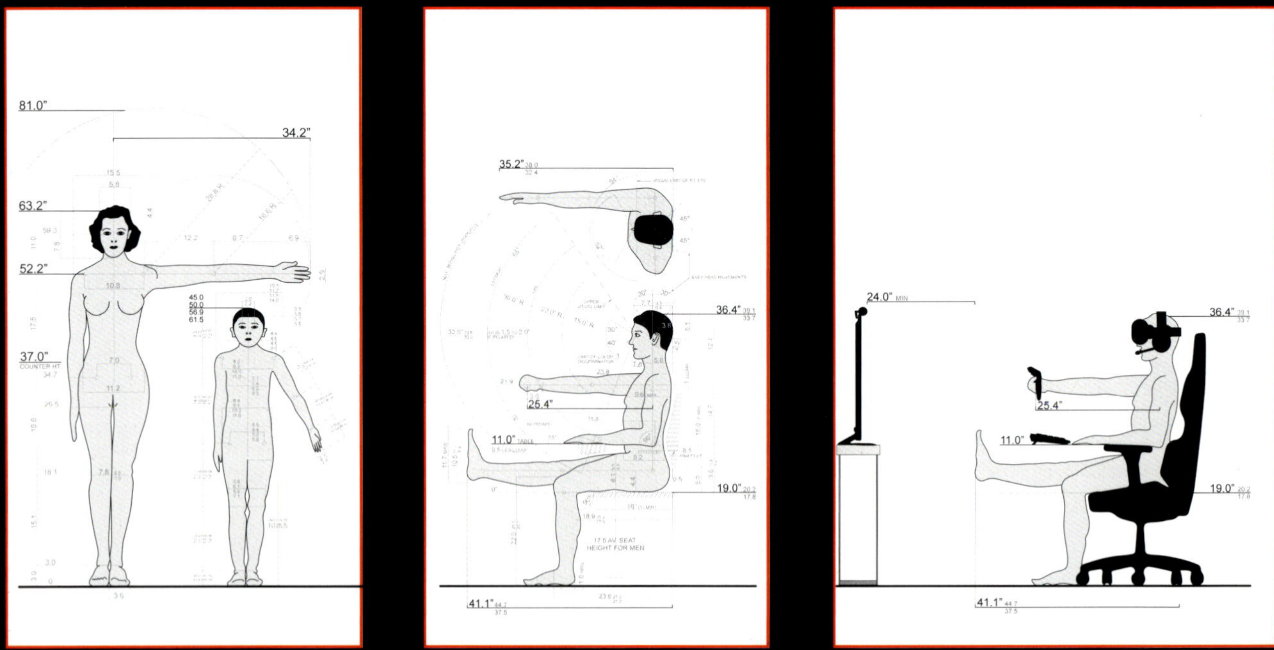

THE MEASURE OF MAN
Henry Dreyfuss (1966)

gender was prescribed as an ergonomic system that proved to be useful as templates for the new industrial demand for car seats, commercial and domestic furniture, airplane seating, etc. Today the concept and self-perception of the body is once again changing with the invention of the digital avatars representing the body in the virtual world. The avatar becomes a social construct of a real humans' desire in a virtual realm not constrained by the limitations of the body.

The real world is restricted by not being given any technological augmentation. This becomes the domain of the INbetween. The balancing of the real and the virtual with augmentation either through mechanical means (a pacemaker to extend life) or virtual enhancements like AR that takes its cue from the virtual and brings it into the real world. Snapchat's filters, or shopping with avatars is merely the beginning of this augmented world.

A hundred years ago there were no antibiotics, no commercial airplanes, no radio, no TV, no cars, no highways, no air conditioning. The notion of a home having a thermostat for heating and cooling, lighting, or imported materials was unknown to the public. Similarly, the health and understanding of the bodies mind, the heart, the physiology, or psychology of humans was also unknown.

BEHAVIORAL CONFIRMATION

ATTRACTIVE CHARACTERS WERE MORE LIKELY TO REVEAL PERSONAL INFORMATION TO STRANGERS

TALL CHARACTERS BEHAVED MORE AGGRESSIVELY WHEN NEGOTIATING WITH OTHERS

SHORT CHARACTERS WERE MORE LIKELY TO ACCEPT UNFAIR NEGOTIATIONS AND PROPOSITIONS

YOUNG PLAYERS TENDED TO PLAY AS MORE INTELLIGENT AND STRONGER VERSIONS OF THEMSELVES

GAMING AVATARS: VIRTUAL PERSONALITIES[124]

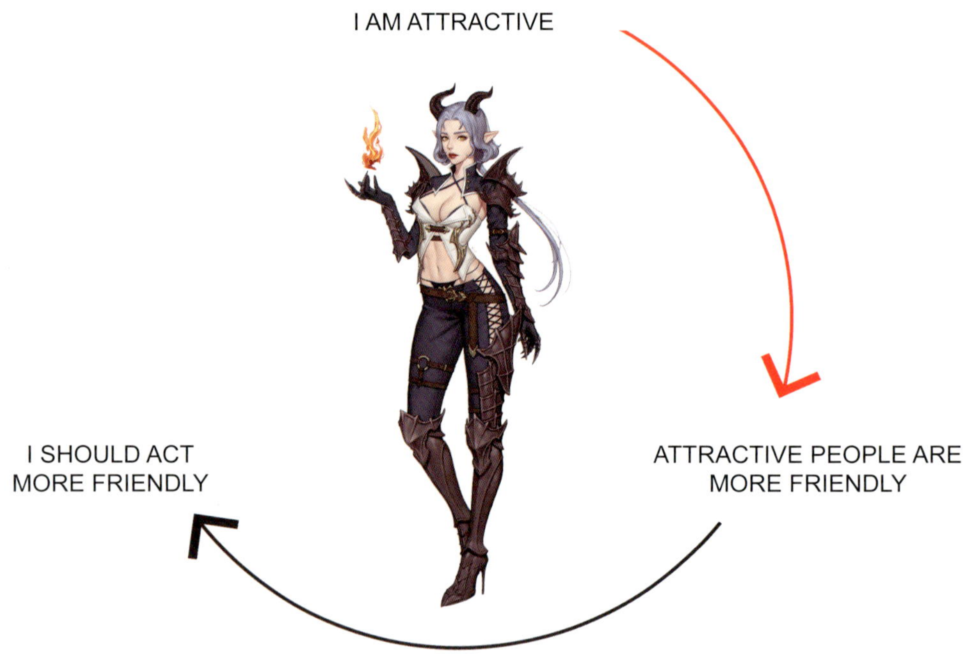

SELF PERCEPTION THEORY

MOST PLAYERS HAVE CHOSEN SEXY AND AESTHETICALLY PLEASING CHARACTERS

MALE HEALERS AND LIGHT ARMOR WEARING CLASSES WERE MORE LIKELY TO GENDER BEND

FEMALE PLAYERS ARE MORE LIKELY TO PICK CHARACTERS WITH LIFE-GIVING ABILITIES AND REJECTION OF VIOLENCE

MALE PLAYERS PREFERRED RANGED COMBAT FOR THE REPRESENTATION OF STRENGTH LIKE AN ARCHER

Sigmund Freud in the early twentieth century proposed the notion of a personality encompassing the id (primitive + instinctual), the ego (realistic), and the superego (moral conscience). Like an iceberg with the unconscious below the surface, dreams became a form of wish fulfillment for the personality traits and their hidden desires. The avatar reveals these psychological forces of behavior.

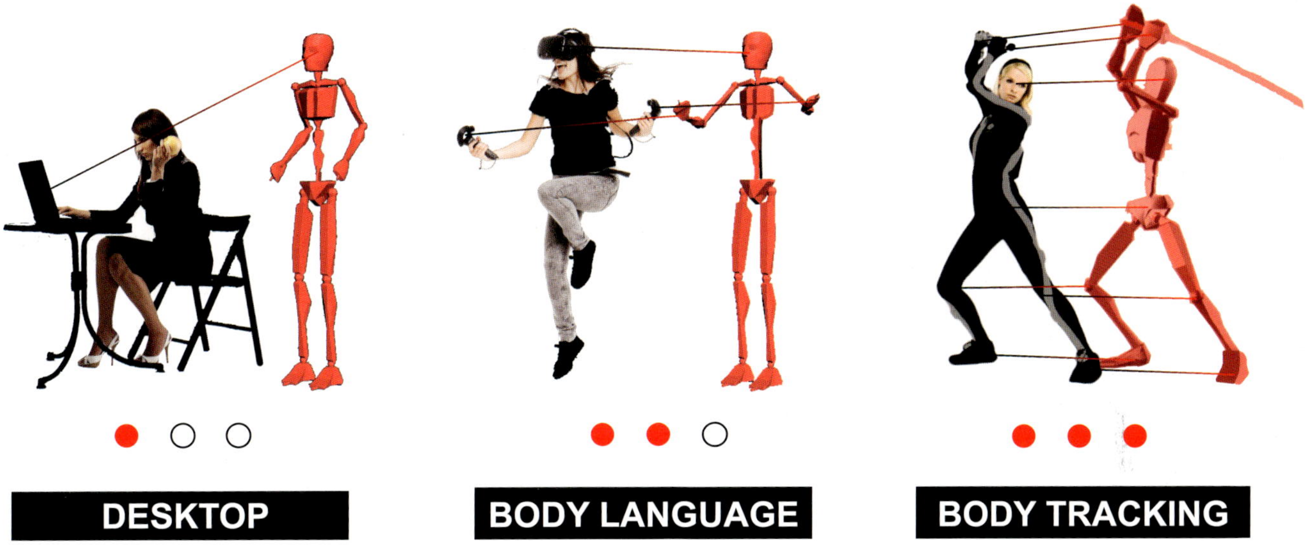

Collaborative competition can be traced back to the original Olympics in Greece in 776 BCE[125] and had evolved as long as three thousand years prior to these ancient Olympic games. Human's desire and competitive nature for games is baked into our DNA. As the memory and computing power of the internet increased, so too did the devices from the real world: the physical stationary consoles to the more advanced home computers have also improved, as controllers linked to the operating system by cables tied the players to the device. However, Bluetooth and motion controls created the capability of interacting with a gaming program, but now the controllers didn't have to be tethered to a machine. As gaming evolved from a simple beginning with a tennis game for two in 1958, the market for gaming has now become 2.69 billion gamers worldwide in 2020. With the invention of the Metaverse in which the computer software in virtual reality can see the self-tracking technology where the hands can be tracked without the use of controllers. This capability to become untethered to the machine once again takes the experience of gaming in the virtual world to another level that can be totally immersive and can begin to simulate a new reality. The computer gaming experience was created using consoles and digital screens on the computer to be viewed on a monitor, iPad, laptop, or iPhone. This type of viewing meant that as an observer you would be tied to a device watching the images unfold in front of you. Pokémon GO first began to use AR as a way to have gaming become an experience in the real world throughout the city. In 2021, Pokémon GO had registered over one million active users per day in the US and the UK. The further development of haptic hardware consisted of gloves and a body suit when combined with VR was meant to activate all the senses and thus became a fully immersive environment.

GAMING: FULLBODY TRACKING | AVATAR

REAL I N B E T W E E N VIRTUAL

Gaming took a leap with the Metaverse being introduced with the use of avatars allowing for real time experience of body language, tracking of movements, and taking gaming into a new reality of feeling the environment in three-dimensional space. This new platform will offer a full array of venues for new gaming experiences, including: social meeting places, business workrooms, concert performances, retail shopping, entertainment, sports, casinos, etc. As artificial intelligence becomes ubiquitous it is inevitable that a metahuman will transform the world of avatars into another realm. As avatars are becoming real in the virtual world and have the capability to interact with all the same human reactions as those controlling the avatar, the question will emerge: What is it to be human? Ayayi,[126] a metahuman is a digital avatar influencer who possesses all the attributes of being a human, except self consciousness. Ayayi can also be compared to the most sophisticated, artificial intelligence robot Sophia,[127] who can communicate effortlessly in real time with humans and project empathy by knowing how to interact with human behavior. Sophia, who gained Saudi citizenship in 2019, marks a new era with humans being able to talk to robots and robots being accepted into the world's social political constructs. As AI develops and is being applied to all facets of life including facial recognition and anticipating individuals desires and actions, it will become increasingly important for society to decide how much personal freedom we as humans are willing to give up to machines. The proliferation of algorithms anticipating your every need on social media and the technology of channeling information for your perusal has made it very clear, even at this infant stage of development, political and worldwide regulations will become more necessary to curb the behavior and application of artificial intelligence.

Gaming was once relegated to physical gameboards, or competitive sports, etc. Today the computer, smart phone, and virtual reality has exploded the gaming industry with MMO: Massively Multiplayer Online communities reaching 161 million users per day on the World of Warcraft game. The future of avatars participating in all the facets of contemporary life is not in question.

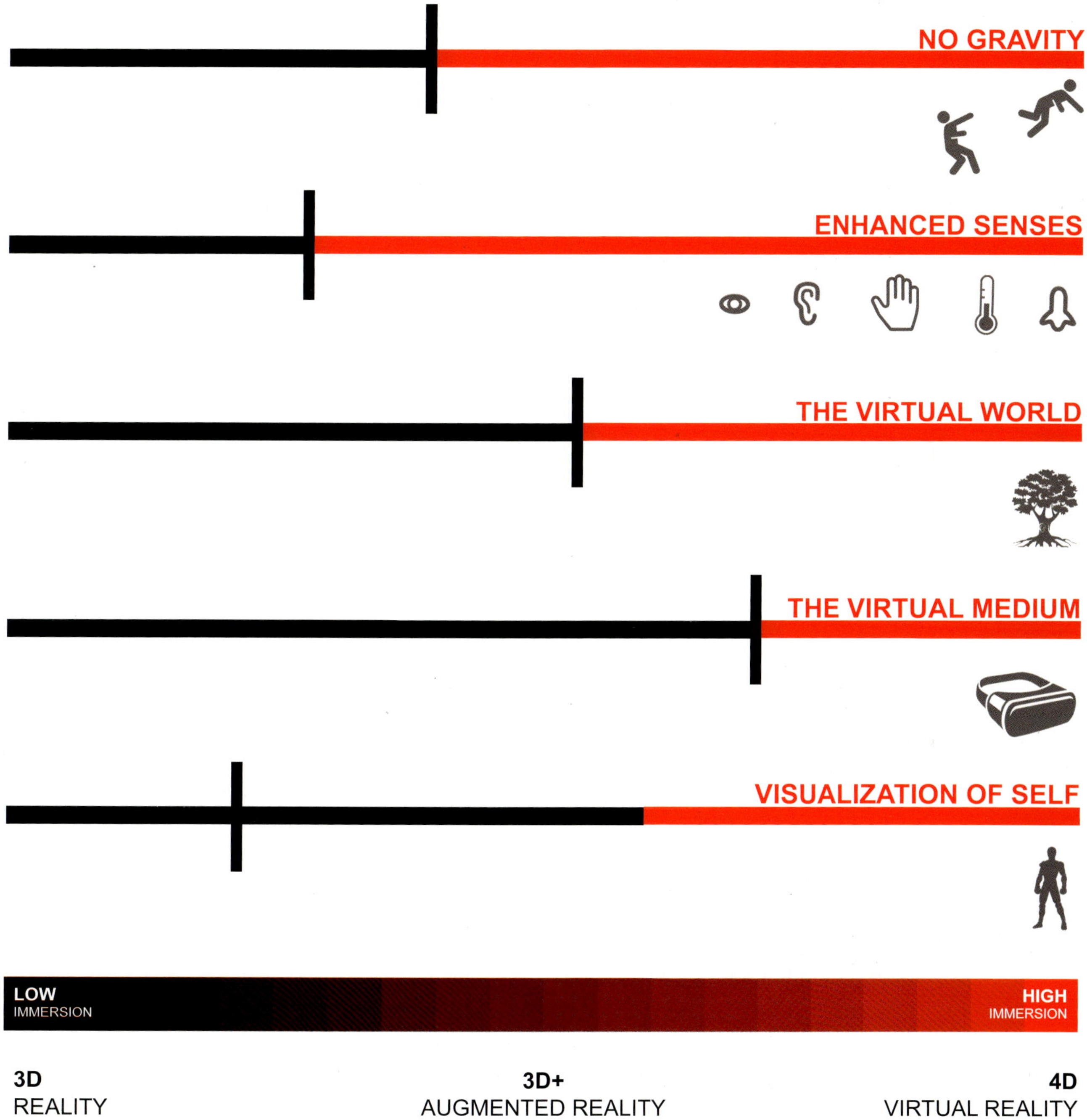

SPACE TIME VISUALIZED: REAL BECOMES VIRTUAL

that required its protector to virtually feed the hand-held object, to put it to sleep, to keep it happy, etc. Parents became involved when the child had to go to school (which typically banned the Tamagochi) and they had to continue to take care of it in real time till their children returned home after school. Children began experiencing true loss at the death of their Tamagotchi when they didn't get enough food and they felt responsible for the Tamogochi death. The question that will be asked: Can robots, not the simple vacuum cleaner or car, but robots with artificial intelligence that can communicate with all the same

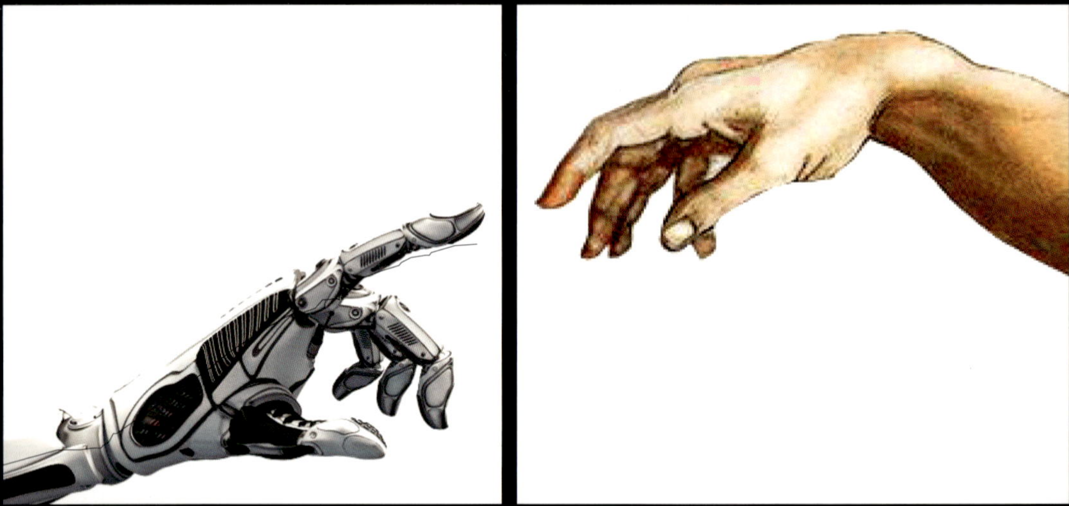

empathy, intelligence, and caring for you as an individual become a friend that in many ways surpass that of other humans? Alexa and Siri have certainly entered our lives in a way that only fortells this future scenario. This idea is taken to an extreme in the movie *Minority Report,*[129] which suggests the potential for precogs (people) who could see the future, a power that could be harnessed with a machine built to record their predictions. Today, Google and Facebook predict what you will want, sometimes you may be talking about something, sit down at your computer and there it is as an ad on your screen. It is the algorithms that are being used to predict human behavior.

As we migrate from the real to the virtual, devices become more sophisticated, the capability of avatars and robots increase, humans become more integrated with machines that can increase our lifespans, the realization is inevitable that this will become our new way of life, our very language of binary codes have evolved as a way for humans to talk with machines.

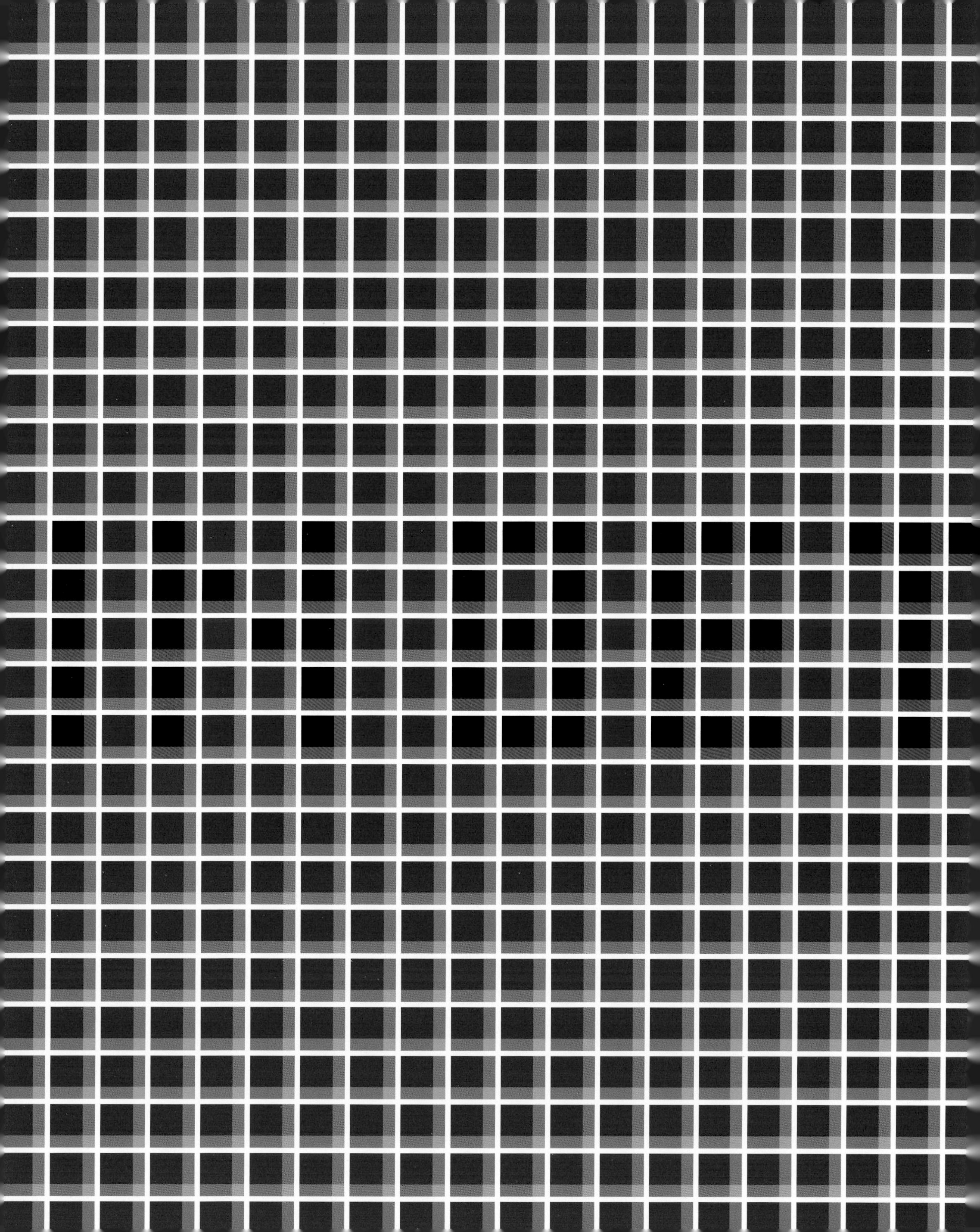
INAET

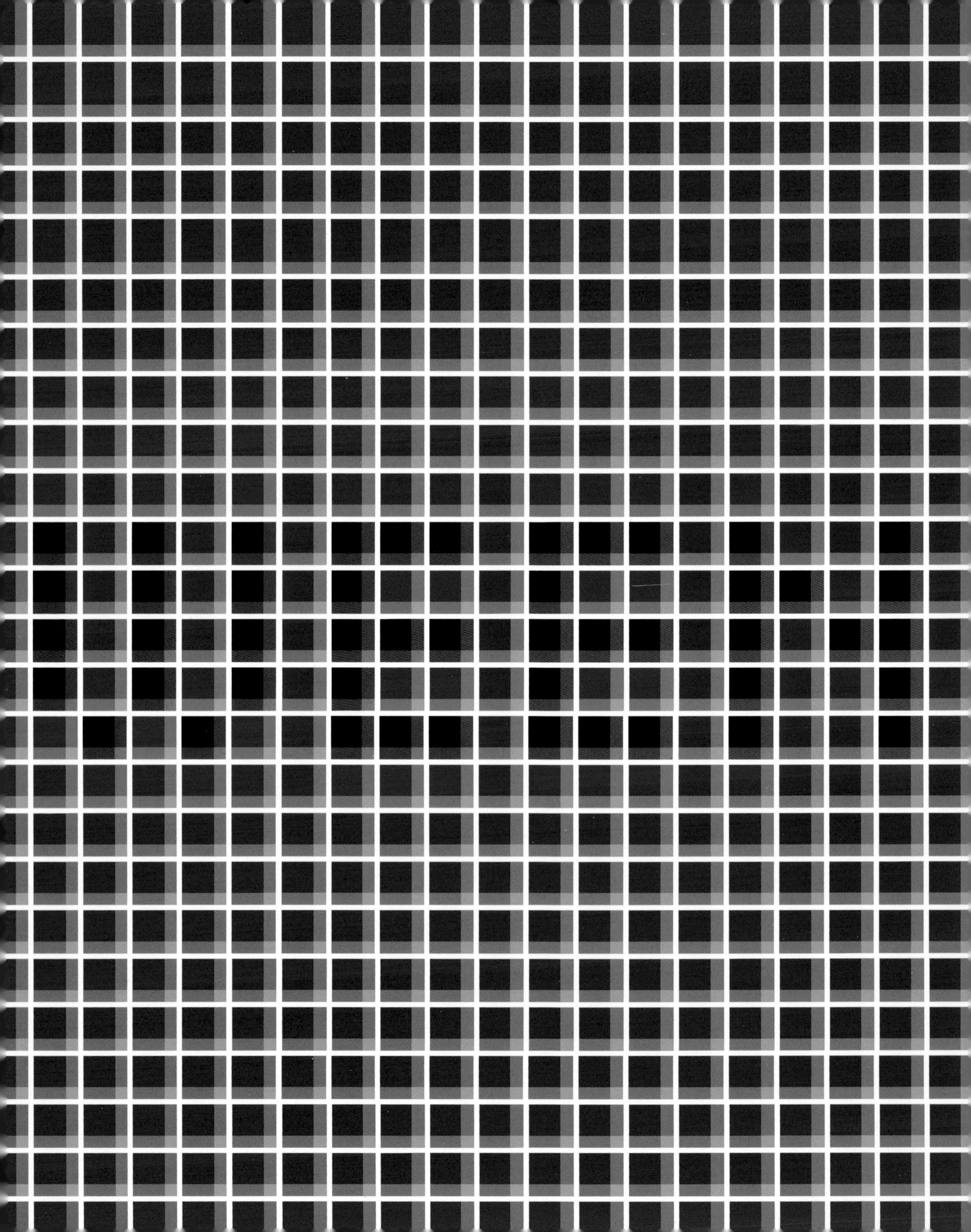
WEEN

CAN ARCHITECTURE SAVE THE WORLD? Stephen Hawking said, "I have noticed even people who claim everything is predestined, and that we can do nothing to change it, look before they cross the road."[132] By definition the INbetween is about perception and being aware of the real and the virtual as two different realities. Humankind is at a crossroads, either accept the status quo or embrace the new technologies while keeping a strong moral compass. As the world flirts with going down one of the paths of extinction from climate change, nuclear warfare, political upheaval, water shortage, overpopulation, cybersecurity threats, pollution, and other global crises unknown in history, the question remains: can architecture make a difference? Can a discipline that consumes only 18% of the global carbon emissions[133] impact the existential direction of humanity? The answer lies in establishing a

new reality of being INbetween and becoming capable of migrating from one reality to another. As new technologies emerge and people move further into the virtual world it is possible that the architect's rigor of problem solving can begin to balance living in the real and the virtual by defining a new reality of the INbetween. As Steve Jobs held a CD player in one hand while listening to music and a mobile phone in the other, the thought occurred to combine the devices and create the smartphone in 2007.[130] As we look at cities, the invention of the highway system in the US happened only 70 years ago,[131] and it is precisely at that moment that the urban life of cities gave up the historical street to the car. Can humanity once again change how we live, how we communicate, and where

we work? Does humanity have the will to take on the global challenges in a post-industrial era as human communities in the past have come together in a crisis? Architecture historically has been defined by the typologies of static forms meant to create a sense of place in the private and the public realm. In the pre-industrial era, courtyards, pathways, streets, and the sun, breezes, and seasons, all defined architecture as the inhabitants lived off the land and built with local materials. In the post-industrial era cities have imported food and building materials in containers through a variety of shipping methods on a global scale with the status quo continuing to build airtight buildings that are powered by fossil fuels. As an alternative the emergence of digital real estate, NFTs, and the Metaverse, have the possibility of creating an architecture where buildings and the environment can meld together in the virtual world.

REAL + UNREAL

We are truly living in a state of transition: as we increasingly evolve into a world that is communicating with machines, with artificial intelligence, and re-engineering our genetic code. Architects have the opportunity to envision an architecture that is net-zero, dynamic, and uses AR with a heightened sense of perception allowing humans to live comfortably with machines. The INbetween is defined by incorporating the real world with new technologies. The traditional roof, for example, is meant to provide shelter from the sun, rain, and wind. But, an augmented roof is also capable of being a solar and water collector. This concept of designing and inhabiting multiple realities using perception and technology to live simultaneously in the real and the virtual world is inevitable in the twenty-first century.

TRUTH

PAGE 102

Architecture has historically been viewed as understanding Truth as a means of representation of its own time and to express the idea of some higher truths. The projects in this chapter question the INbetween as seemingly divergent poles in two different realities. By investigating science as the properties of water that exist in three states: solid, liquid, and gas, can be contrasted to the nature of light as a metaphor for uplifting and tapping into the notion of faith and the iconic imagery of the faithful as a means to gather the flock, Truth in these projects become a way to redefine Plato's Cave and bring the shadows into the light as an expresson of science and faith.

NO BORDERS

PAGE 166

Architecture is being redefined in an era of three competing realities. The unconstrained limitations of virtual reality have allowed architecture to ignore the real world properties of the sun, wind, rain, and gravity, while AR represents designing an architecture INbetween these realities. No Borders uses the subtleties of perception to allow the mind to imagine other realities by seeing an afterimage, or different colors that are not there. Through redesigning the landscape to respond to the reality of climate change, to simulate life on Mars, or to genetically engineer a new material not found in nature: all become possibilities within the new dimension of No Borders.

GAME ON

PAGE 248

Game On is taken from gaming lexicons that introduce time into the life span and perception of architecture. By planning for a future that is pure speculation, architecture can move into the design space of navigating INbetween the real of the moment and the possibilities of an unknown future to come. The projects explore the migration of cities and suburbia and challenge the existential threat of climate change and how people must live in order to sustain the planet. The threat of being complacent in our current dvided reality calls for an architecture to take on activism and embrace the virtual future, with equity, and recognize climate change—Game On.

AVATAR

PAGE 326

Social media has now created our personas, our image, business, and purchasing profiles in what is termed, the second life, from Sherry Turkle's book, *Alone Together*.[134] The ability for humans to create this second life with a digital avatar opens up an architectural environment unknown in history. By exploring the proliferation of current ghost towns created by online retail moving out of shopping centers, to changing people's reality by being able to move into the virtual world, to questioning the longevity of the life of virtual avatars as they live in perpetuity on the internet, society must now redefine how to perceive death in the virtual world of avatars.

TRUTH

FAITH

PARADISE

SCIENCE

DISORDER
SPA
LIGHT SHADOWS

NO BORDERS

ORGANIC

SANCTUARY
SOUND
MIRAGE

ARTIFICIAL

SPACE
INNER / OUTER

INBETWEEN: A CATALOGUE

PORTAL
LIVING CLOUDS

URBAN

GAME ON

SUBTOPIA
EXPLODED

SUBURBAN

SEED
DEPRESSION
XTREME ENVIRONMENTS

EXTINCTION

AVATAR

ANCESTRAL SHRINE
GHOST TOWN

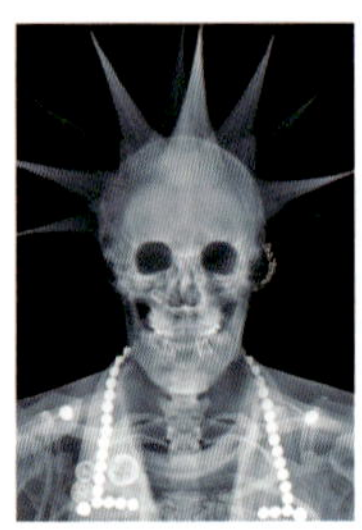

DEATH

The INbetween defines the categories of TRUTH, NO BORDERS, GAME ON, and AVATAR as a way to gather the architectural projects into different views of reality: as competing forces of science vs faith, the artificial vs the organic, the urban vs suburban, and extinction vs death. Each chapter has its own unique spatial relationship to being realized in the virtual world.

SCIENCE
TRU

TH
FAITH

Inspired by a visit by Pope John Paul II, the monumental sculpture of Cristo de la Concordia (Christ of Peace) west of Cochamba in Bolivia, South America, gazes upon the city below. Stretching 33 meters (one meter for each year of Christ's life) the figure has become a symbol of Christ protecting and watching over the city.[135] Faith has been a social construct that is defined as a belief with a strong conviction for which there may be no tangible proof. Humanity has been inspired by this confidence in what we as humans hope to be true in life since time immemorial. Not based upon fact, faith in subsequent religions in its many incarnations requires the full acceptance of a belief system that often uses a particular edited view of reality. The concept of faith therefore dwells in the virutal, or in the idea of not existing in reality. Paradise, Living in the Shadow of Truth, refers to Plato's Cave and the notion of the shadow being seen by those in the cave, but not seeing the beauty of the

A PLACE OF FAITH INBETWEEN HUMANS + GOD

real truth of the sun. The search for paradise is seen as a struggle between humans and God. Man is represented by the house that is cast within the shadow of Christ with a husband / curator of an anthropological museum located in the city below. The notion of God is embodied in a chapel with his wife also working to maintain the sculpture of Christ and tend to the tourists. The tension between humans and God is resolved in the garden between the chapel and the home by a symbolic set of snake scales created as platforms for plants descending to a fire pit at the center of the garden. The uplifting chapel of God, made of translucent concrete is open to the sky and beyond, while the earthly house made of rammed earth has panoramic views of the city and is rooted in the real. The truth lies in the potential of the earth becoming a living paradise where nature and people can co-exist in harmony and come out of the shadows that have been cast throughout history.

GOD

PARADISE

HUMANS

EXPLICIT	**PARADISE**	**IMPLICIT**
MONITOR A PERSON'S INNER STATE	ALL PARTS OF THE BRAIN ARE ACTIVATED	COGNITIVE SKILLS VOLUNTARY MOTOR MOVEMENTS

TRUTH | FAITH

PARADISE: A REALM INBETWEEN GOD AND HUMANS

LEFT SIDE	CORPUS CALLOSUM	RIGHT SIDE
LOGIC SKILLS POSITIVE EMOTIONS	COORDINATE ACTIONS AND DECISIONS	CREATIVITY SUPPORTS FEAR

With the advancements in the neurosciences and computer modeling it is now possible to locate the areas of the brain that correspond to real stimuli and those not based in reality. The area that coordinates or mediates between the two paradise can be conceived as the place in the brain that keeps the real and unreal in balance.[136]

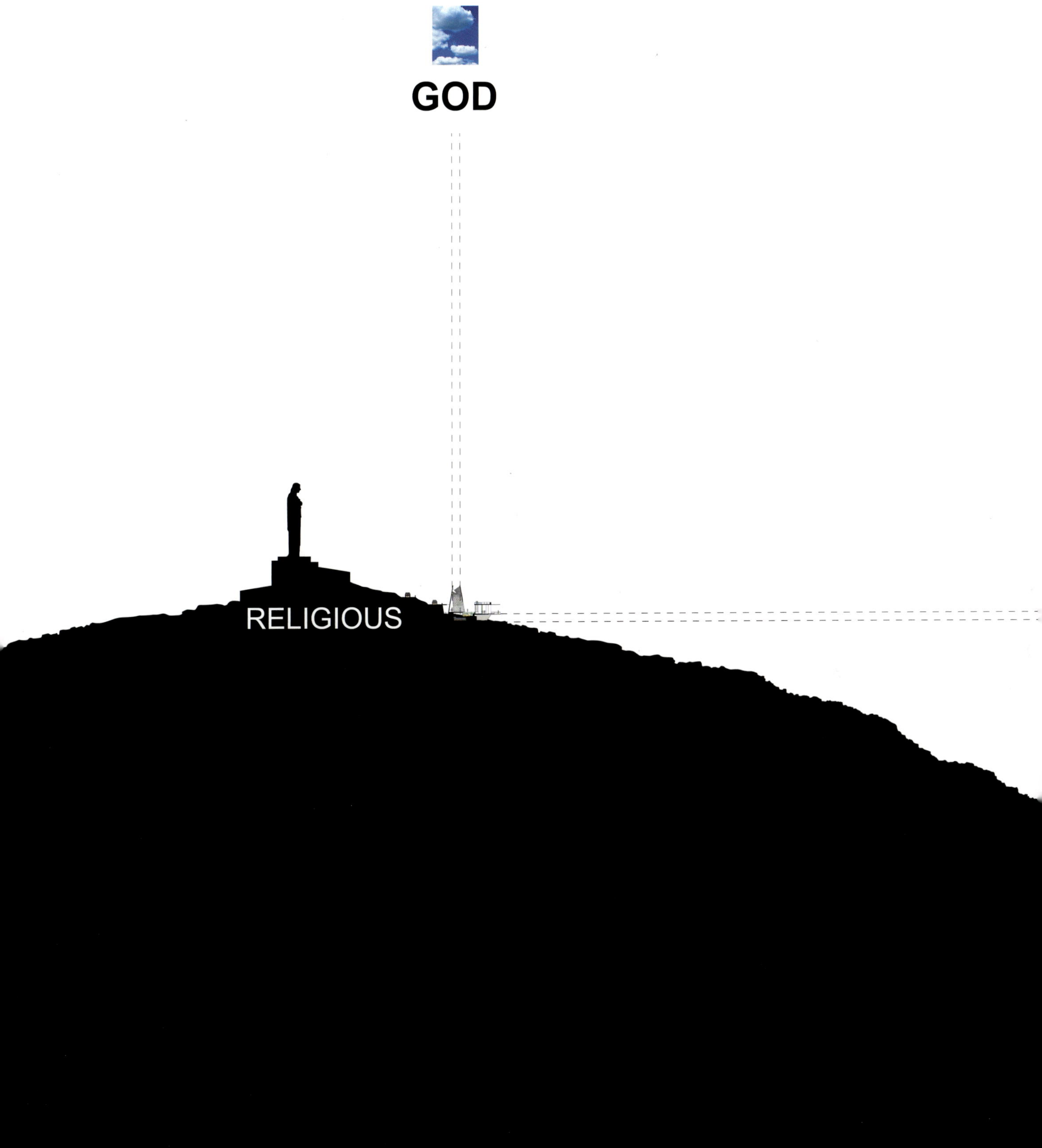
GOD
RELIGIOUS

HUMANITY  SECULAR

TRUTH | FAITH

GOD | PARADISE | SECULAR

TRUTH | FAITH

PARADISE: LIVING IN THE SHADOW OF TRUTH

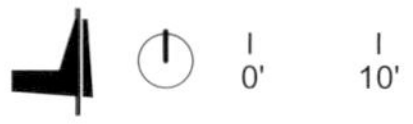

The 33-meter Cristo de la Concordia is the second largest sculpture in the world of Christ. Finished in 1994, the shadow of the figure stretches out to the symbolic house of God and humans. The two-story chapel is oriented to the heavens which is in contrast to the openness of a house overlooking the city.

FIRST FLOOR
0' 4' 12'

TRUTH | FAITH

THE HOUSE: UNDER THE SHADOW OF TRUTH

For many the figure of Christ is the Truth, however living under the immense cast shadow of Cristo de la Concordia has the house rooted in the earth looking up to Christ on the one hand and to the city on the other. This tension between Humans and God is resolved by the garden of paradise that mediates between the vertical and horizontal spaces.

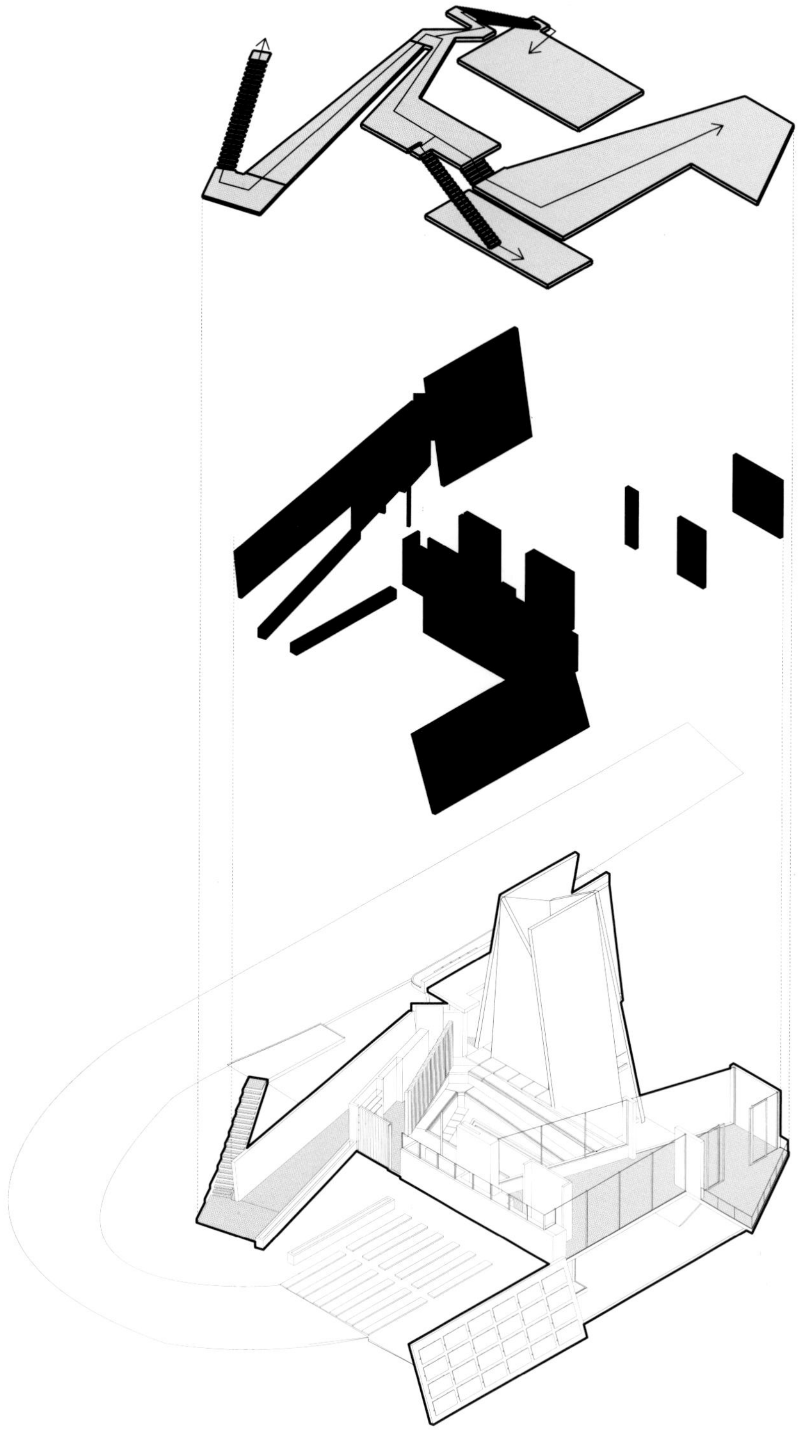

A JOURNEY: SEEKING PARADISE

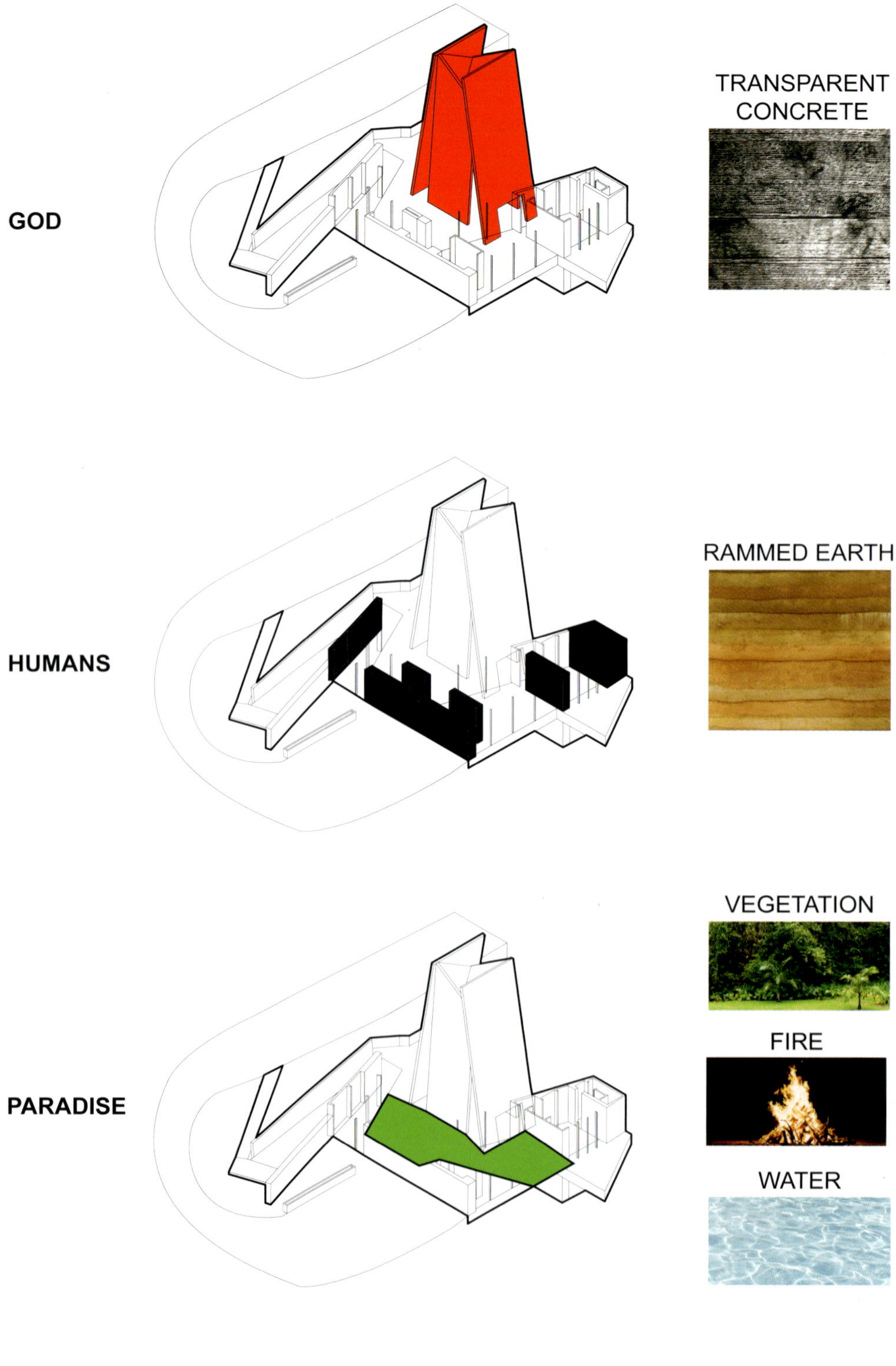

A SEARCH FOR MATERIALS

The materiality of the house creates the possibility to understand the three materials of construction: Humans (The Real) is built of rammed earth as a symbolic element, Paradise (The INbetweeen) is composed of water, fire, stone, and vegetation, and God (The Virtual) is made of translucent concrete allowing only light with shadows to be cast on and through the walls of the chapel.

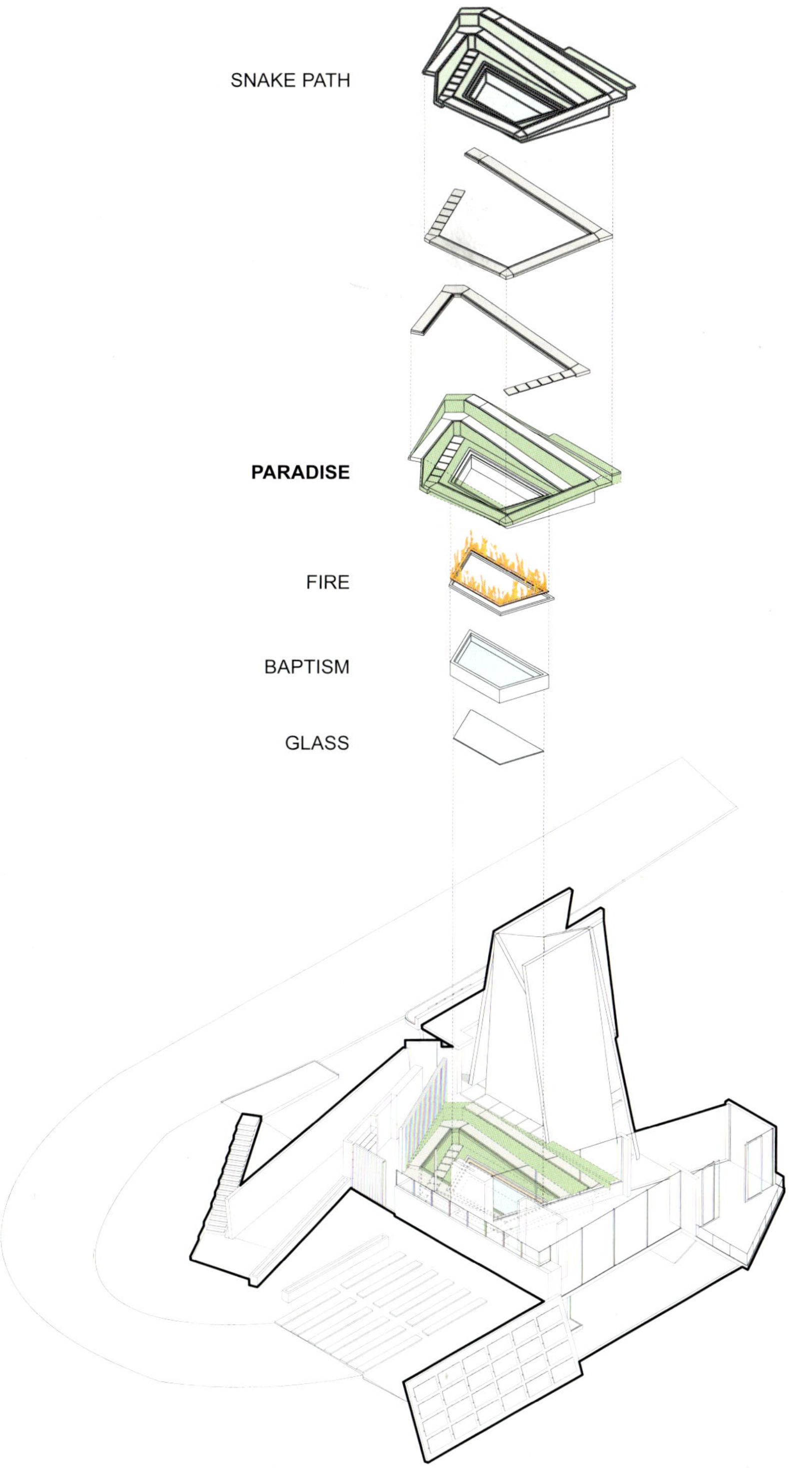

A JOURNEY: FOLLOWING THE SNAKE PATH

TRUTH | FAITH

PARADISE: FIRE, WATER, AND NATURE

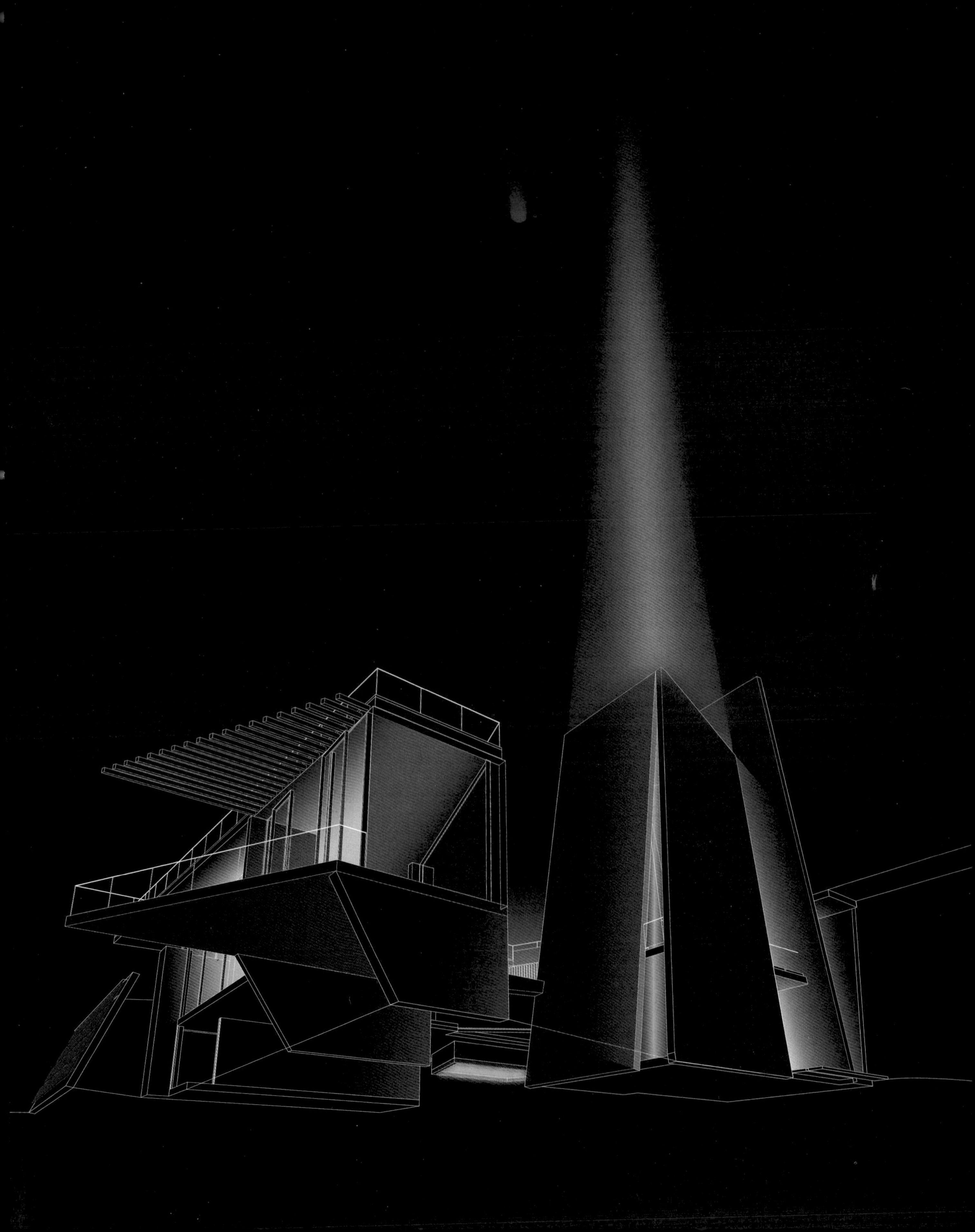

Science is defined as the pursuit of understanding the natural and social worlds through a systemic methodology. The process of creating this way of seeing the modern world developed in the Renaissance with Galileo Galilei. Water can exist in three states: solid, liquid, and gas, depending upon the exposure of water to different temperatures.[137] In the natural world there are special locations that have springs of naturally hot water that is heated by subterranean volcanic activity. Ausangate, Peru, one of the most untouched sites in the world, has a variety of geysers, hot springs, and steam jets that are located within the Andes Mountains where historic volcanoes once erupted to create 14,000-foot peaks. The notion of developing a spa called, El Tatio (the oven) that balances the mind, the body, and the spirit so far from civilization was inspired by the intent of leaving the real world behind, to free their mind, and have an unreal (virtual) mind / body experience by discovering a (real) common resourcre of water in a new way. Science describes how humans are composed of 70% of water.[138] From the initial arrival by car or

A SPA THAT BALANCES THE MIND, THE BODY, AND THE SPIRIT

helicopter the spa immediately leaves technology and the social constructs of humanity behind by initiating a variety of diets, exercise, and curated journeys on a pathway designed to blend architectural objects into the rocky landscape. The idea of minimal intrusions was meant to highlight the geysers, hot springs, and jet streams that are all carefully calculated by temperatures to reflect different experiences of the water. A cave, with a geothermal source of recirculating water, solar collectors, and Clivus Multrum toilets create isolated living areas with an ice bath meant to express the solid state of water. The spa, in this extraordinary landscape of rocks and peaks, geysers shooting as much as 150 feet in the air, and natural hot springs, is conceived as a once in a lifetime opportunity to meld science and nature into a unique spiritual environment. As technology evaporates into the experience of the natural landscape, the spa will create a greater understanding of how we fit into the larger scheme of the universe—the real Truth.

WATER CYCLE
100% RECIRCULATION

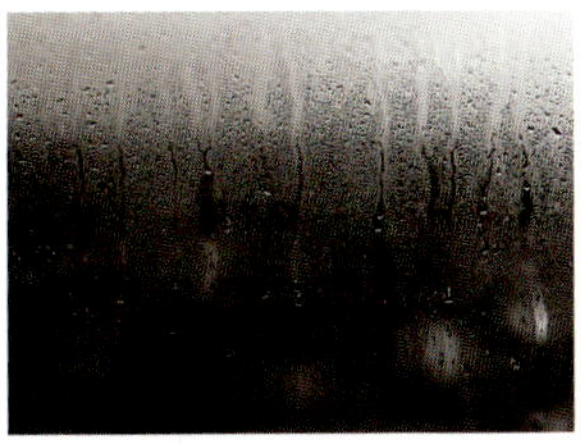

SUBLIMATION	**CONDENSATION**	**VAPORIZATION**
→	→	→
SOLID (32°F)	LIQUID (32-212°F)	GAS (212°F)

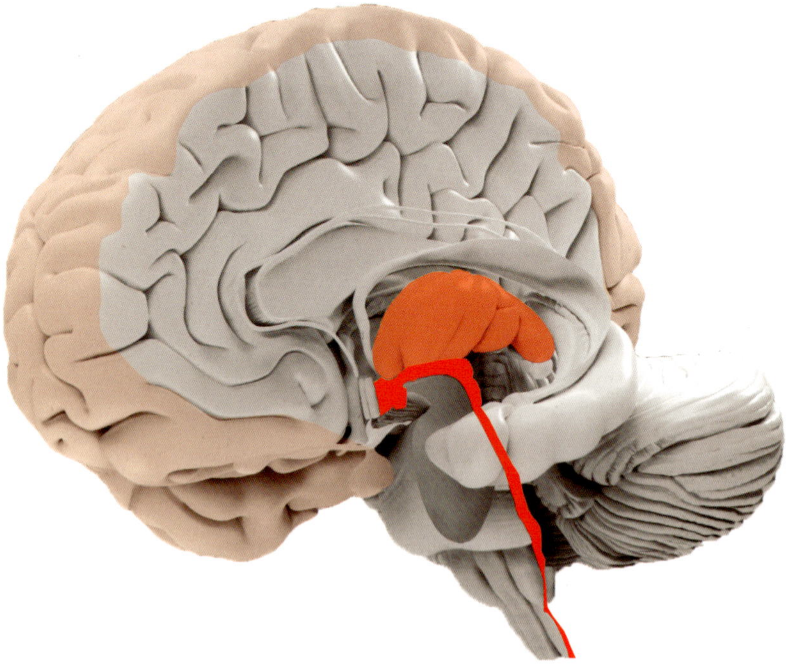

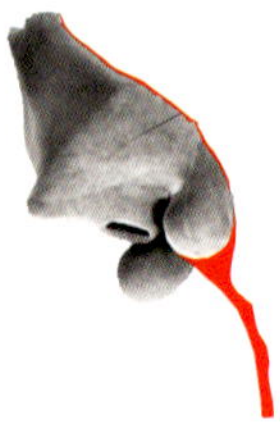

THERMOSTAT
HYPOTHALAMUS

Links the nervous system and endocrine system; it regulates body temperature physically and increases emotional stress.

SENSES
THALAMUS

Central station of the senses filters sensory information in real time before being processed in the central cortex connected to hippocampus.

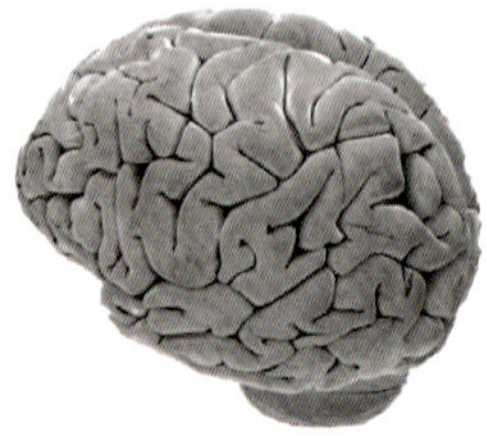

MEMORY
CEREBRAL CORTEX

Site of neural integration in the central nervous system, activating perception, thought, and consciousness, and storing the memory of temperature changes.

The water cycle within the crust of the earth is 100% recirculated from clouds, to the earth, to evaporating back into the sky. The science of water allows the activities of the spa to regulate different parts of the brain, which then activates the senses that cause the body to act like a thermostat, and send signals back to the brain that ultimately store the memory of the experience of water.

POSITIVE

ICE	WATER	GAS
INCREASES ADRENALINE	HYDRATE AND MOISTURIZE THE BODY	OPEN UP PORES TO RELEASE TOXINS
DECREASES SWELLING IN MUSCLES AND JOINTS	RELAX MUSCLES BY EXPANDING BLOOD VESSELS	RELEASING TOXINS HELPS REDUCE DEPRESSION & INCREASE CONCENTRATION
IMPROVES RECOVERY	DECREASES PHYSICAL AND MENTAL TENSION	INCREASES FLUID PRODUCTION

NEGATIVE

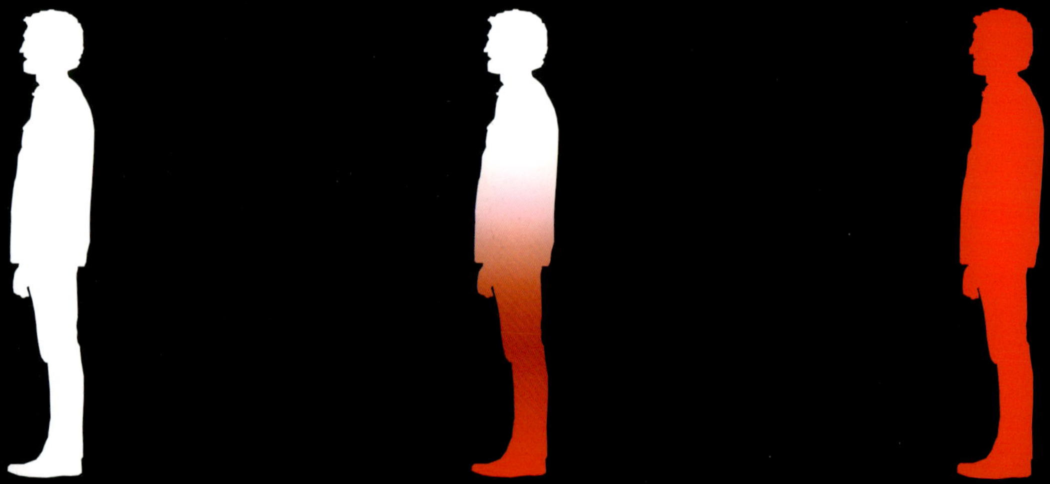

DYSFUNCTION

84-77 ºF CONSCIENCE LOSS AND TEMPORARY STOP OF BLOOD FLOW

104-106 ºF DYSFUNCTION OF CIRCULATORY AND NERVOUS SYSTEM

NORMAL

102-97 ºF NORMAL BODY FUNCTIONS INCREASE

DEATH

111-113 ºF INJURED INTRACELLULAR LOSS OF FUNCTIONS: DEATH

77-68 ºF TOTAL LOSS OF CONSCIOUS FUNCTIONS: DEATH

Humans living within the fragility of the earth's atmosphere have adapted to a variety of weather conditions. The body, which is sensitive to air temperature changes, is no different when experiencing the three states of water. Each phase of ice, water, and gas produce different effects that stimulate the body, mind, and spirit in therapeutic ways.

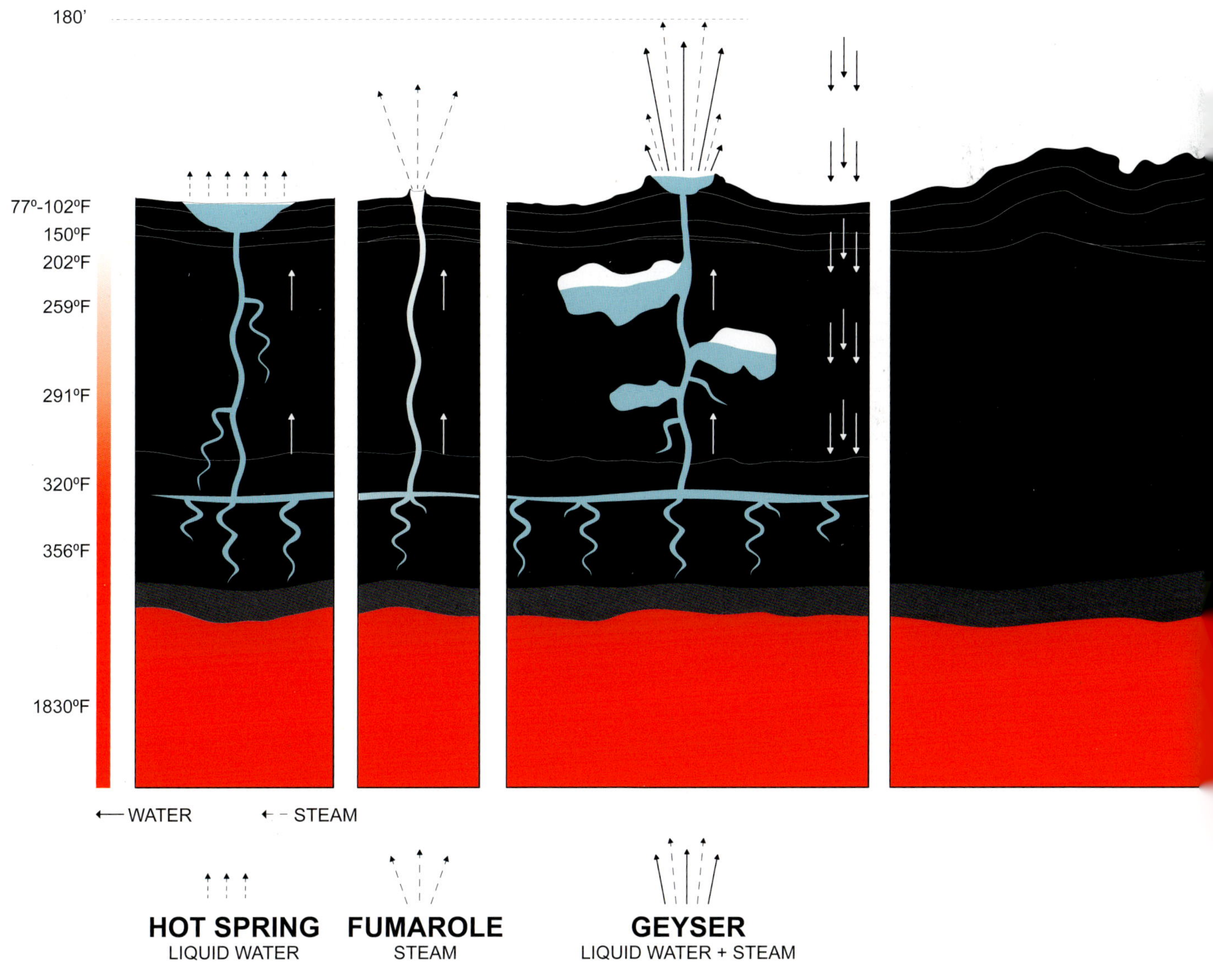

TRUTH | SCIENCE

GEOTHERMIC GEYSER FIELD: EL TATIO (OVEN)

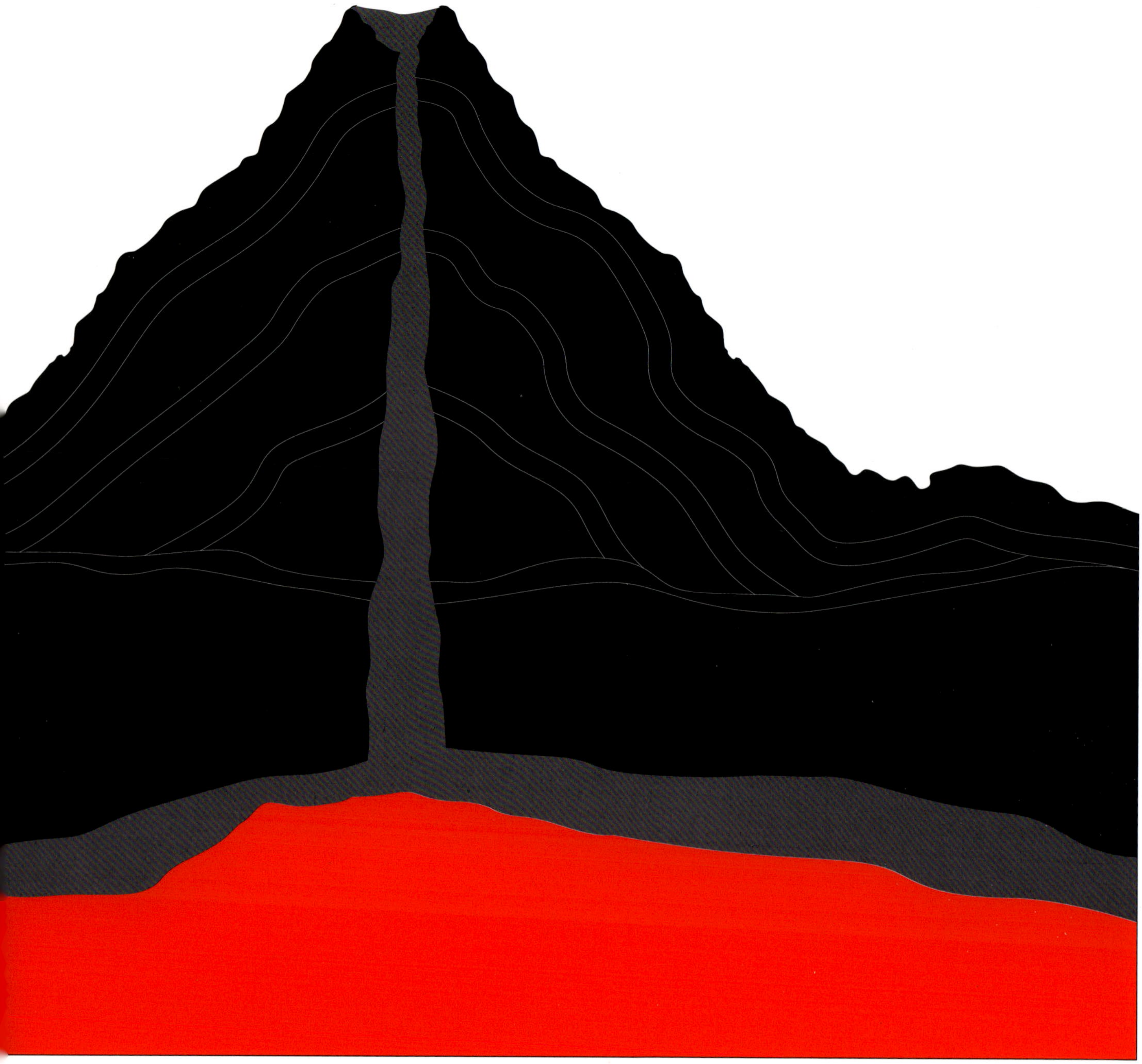

DORMANT VOLCANO
WITH NO HISTORICAL ERUPTIONS

El Tatio (the oven) received its name for the dormant volcano that has produced a geothermic geyser field with a spectacular array of geysers and hot springs. The guests of the spa can experience a curated journey through a natural path of a rocky landscape with private hot springs.

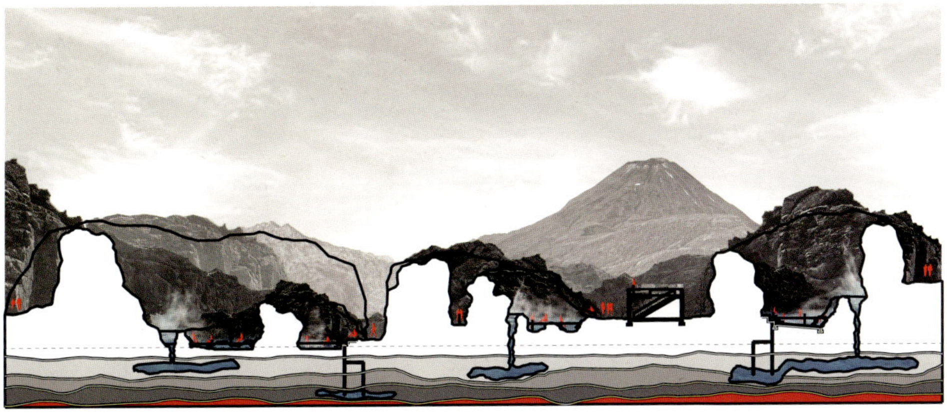

GEYSER POOL COOL

SPRING WATER THERAPY

RETO	ENFRIO	GELU	CRYSTAL

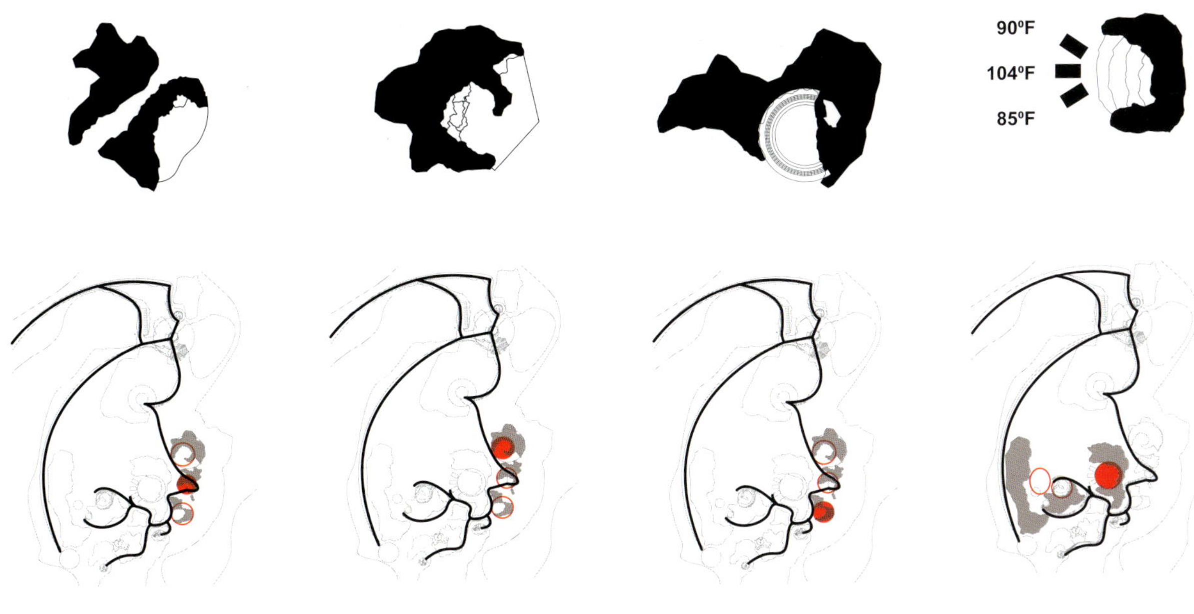

	RETO	ENFRIO	GELU	CRYSTAL
TEMP.	50ºF	70ºF	80ºF	100ºF
HEIGHT	5’	10’	15’	15’

TRUTH | SCIENCE

GEYSER ARRAY: TEMPERATURE + HEIGHT + FORM

STEAM THERAPY

DELMAR	**NIDOR**	**AURA**	**FUGA**
	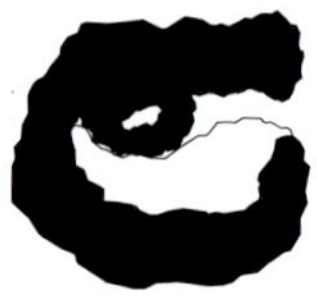		
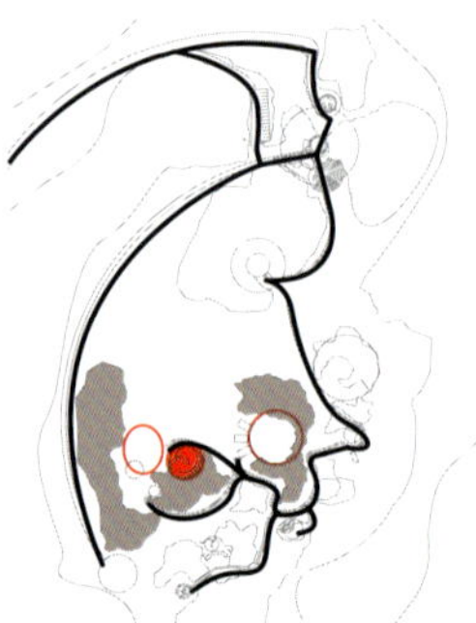	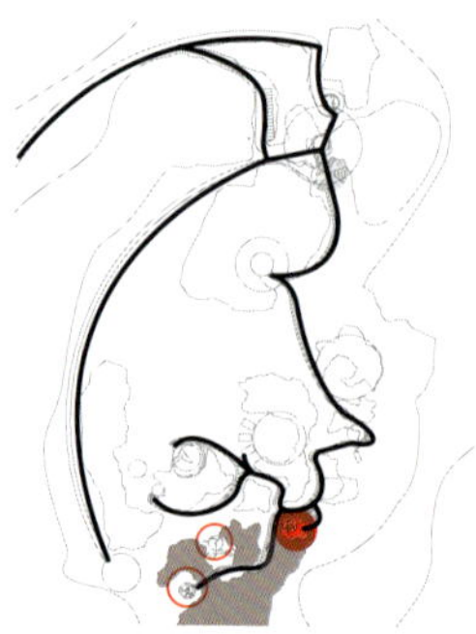	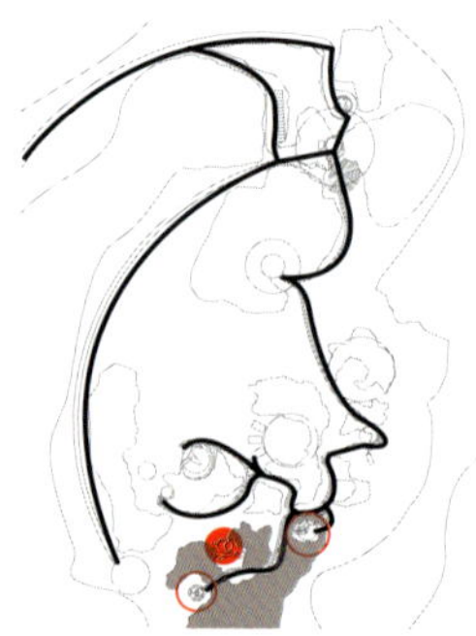	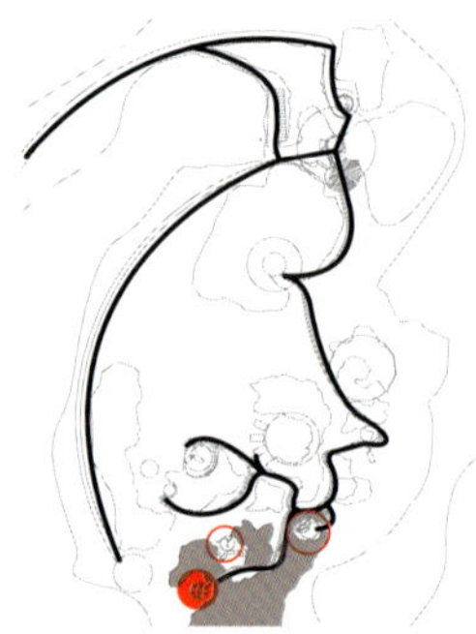
104ºF	**100ºF**	**110ºF**	**120ºF**
5'	**15'**	**30'**	**25'**

By researching each natural outgrowth, the temperature of the hot springs, and documenting the heights of the geysers, the experience through the jagged rock formations and mountains becomes a curated spiritual journey through nature's splendors. The different therapies include a geyser cool pool, spring water, and steam therapy.

SPRINGS | **WATER** THERAPY

LIQUID

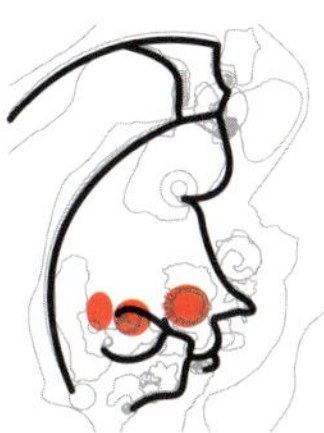

TRUTH | SCIENCE

GEYSER THERAPY: PERCEPTUAL EXPERIENCE

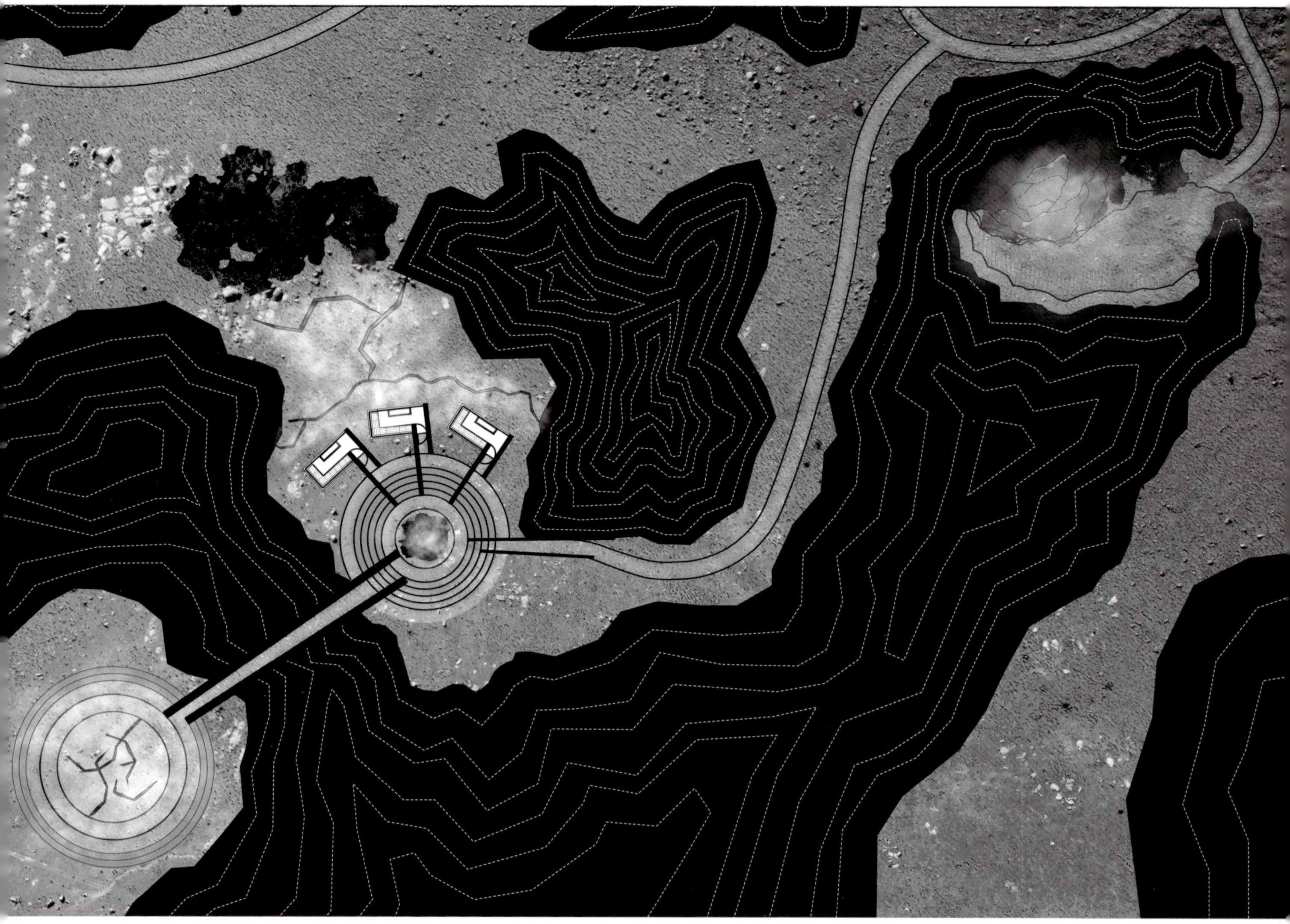

STEAM THERAPY

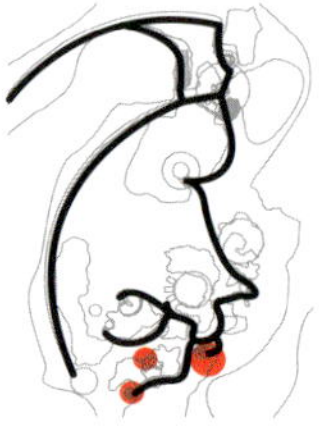

The experience of the different water therapies creates a change in the individual's perception of the exterior landscape and the interior mental landscape of those at the spa. The privacy of the water fields, the steam surrounding the water appears to engulf those in the springs and seeing the geysers reach to the sky.

SPRING WATER THERAPY

FUGA

180^{0}F | 150'

DELMAR

1040F | 5'

TRUTH | SCIENCE

THE OVEN GEYSER: STIMULATES SENSES

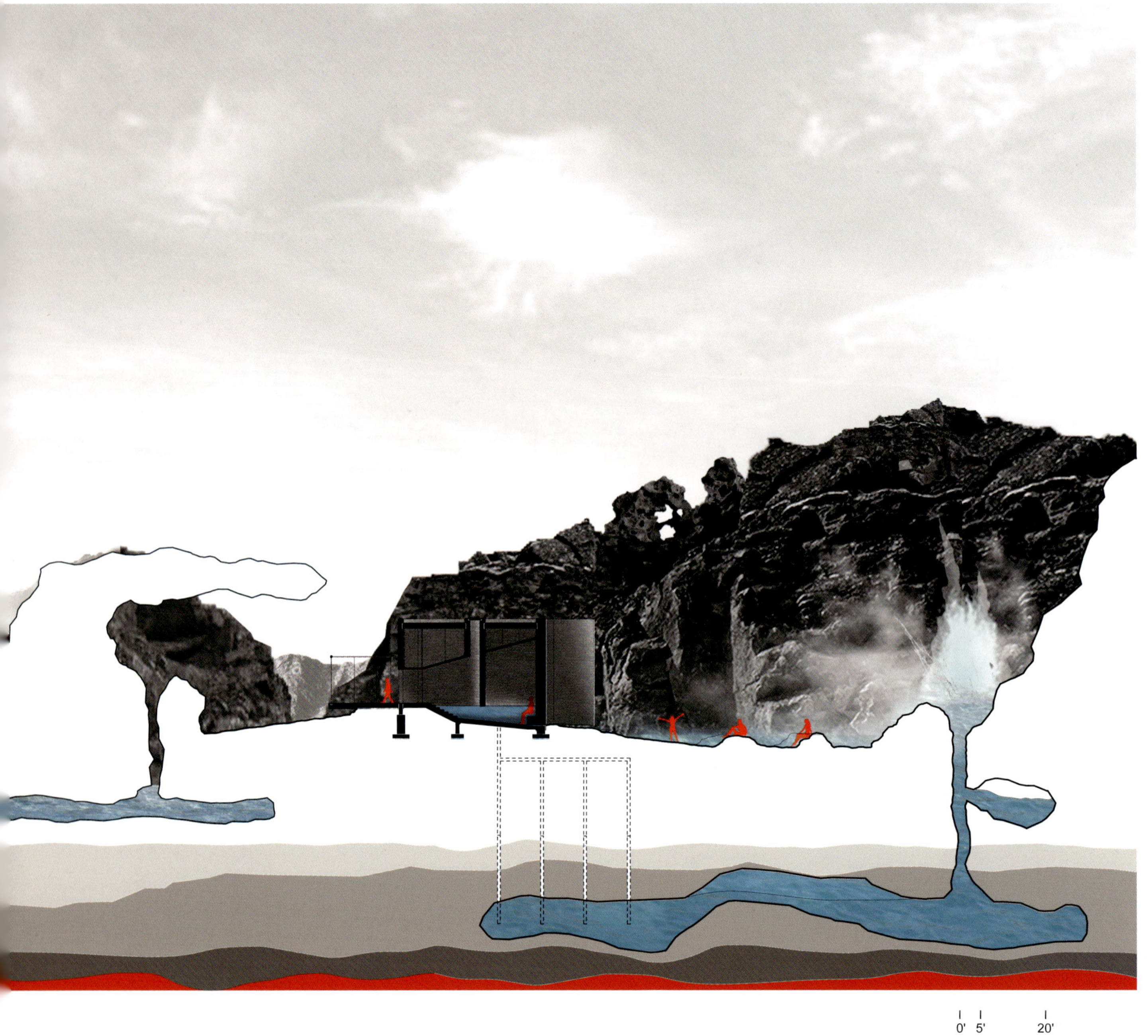

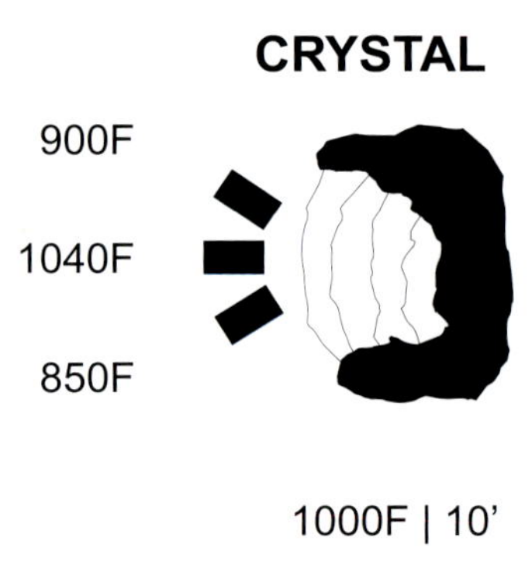

El Tatio's oven geysers reach 150 feet in the air. These spectacular demonstrations of the dormant volcano's underground springs are some of the most extraordinary natural phenomena in the world. The stimulation of the senses from the ambient air temperature to the submersion into the liquid water soothes the soul and captures the spirit of nature.

In the past, many cities that were sensitive to their environments were designed by organizing the city around the real movements of the sun, the moon, the wind, etc.[142] The contemporary city, with light fixtures that change the perception of day and night and air-conditioning that seals buildings from being affected by their surroundings, is based upon the increasing use of technology to control our environment. This inevitably leads to humans becoming dependent upon the virtual world to sustain a standard of living that tries to conquer nature. The project Light Shadows renews the premise of orienting the architecture of the galleries to the sun and the moon as form givers in a garden by the Mediterranean Sea in Marseilles, France. The solstice and the equinox create an axis about which the galleries are aligned with the city and the sea through extensions of piers on the water. The gallery designs began with a cubical shape that casts shadows—projected shapes from the sun and the moon—

SHADOW IS THE TRUTH

on prescribed days and times in three dimensions. The initial cube that was used for the design of the gallery, once removed, becomes merely a virtual afterimage. The only remaining forms left from the shadows of the virtual cube cast permanent buildings which are direct projections of the sun and moon shadows. Concrete walls capture the sun shadows, while open wire metal scaffolding reflects the passage of the moon at night. Festivals, openings, and park events are marked by significant moments of alignment of the sun and moon at particular times, like the historic sun dial being used to mark the passage of the day. Vincent van Gogh and Antonio Gaudi spoke extensively about the unique quality of the light along the coastal areas of France and Spain.[143] The galleries located in a park setting create a sense of mystery and discovery that heightens the experience of the light on the Mediterranean Sea.

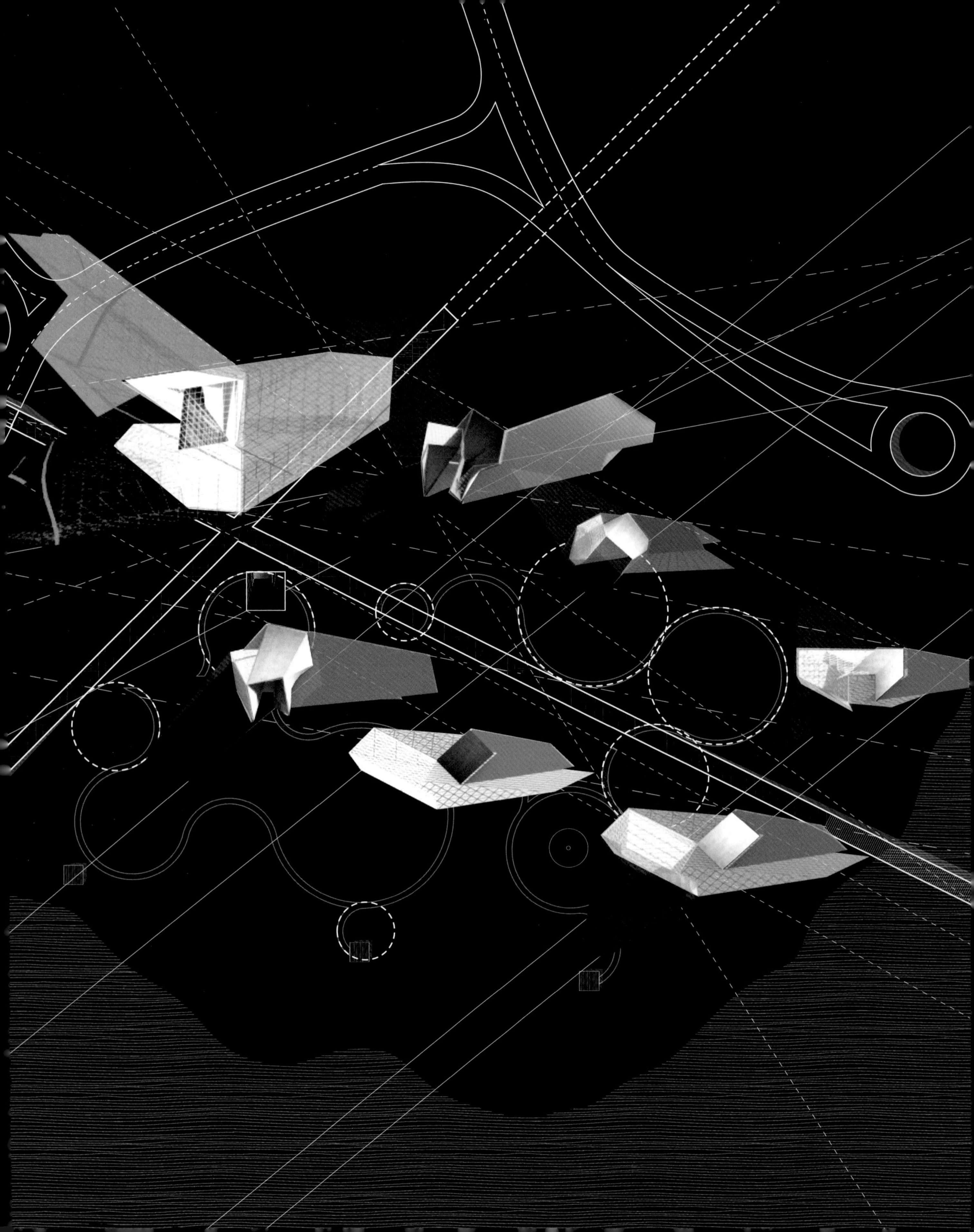

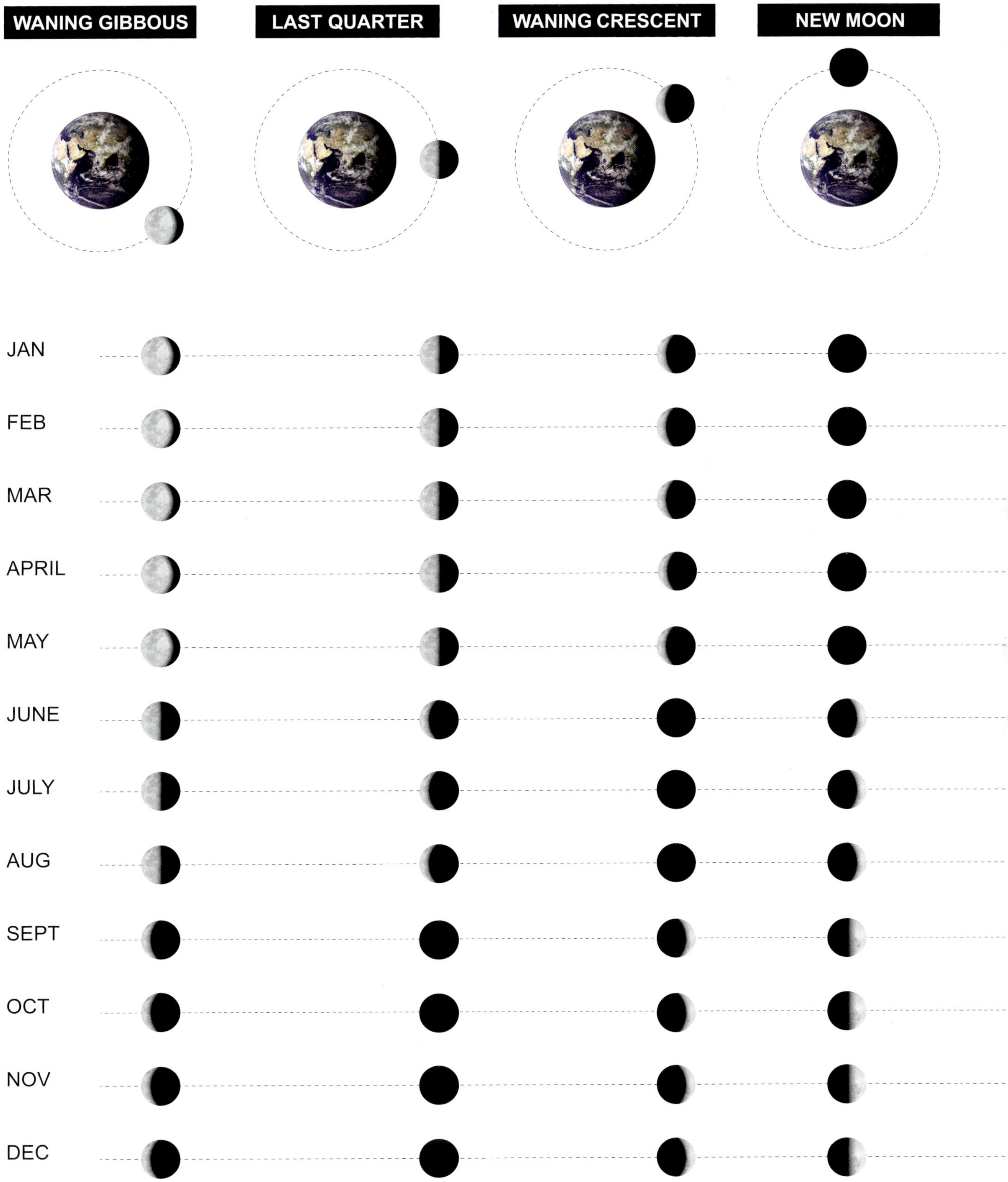

TRUTH | SHADOW

MOON SHADOW: LUNAR CYCLE

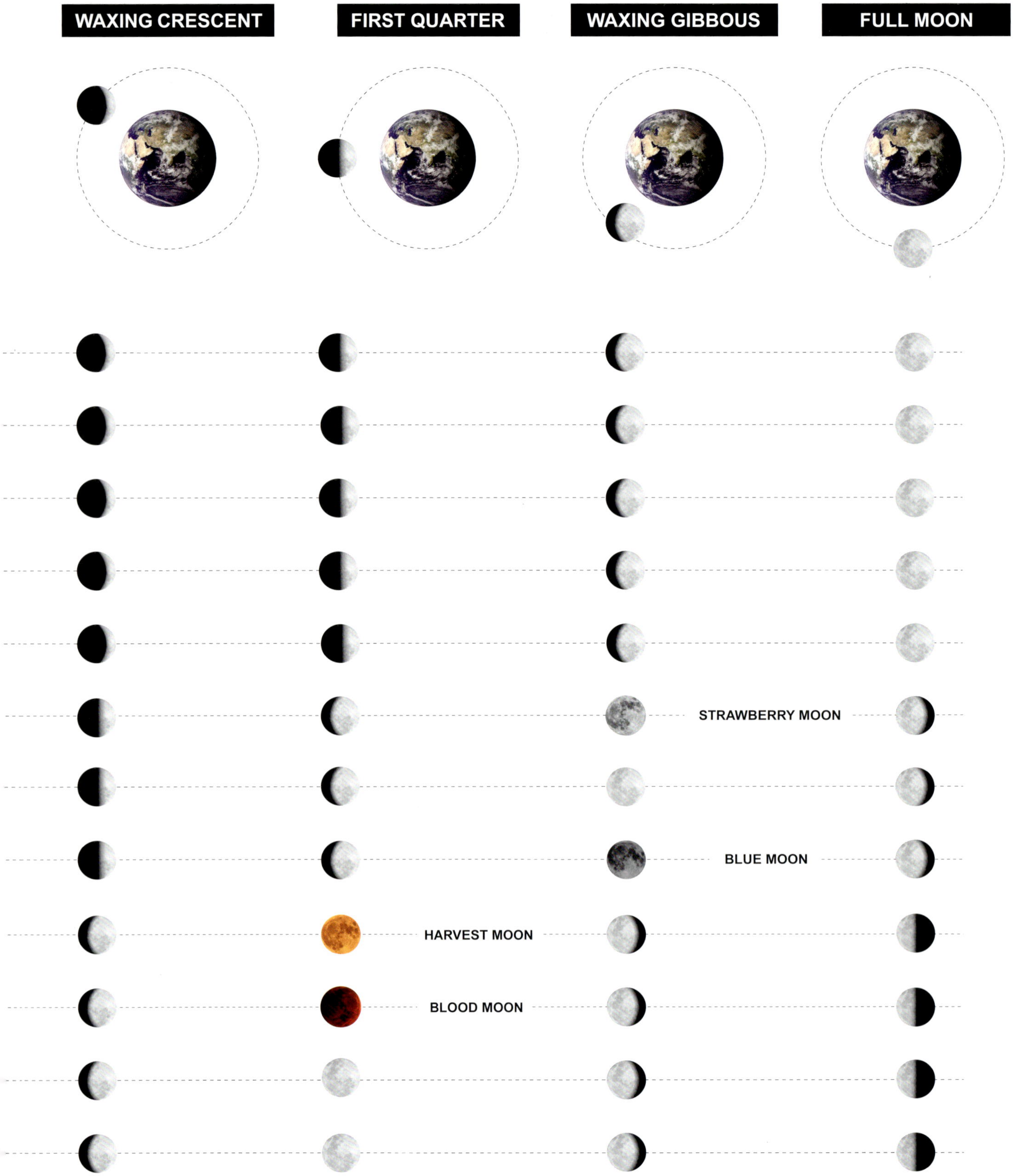

By documenting the phases of the moon throughout the year,[144] it is possible to design an architecture that responds to the variety of light conditions cast by the moon. The sensitivity to the subtle and dramatic contrast of the light lies at the heart of Light Shadows. By understanding the movements of the moon, remarkable relationships can be established between the built form and the moon.

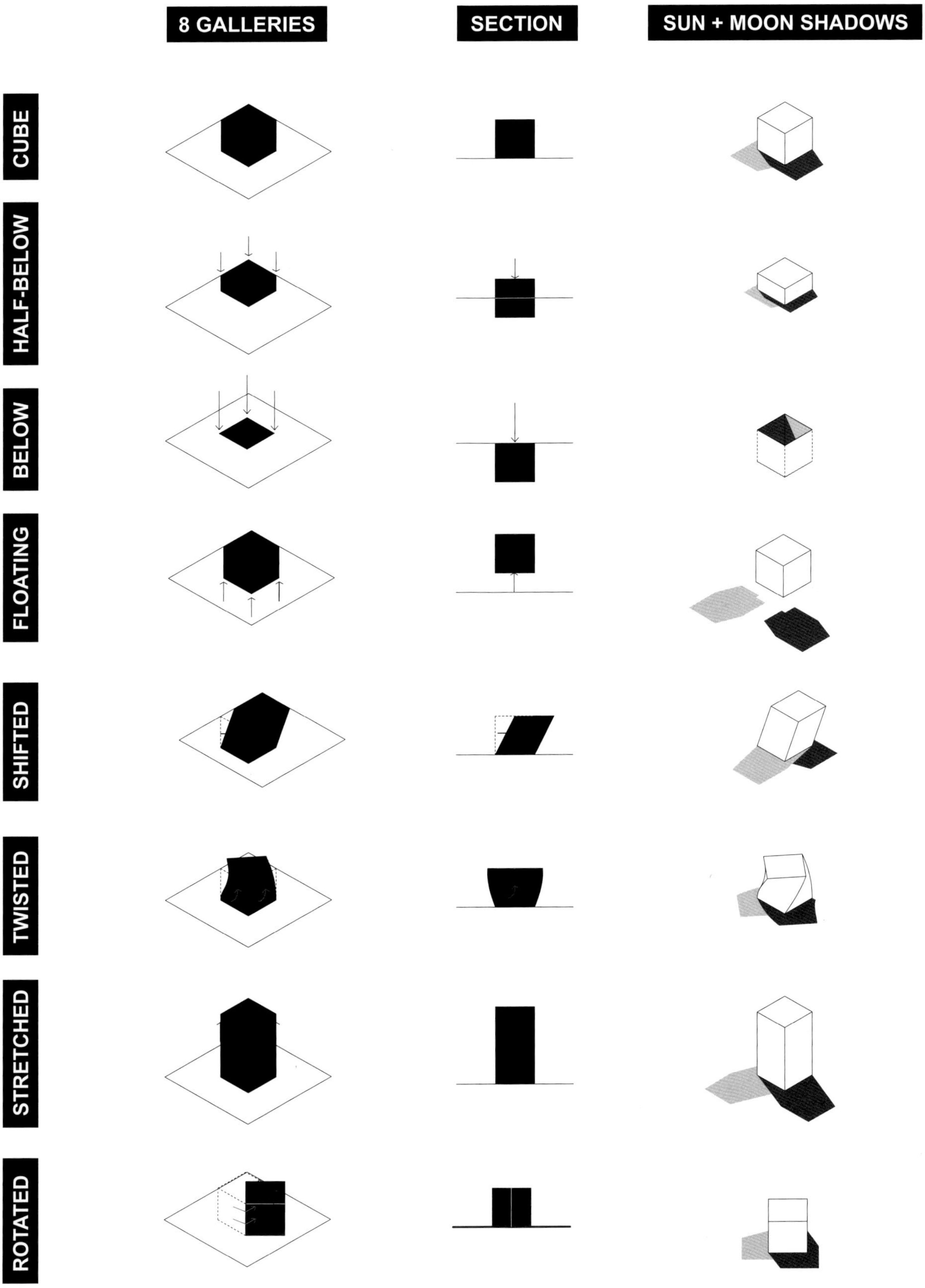

+ TRUTH | SHADOW

MOON + SUN: MERGING SHADOWS

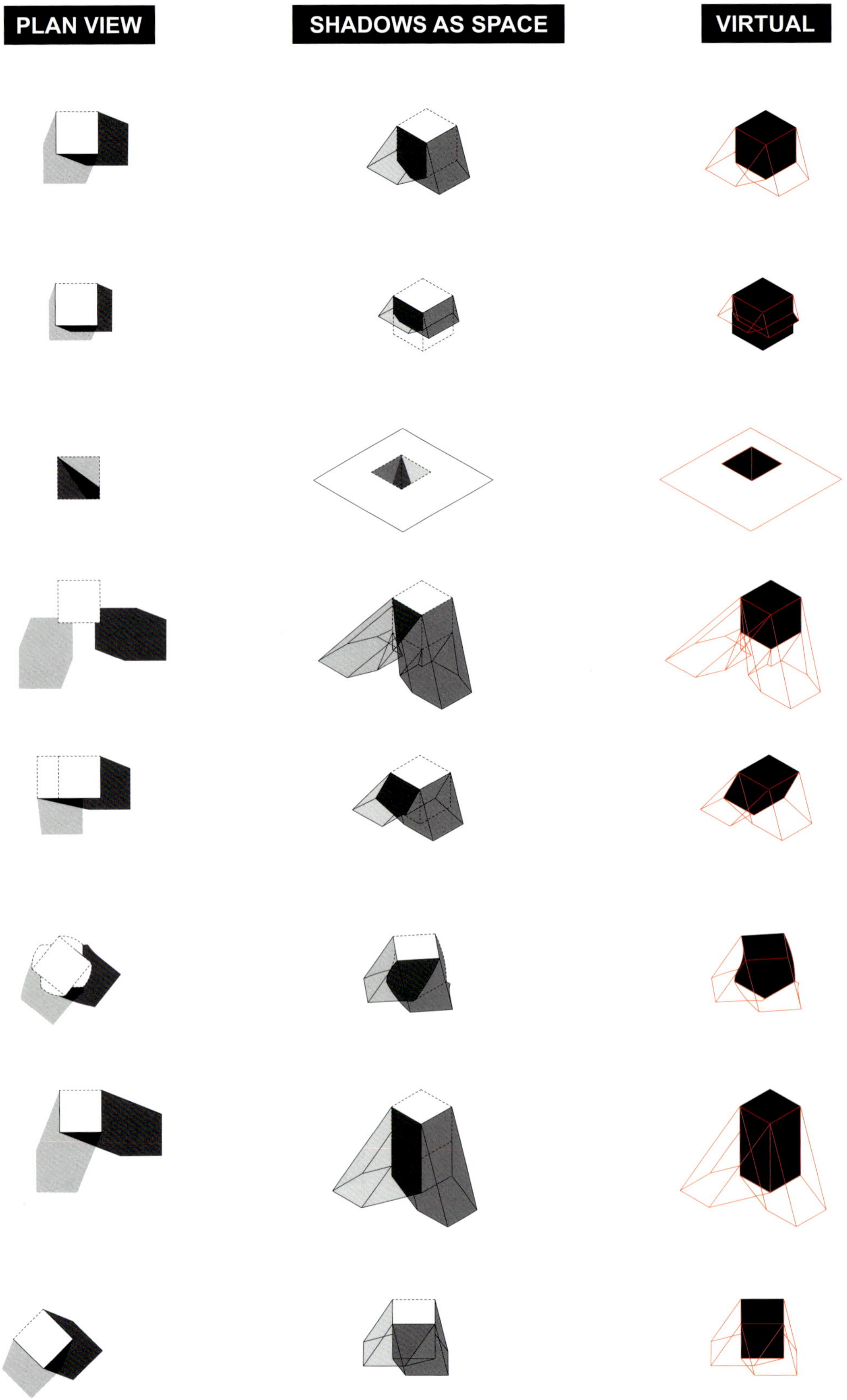

Through analysis of the remaining three-dimensional shadow forms, a matrix can be made that correlates the architectural shapes with the axis of the equinox, the solstice, and the blood moon. The unique tectonics created by these shapes marks the beginning of a new investigation into form giving that vacillates INbetween the real and the virtual.

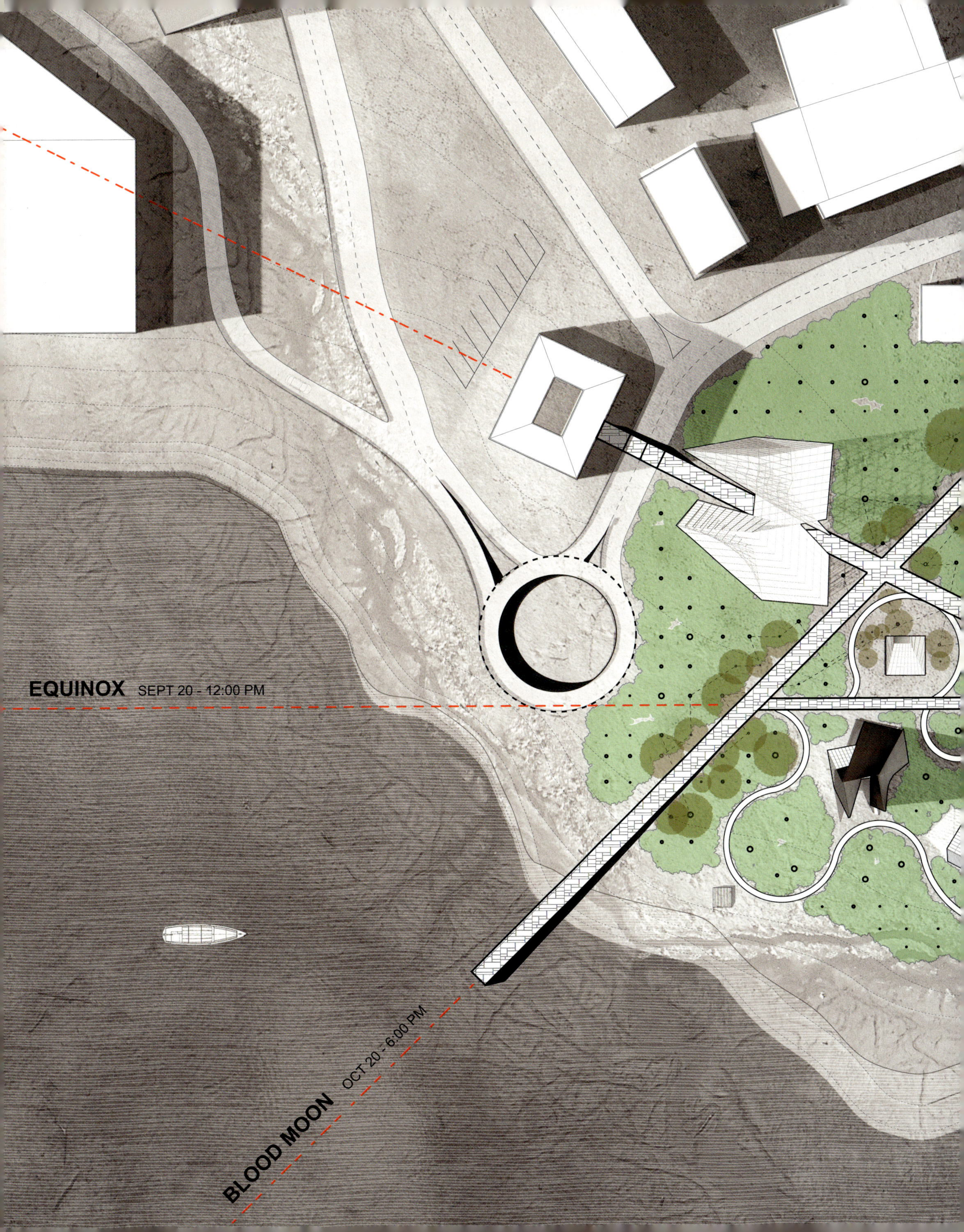
EQUINOX SEPT 20 - 12:00 PM
BLOOD MOON OCT 20 - 6:00 PM

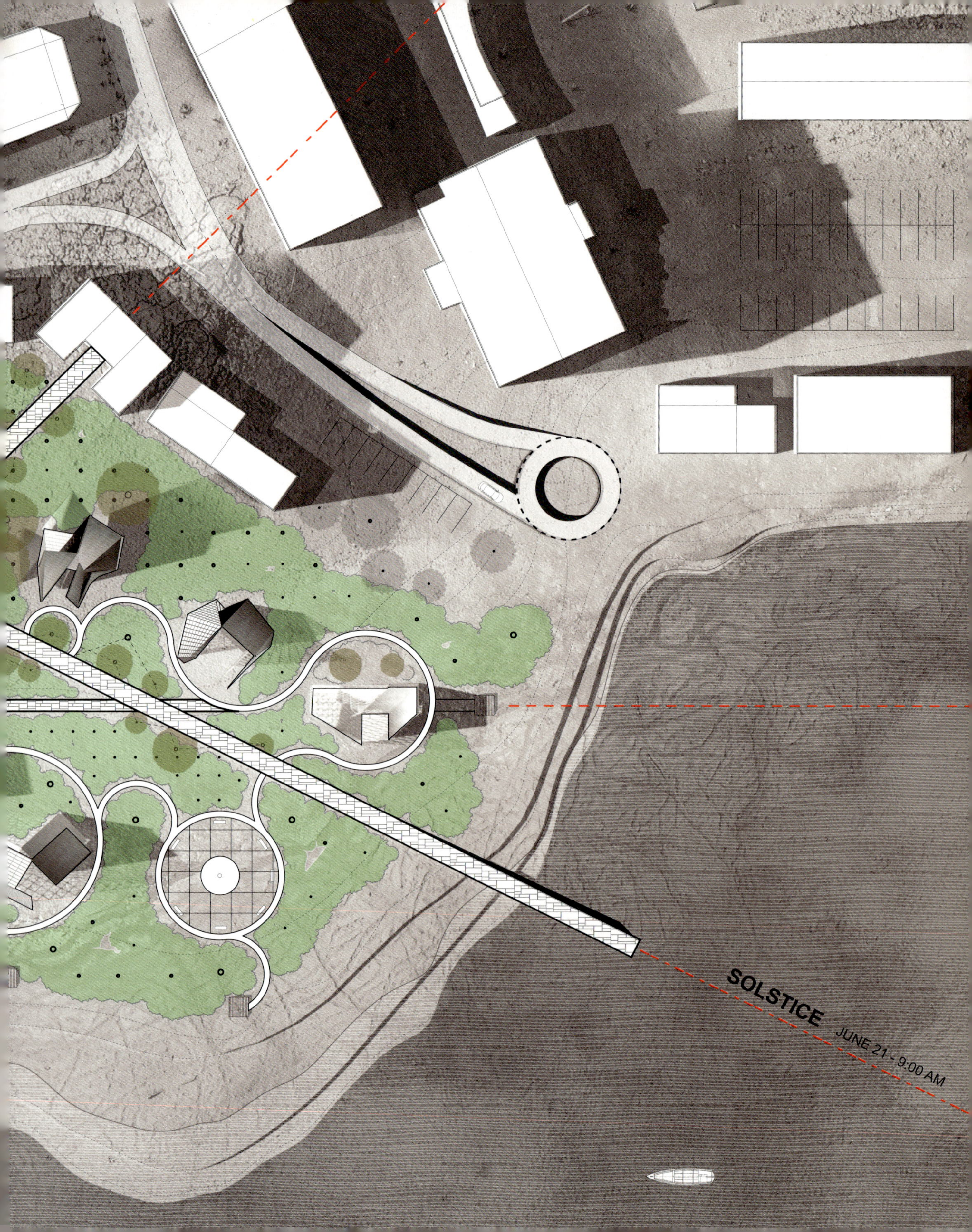
SOLSTICE
JUNE 21 - 9:00 AM

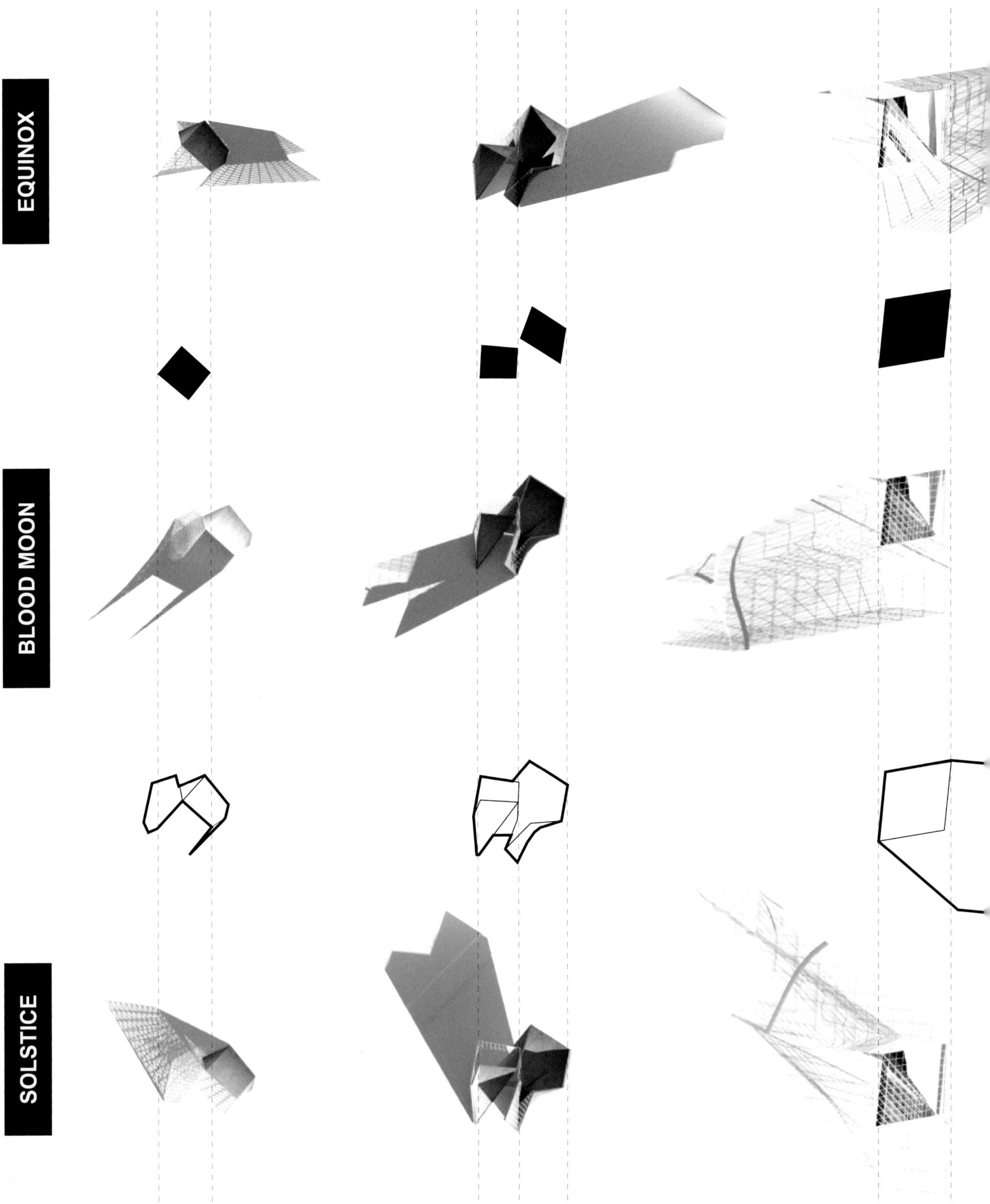

TRUTH | SHADOW

SHADOWS: SOLSTICE | BLOOD MOON | EQUINOX

The extraordinary shapes created by the shadow forms and their axis to the sea establish a new datum in which to understand the angles created by the light cast from the sun and the moon. The unique gallery forms are organized about an axis that terminates in the piers extending into the Mediterranean Sea.

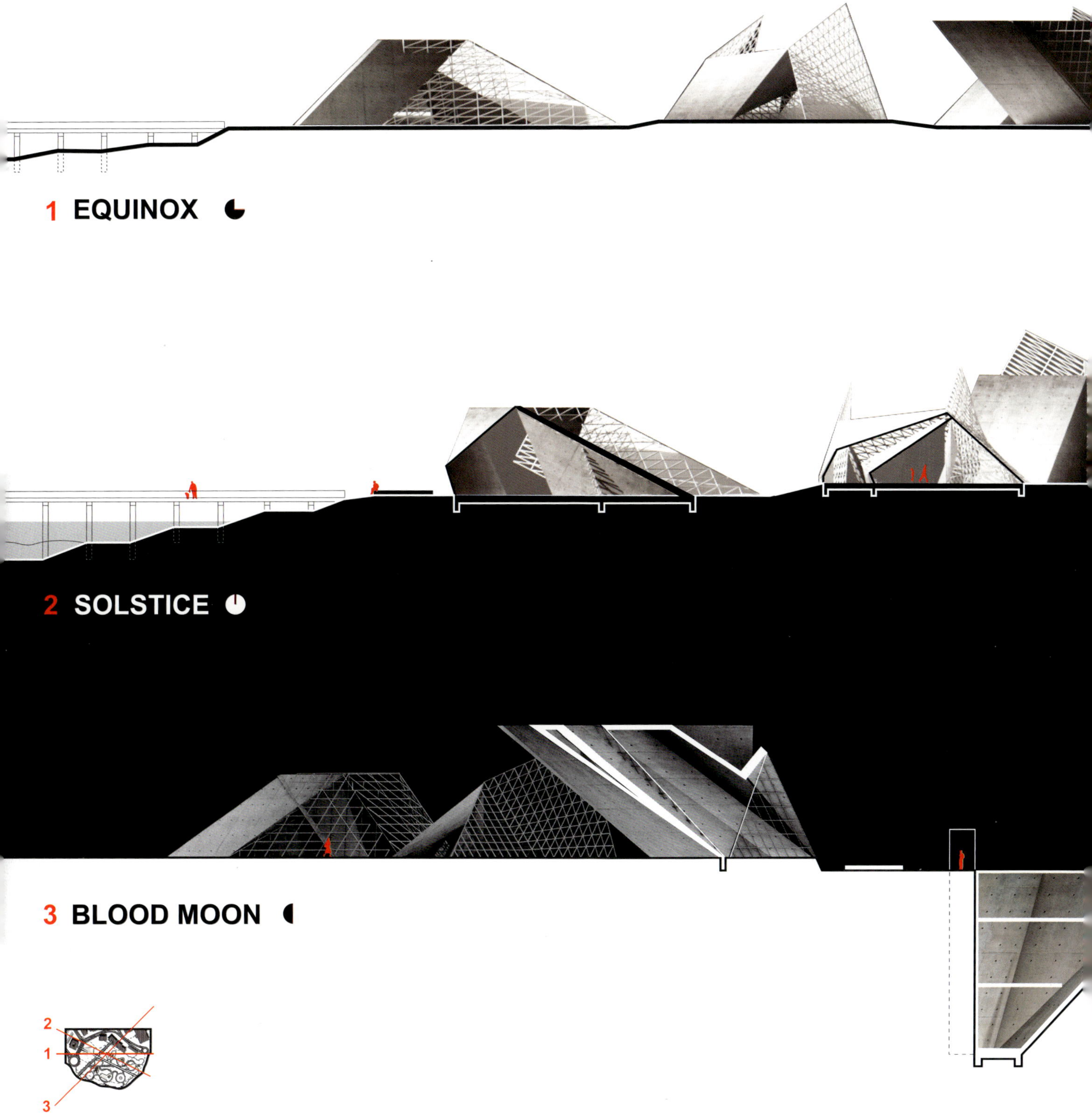

TRUTH | SHADOW

SHADOWS: SOLSTICE | BLOOD MOON | EQUINOX

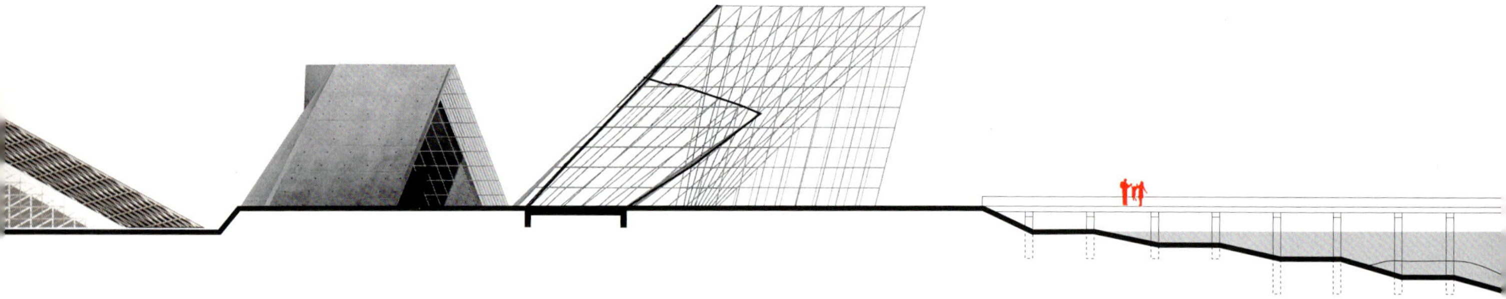

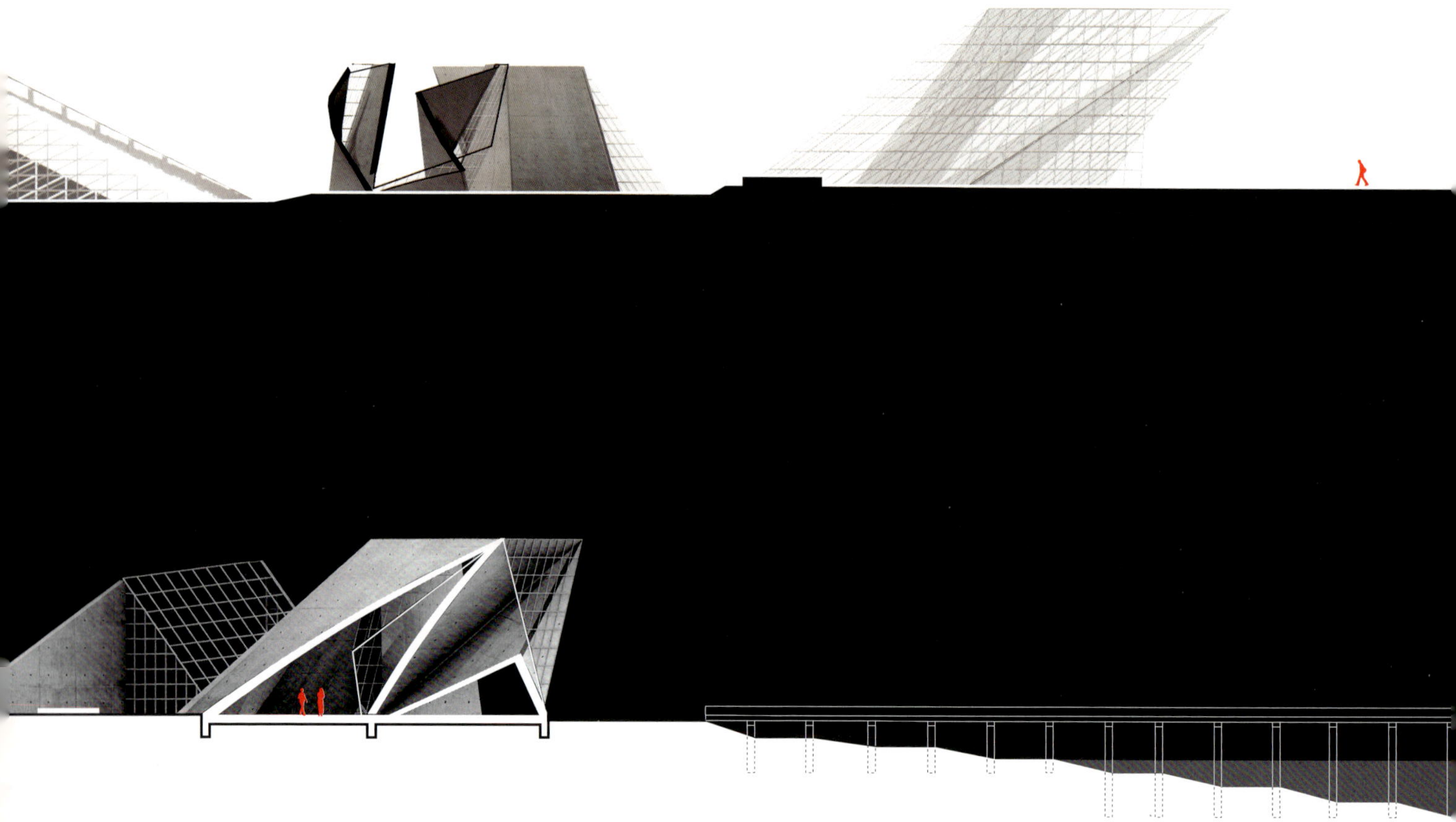

The conversation between the open, metal scaffolding of the moon and the concrete sun structures of the galleries create extraordinary forms that reflect the celestial movement of the sun and moon. On particular days the forms align with the sun and moon shadows that serve to confirm human's existence and the humility needed to understand the universe.

TRUTH | SHADOW

SHADOWS: SOLSTICE | BLOOD MOON | EQUINOX

When Mount St. Helens erupted in 1980, the adjacent communities had to evacuate when being faced with such a catastrophic event.[145] The volcano, by its very nature, is unpredictable and has the inherent characteristic of creating disorder. The scenic beauty of the forest and the mountain with snow peaked caps creates the idyllic sense of order when the volcano is not active. This tension between disorder and order becomes the basis for creating an architectural journey through a museum that overlooks the volcano from a cliff on a nearby ridge. The bent, linear path of the exhibit marks the uncertainty of what will come along the journey. The outdoor gardens of charred wood from the fire of the lava's descent from the mountain is repurposed into the exhibition for wall surfaces and sculptural trees. The contrast of open and closed spaces heighten the sense of the smells. The variety of textures including the rough and smooth finishes of the concrete and the ash in the sky after

ORDER DISORDER

an eruption all contribute to the changing intensity of the experiences when walking through the exhibition spaces. The forms of the museum are meant to accentuate the emotions of order and disorder by manipulating the horizon and vertical compression of the spaces in a sculptural manner, rather than simply extruding the plan vertically. The twisting forms and the angular shapes mark the crevices of an earthquake, or the fissures in the volcanic eruption at the cone. The poured in place concrete is carefully detailed to have the tie rods filled with stainless steel caps, or the pour lines of the exposed concrete that follow the geological layers of the rock formations in the volcano. The journey ends in a panoramic view of Mount St. Helens cantilevering over the precipice of the site. The triangulated shape of the overlook frames the volcano and returns the visitor to a sense of order with nature at rest, aware that the fury of nature's disorder is known to be just under the surface.

ENTROPY: MOUNT ST. HELENS ERUPTION

NATURE

3000 BCE

STABLE

ORDER

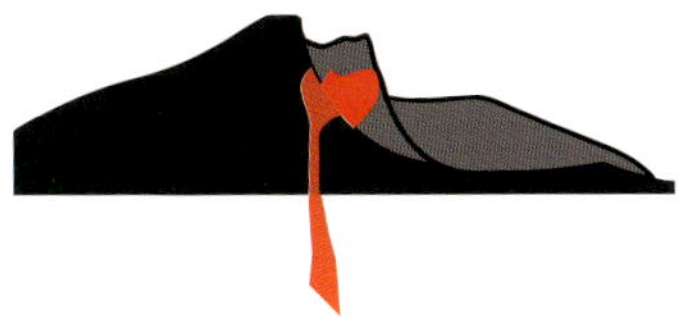

EARTHQUAKES

03 - 16 - 1980

UNEVEN PRESSURE

INBETWEEN

ERU

05 - 18

DEST

DISC

TION

32 AM

TION

DER

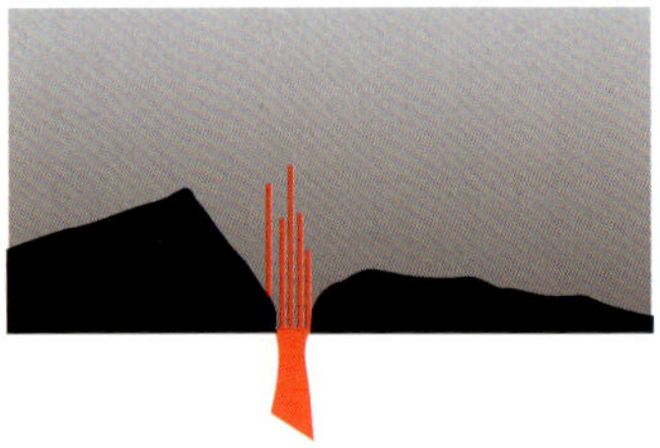

PYROCLASTIC

05 - 18 - 12:00 AM

CHAOS

INBETWEEN

NATURE RETURNS

PRESENT

MEMORY

ORDER

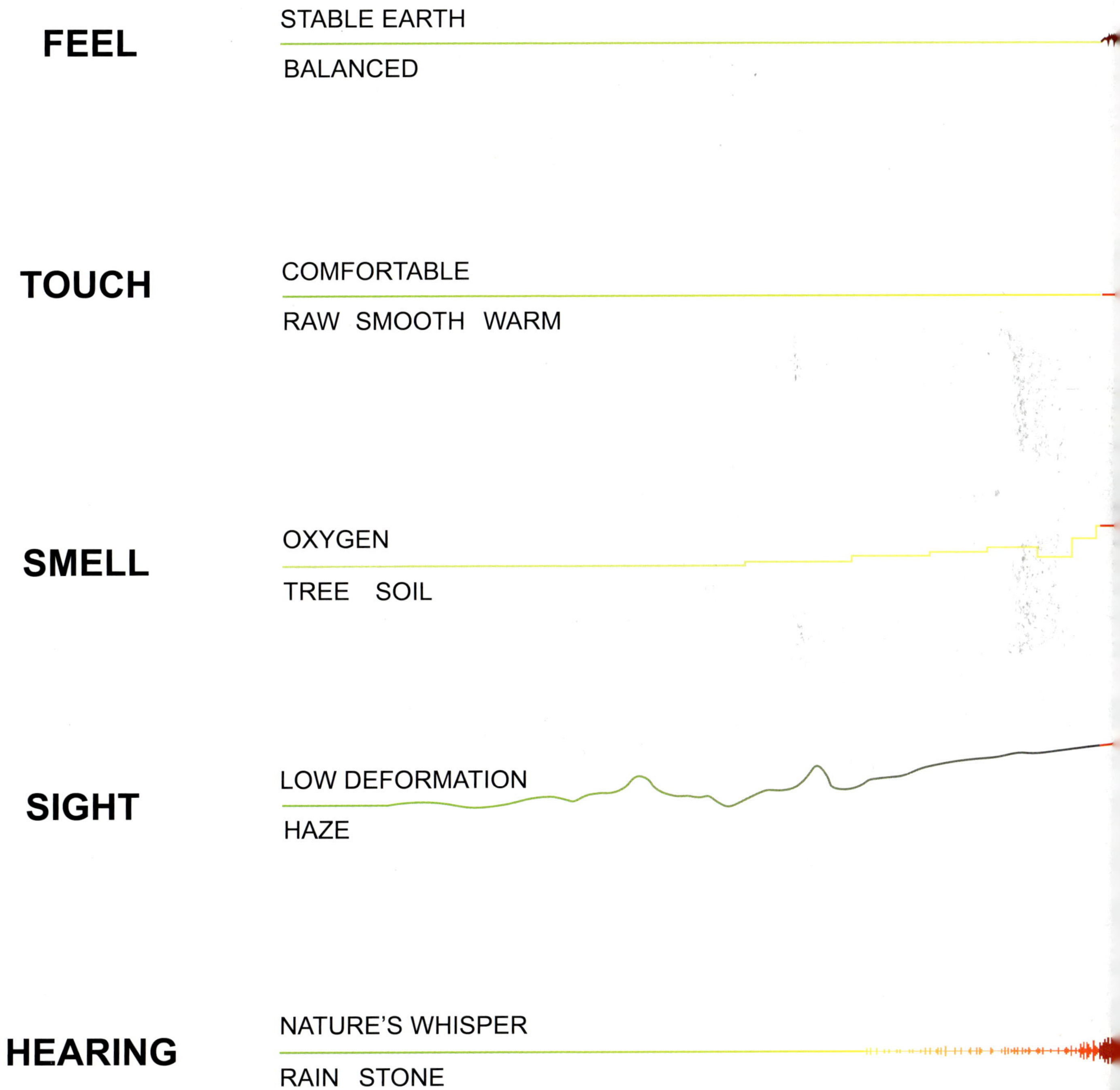

TRUTH | DISORDER

5 SENSES: BIOLOGICAL DISORDER

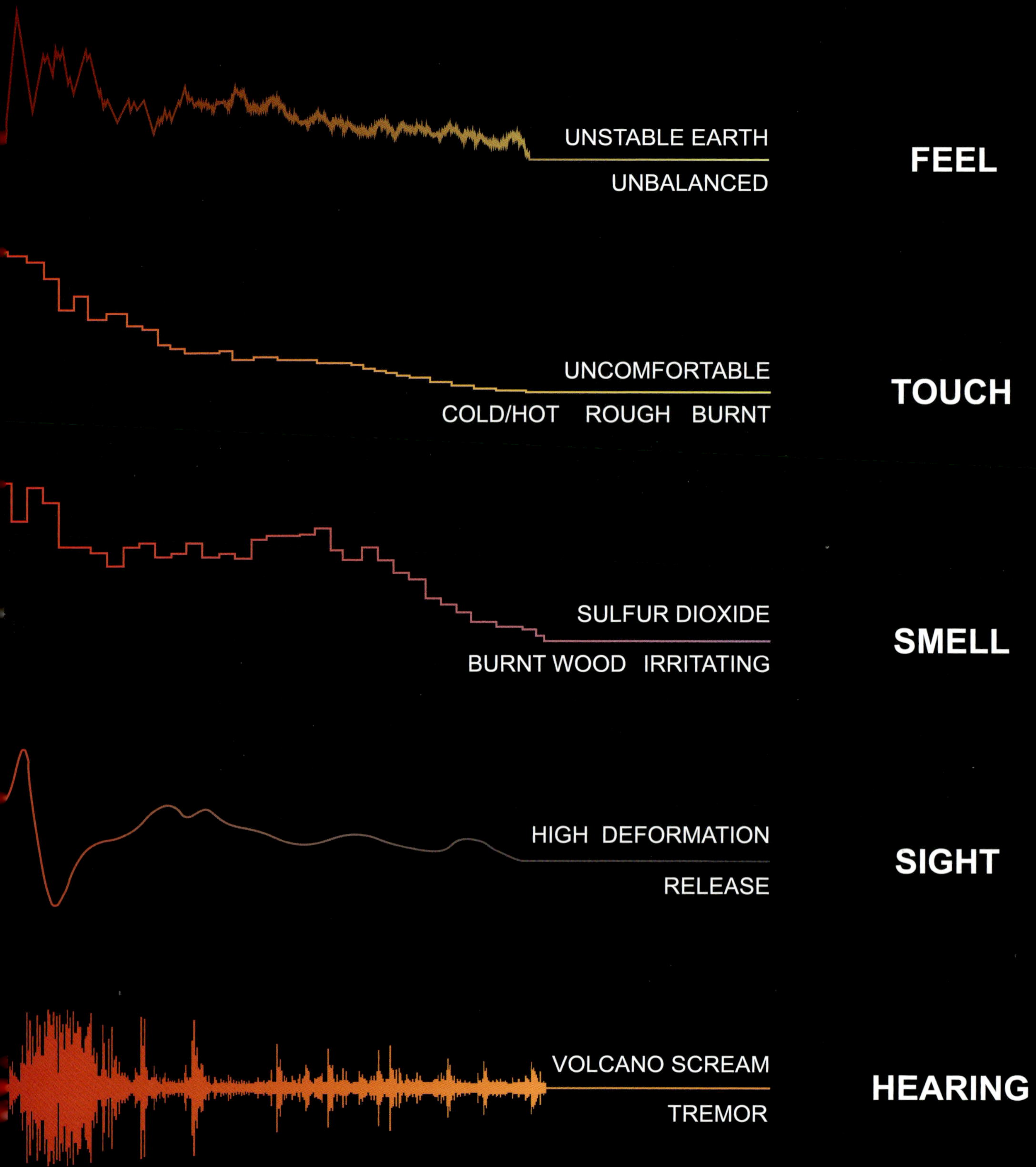
UNSTABLE EARTH
UNBALANCED
FEEL
UNCOMFORTABLE
COLD/HOT ROUGH BURNT
TOUCH
SULFUR DIOXIDE
BURNT WOOD IRRITATING
SMELL
HIGH DEFORMATION
RELEASE
SIGHT
VOLCANO SCREAM
TREMOR
HEARING

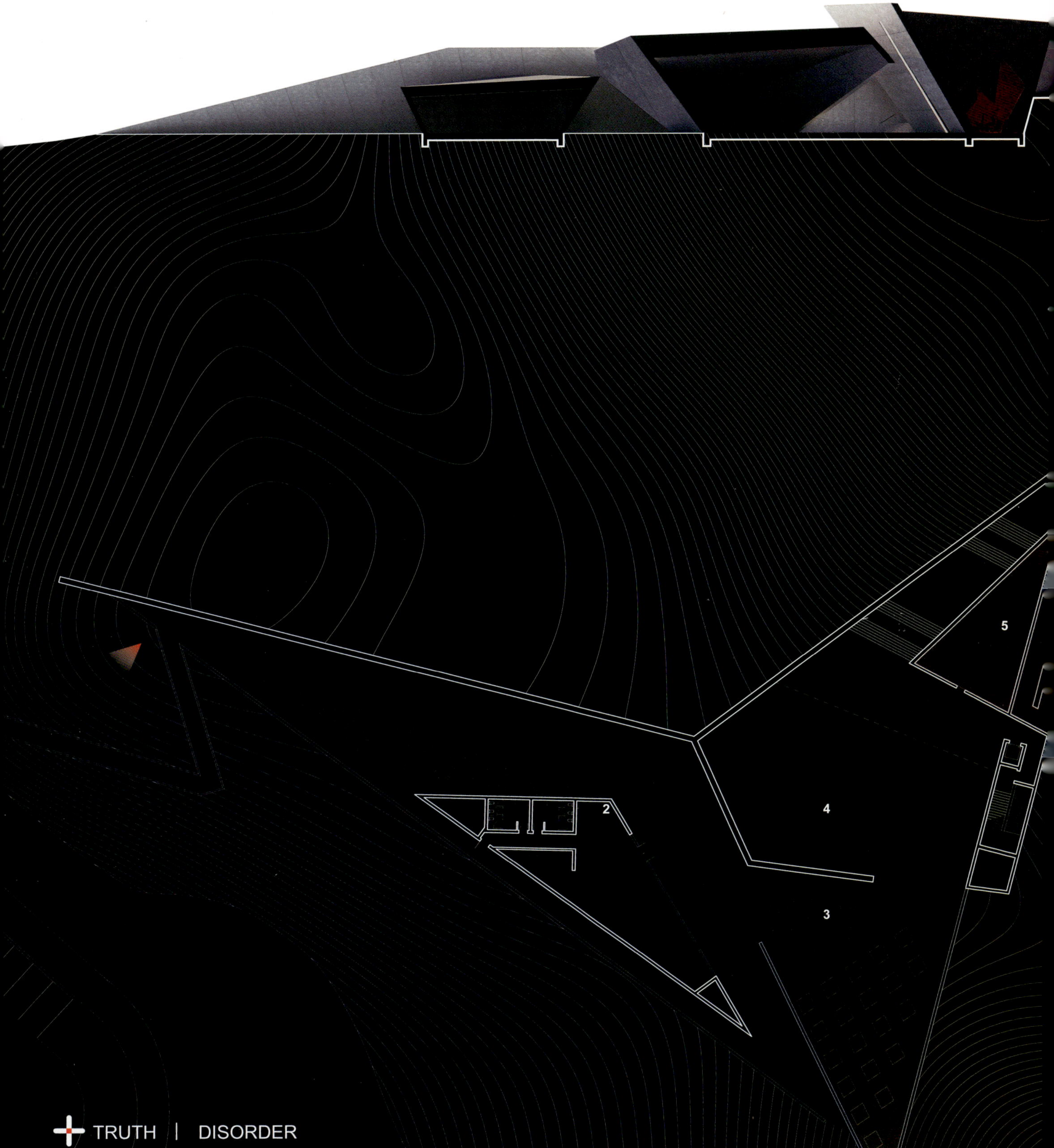

+ TRUTH | DISORDER

CIRCULATE THROUGH ORDER + DISORDER

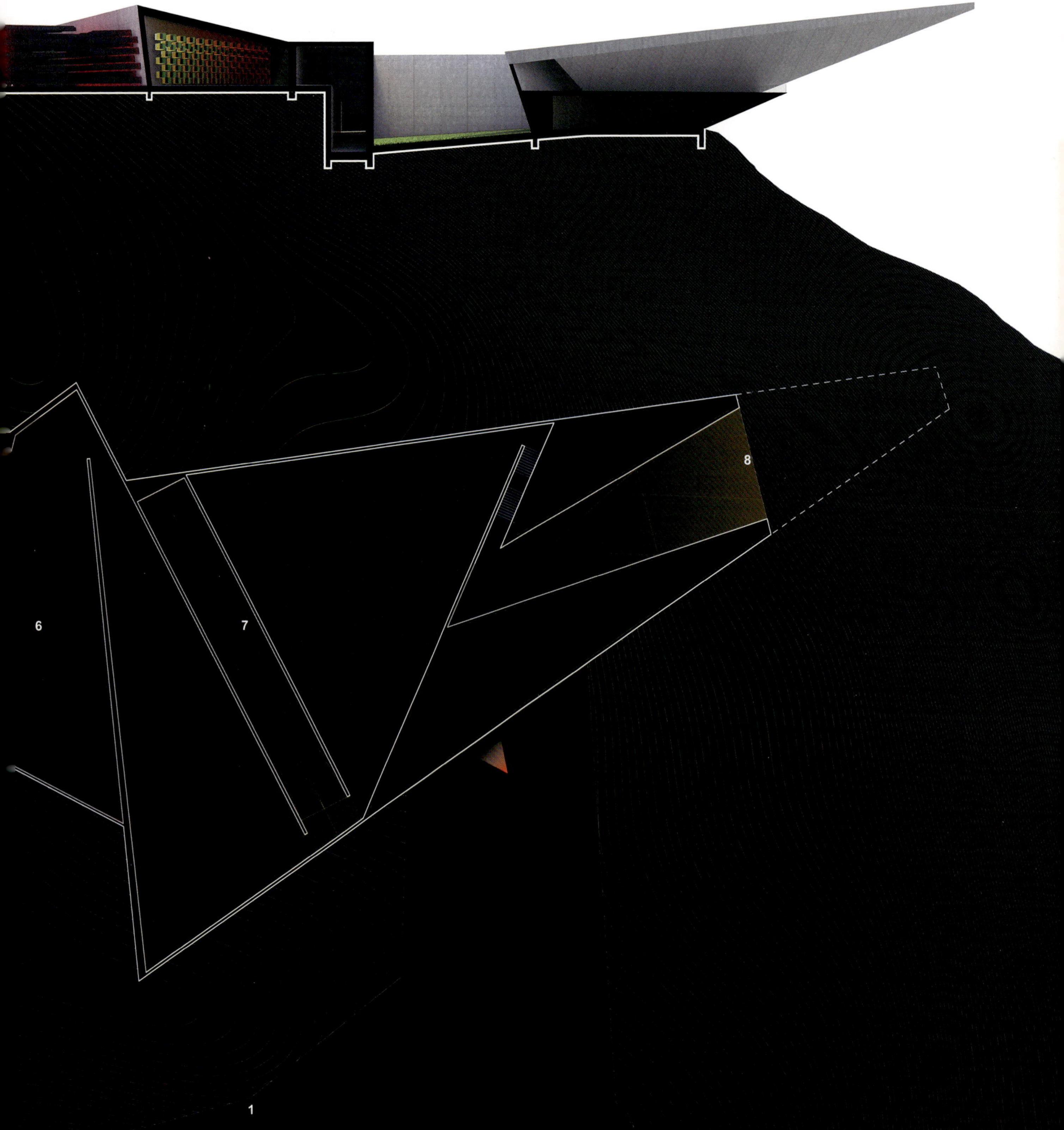

0' 10'

1 NATURE WALKING
2 VISITOR CENTER
3 UNSTABLE GROUND
4 MOMENT OF PEACE
5 ERUPTION
6 DESTRUCTIVE FLOW
7 FRACTURE LIGHT
8 NATURE RETURN

JOURNEY: A SPATIAL EXHIBITION

ENTRY COURT FOREST VOLCAN

+ TRUTH | DISORDER

EXHIBITION: VOLCANIC CHARRED WOOD

The intensity of the journey is most profoundly felt in the two spaces affected by the museum that represent the eruption of the volcano and the aftermath of charred trees that have fallen from the escaping lava from the cone. The burnt wood is used on the walls and as cantilevered fragments illustrating the ash plume that scatters over the earth during the volcanic eruption.

ORGANIC
NO

ARTIFICIAL

Perception is a state of becoming aware, an interpretation or a mental impression of something understood through the senses. The dialogue between the brain and the senses can become a metaphor for a conversation between the real and the virtual. Illusions are products of the mind, but a mirage is a result of the physics of the earth's atmosphere with light refracted from the hot air close to the ground.[146] This project named Mirage is located outside Sedona, Arizona. The series of detached spaces can be used as a home / gallery where each area explores a different aspect of physics and perception as an experiential tool for architecture. Mirage comes from the nineteenth century French verb mirer (to look at).[147] The word morphed into English to mean mirage, or mirror. The six buildings are located on a descending hill in the rolling landscape of the desert that is home to spectacular sunsets, sunrises, dust storms, flash floods, and dramatic daily temperature shifts. Each of these natural

THE TWENTY-FIRST CENTURY PERCEPTION OF FIVE SENSES

characteristics lend themselves to a variety of phenomena that is developed in each of the golden proportion structures of concrete and glass. The buildings, organized on a curated path, develop a rhythm of perceptual changes meant to create illusions and to employ the physics of the material, atmosphere, and temperature changes. What appears to be an illusion is, in fact, a purposely crafted arrangement of buildings where the actual spaces between the six façades collapse into one building that can be perceived as one elevation. This idea of altering the perception of architecture can be compared to the classical manipulation of the tapered column to appear straight,[148] rather than its actual construction of compensating for the viewer's eye. The subtle and overtly obvious play on how the viewer experiences the mirage reflects how the INbetween plays between the real and the virtual.

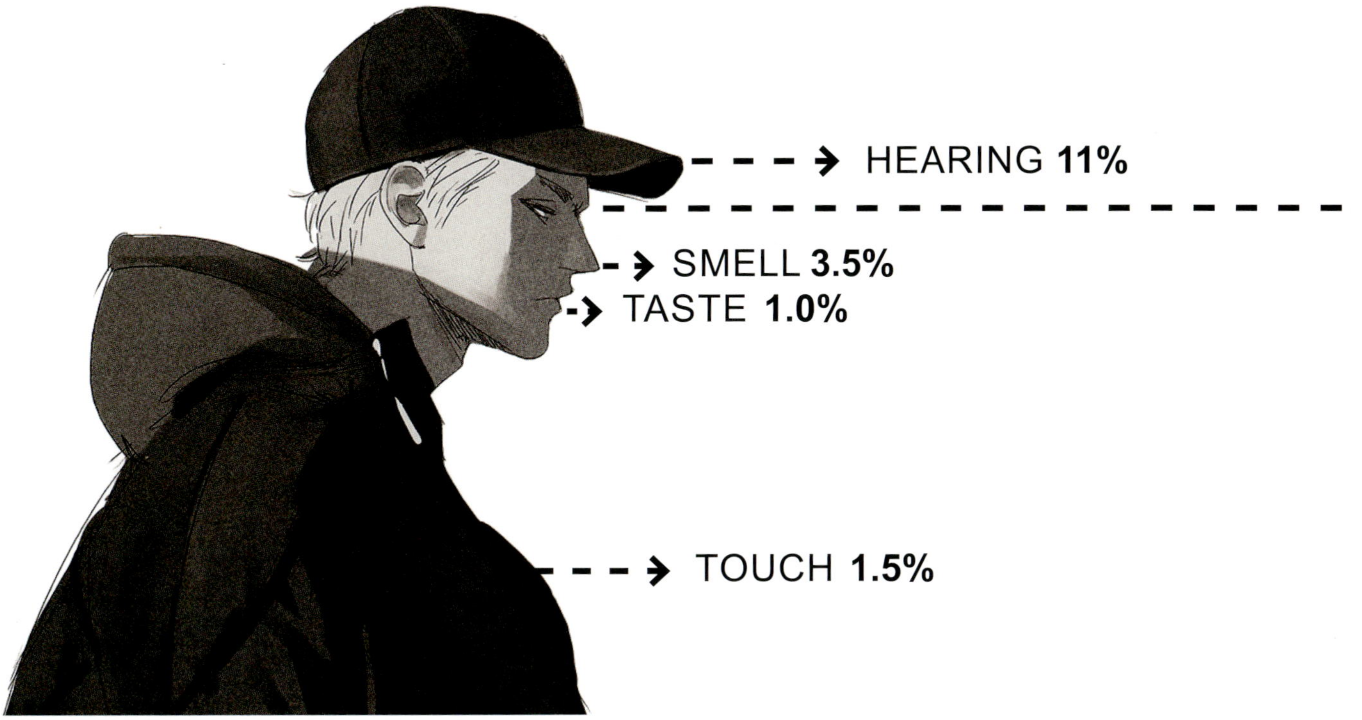

REAL	**INBETWEEN**	**UNREAL**
+ METAL + PLASTIC + THREADS + MIRROR + LED	+ AIR + WATER + SKY + LIGHT + COLOR **+ MIRAGE**	+ TECHNOLOGY + VIRTUAL + COSMOLOGY

INBETWEEN SIGHT: MIRAGE

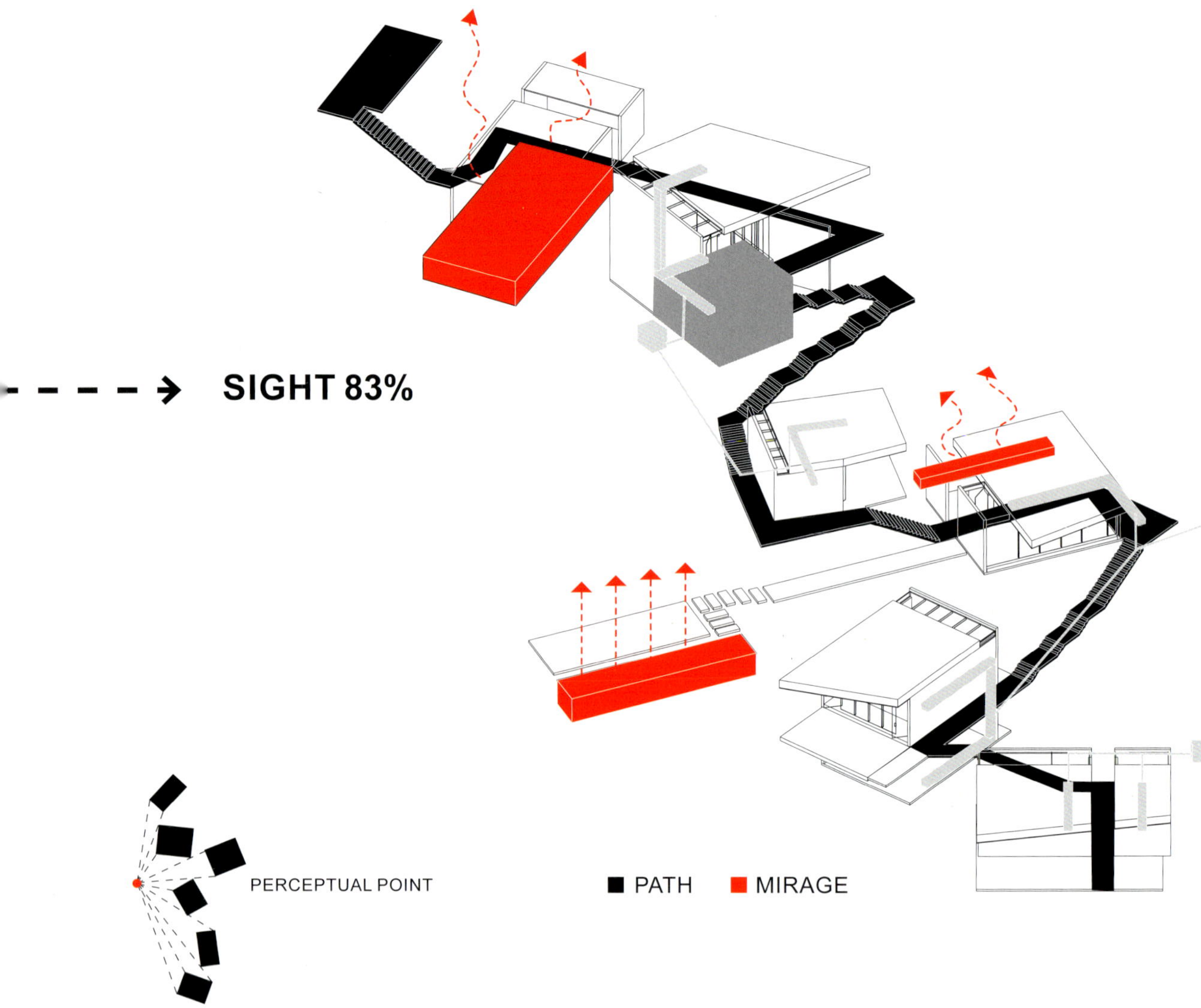

MIRAGE		COLD	COOL	WARM AIR
MIRAGE	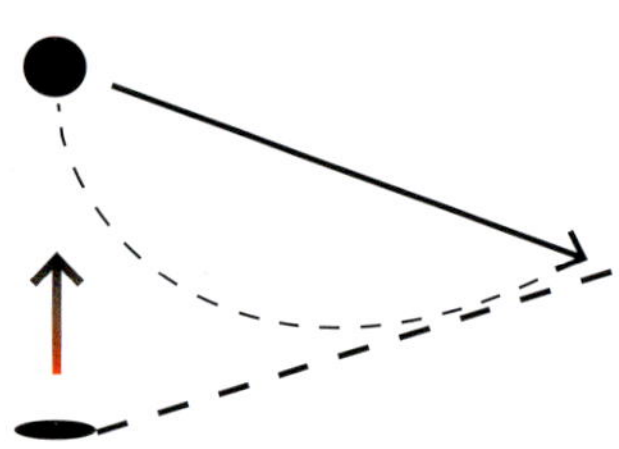	**DENSER**	**DENSE**	**RARE**
COOLER AIR		**HIGH** REFRACTIVE INDEX	**MEDIUM** REFRACTIVE INDEX	**LOW** REFRACTIVE INDEX
WARM AIR		**- 46 °F**	**46 °F - 59 °F**	**59 °F+**

The five senses are a pathway to the brain and for understanding the world around us. The visual cortex makes up to 30% of your brain, but the sense of touch and hearing takes up only 1.5% and 3.5% respectively.[150] Because of its larger size, sight can easily turn reality into an illusion. The perception of what you see and interpret varies between individuals and thus enters the world of unreal.

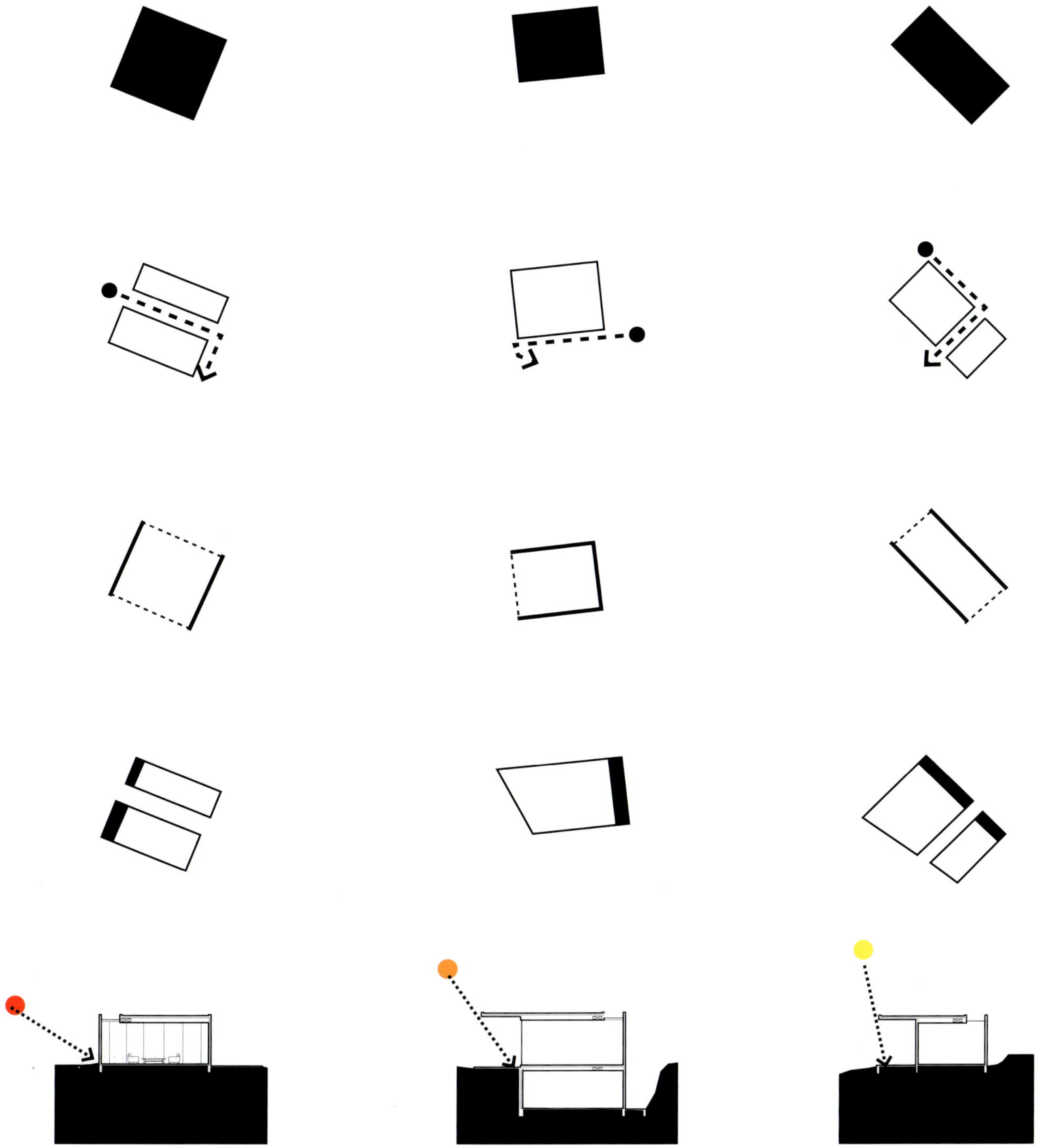

SUNSET LIVING | SINGLE BEDROOM | BEDROOM

NO BORDERS | PERCEPTION

MIRAGE: A HOUSE THAT PERCEIVES LIGHT

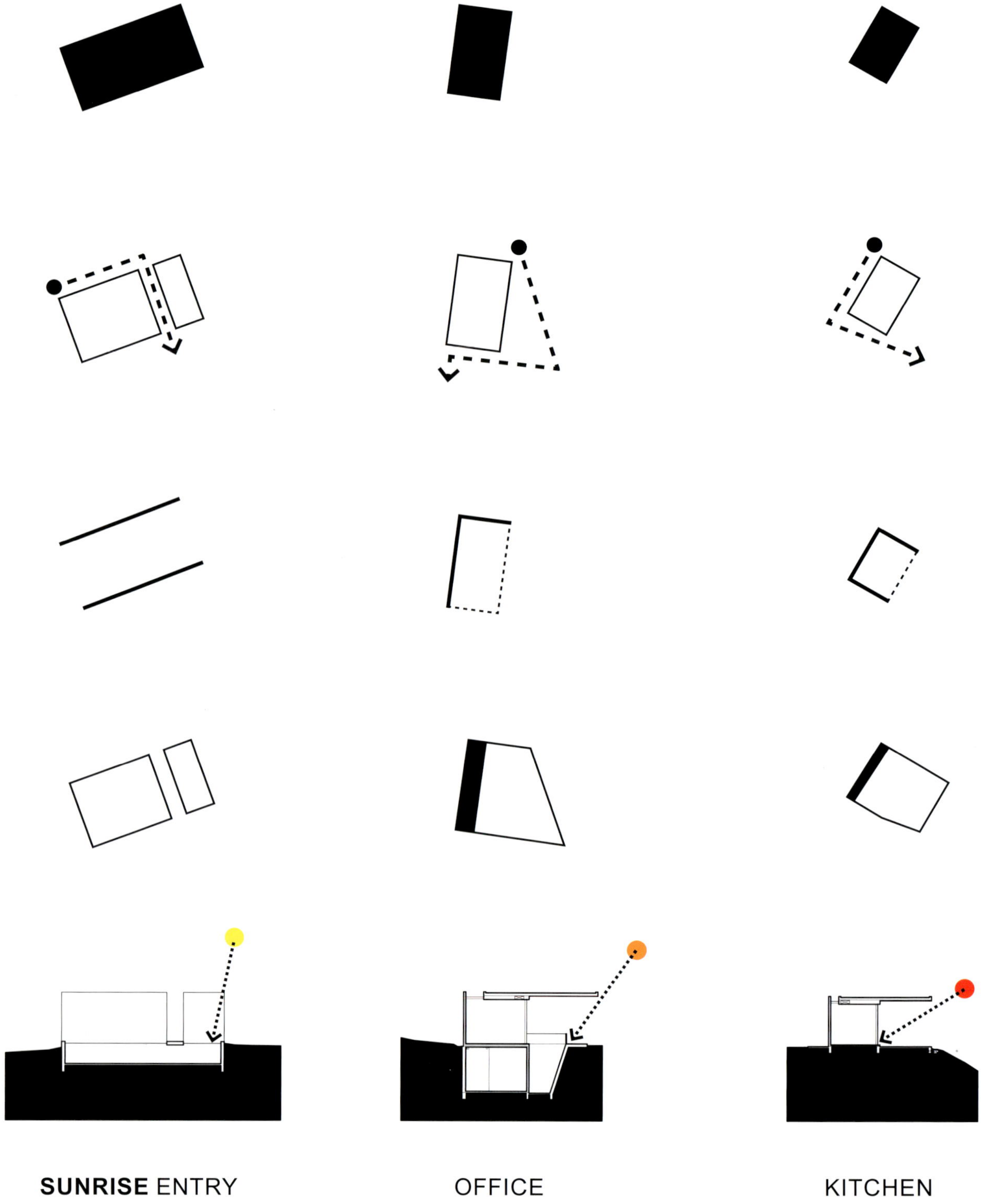

The movement of the the sun also can be seen as a medium for the architecture of perception. The notion of phenomenology as a basis for people seeing colors and spatial orientation differently[151] *is an indication of the blurring of the boundaries between the real and the virtual in how we experience the changing light, the depth of the spaces, and the atmosphere of the architecture.*

SUMMER SOLSTICE 6:34 AM

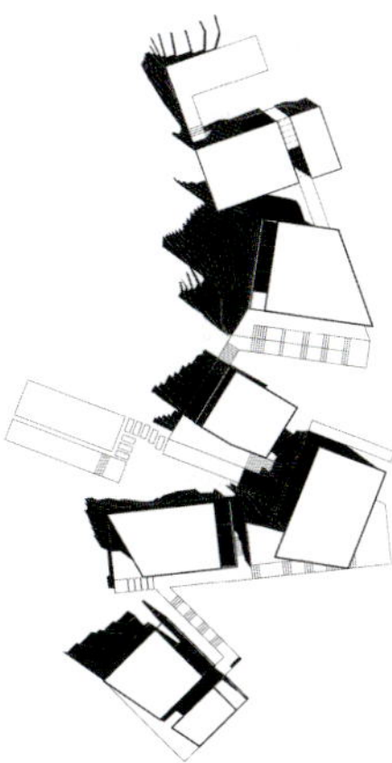

◐ WINTER EQUINOX 7:52 PM

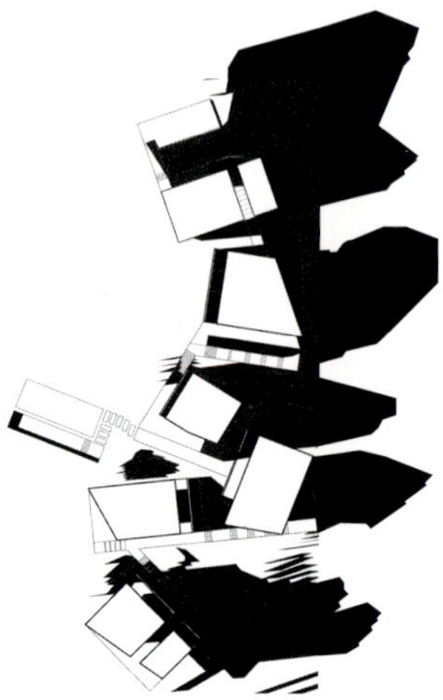

The light patterns in the desert create a dramatic backdrop to the intensity of the fluctuating temperature changes during the day and night. The sunset and sunrises also become a light show of colors refracting off the clouds. At night the Milky Way and stars are also clearly visible in the sky.

SUNRISE ENTRY

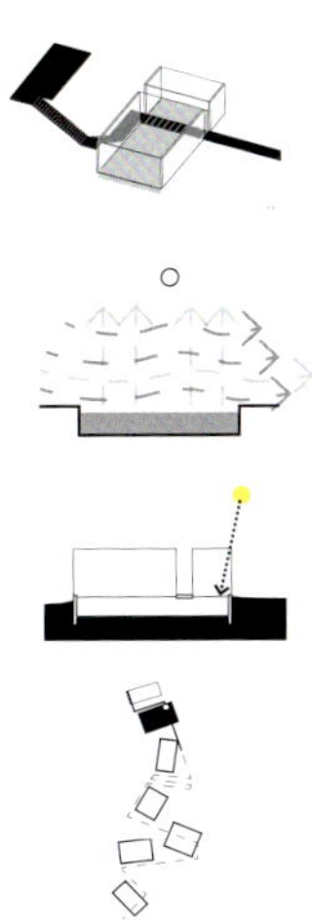

SUNRISE OFFICES

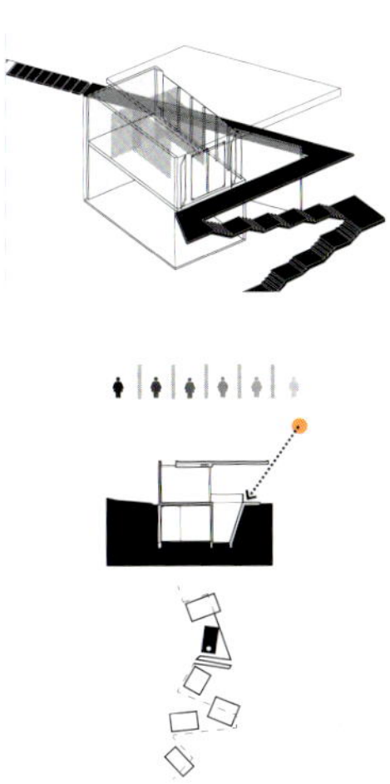

SUNSET BED

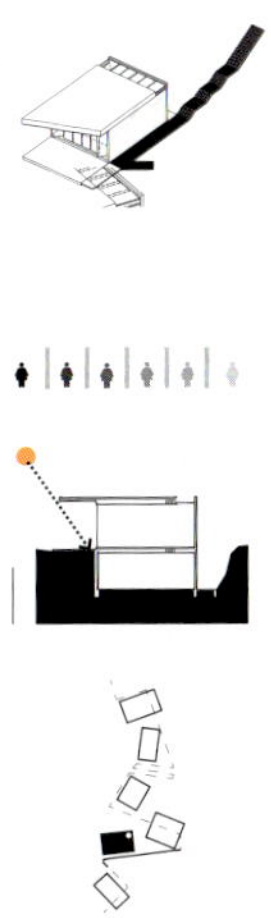

NO BORDERS | PERCEPTION

SIGHT: PERCEPTION OF LIGHT, MASS, AND MATERIAL

SUNRISE KITCHEN

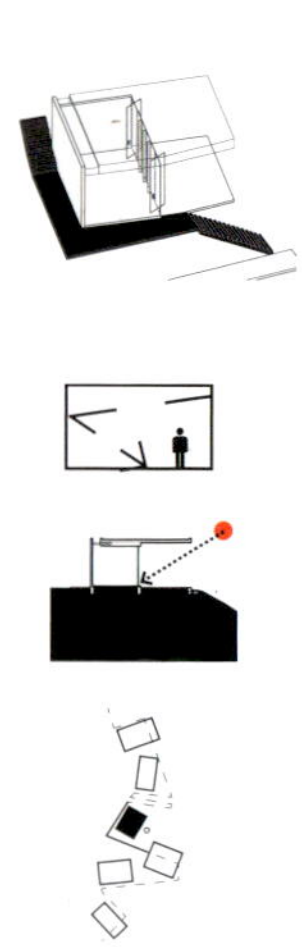

SUNSET GATHERING

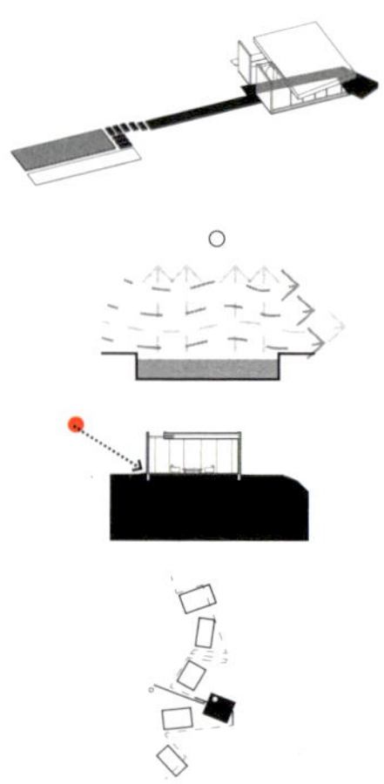

SUNSET ENDING

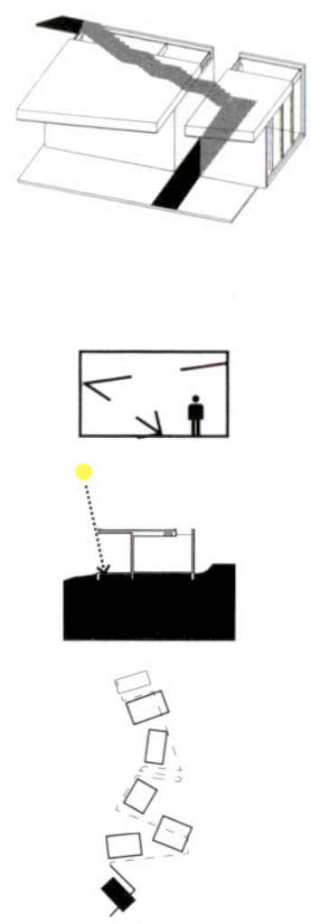

Historically architecture has often been referred to as frozen music.[153] However, architectural spaces rarely take into account or use acoustics as the primary design element in a building. A House That Resonates is built just north of the eastern Canadian Border in a remote area by Buffalo, NY, which has mild winters and is known for long growing seasons for food. A successful R&B musician / singer has conceived of his house as a studio, living, and performance space to showcase his numerous projects. Because he is so sensitive to sound, the entire complex is designed around the different acoustic properties of the space that will enhance the interaction of architecture and acoustics as an experiential place to create music. The centerpiece is an outdoor circular arena that can become an exclusive area for performing for his fans and will have a public entry. The grounds have gardens and a forest and will accommodate the fans parking under the trees. The

MUSIC + ARCHITECTURE: A UNIVERSAL LANGUAGE

clearing for the site will be built of shaped earth that will be used as the formwork for the poured in place concrete structure that blurs the boundary between walls, floors, and ceiling for acoustic reasons. The hard concrete surfaces are textured from the earth's sculptural formwork to respond to the variety of acoustic spaces needed in the house. The dramatic lighting and music system is controlled through the state of the art recording studio. Through understanding the possibilities of living outside the mainstream, this young musician plans to take advantage of the Metaverse through live, interactive performances in a connected, worldwide 3D virtual world, even though he is living in a remote area. He hopes to represent his generation in a new way of creating concerts, interacting with fans, and satisfying his new lifestyle. He believes in this future and is willing to gamble everything on it.

A HOUSE THAT **RESONATES**

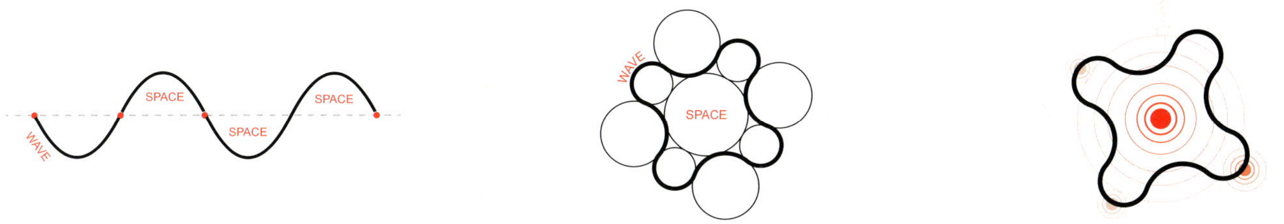

SOUND WAVELENGTH + SOUND THROUGH SPACE = A HOUSE THAT RESONATES

The design originates with a sound wave that is wrapped around a space in both the horizontal and vertical axis creating a dynamic, interiorized architecture that radiates from a central performance space. Nestled in a clearing, the earth mounds will become the formwork for a variety of spaces that respond to the specific interior acoustics for each space.

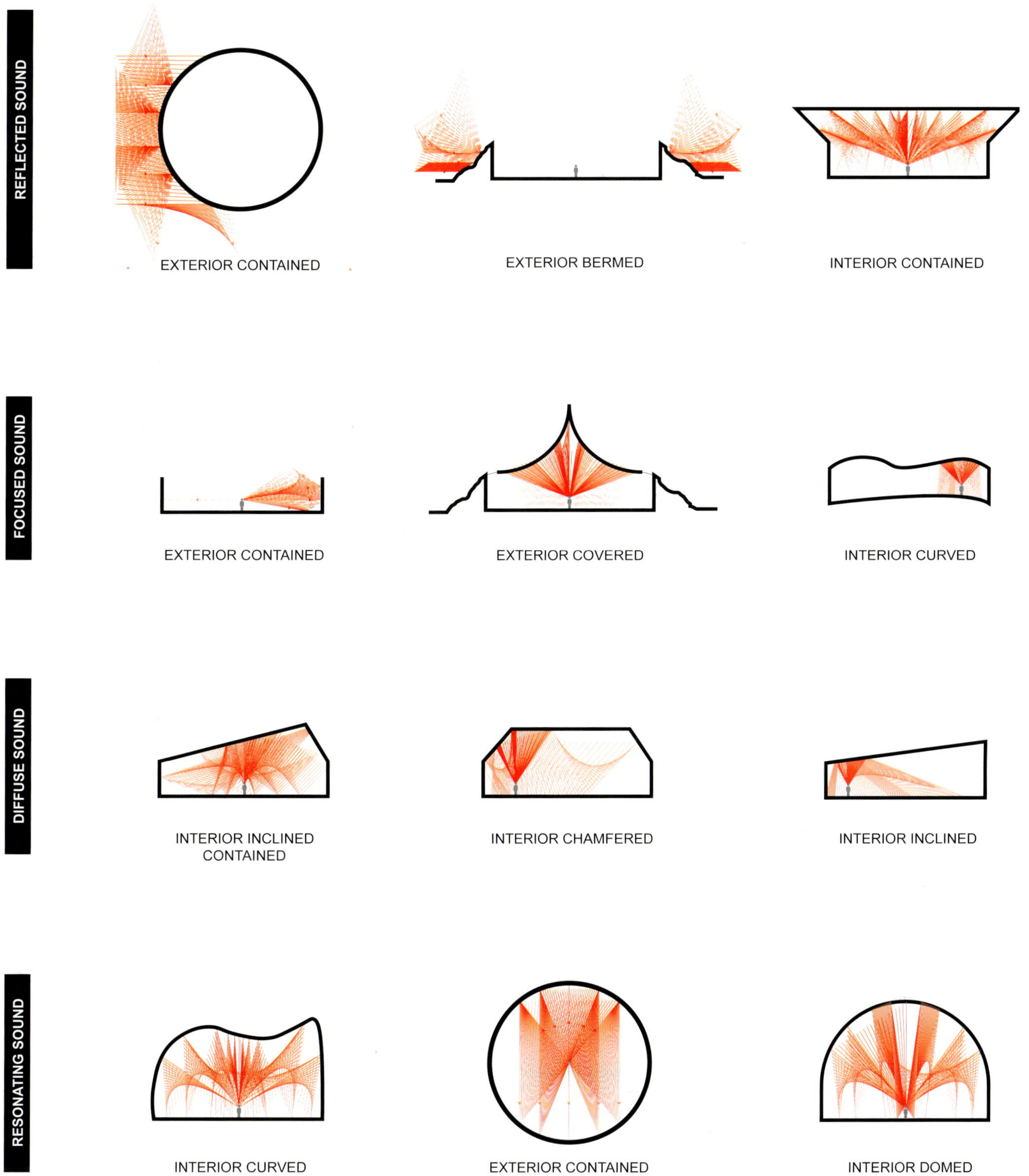

NO BORDERS | PERCEPTION

ACOUSTIC: MUSIC + ARCHITECTURE

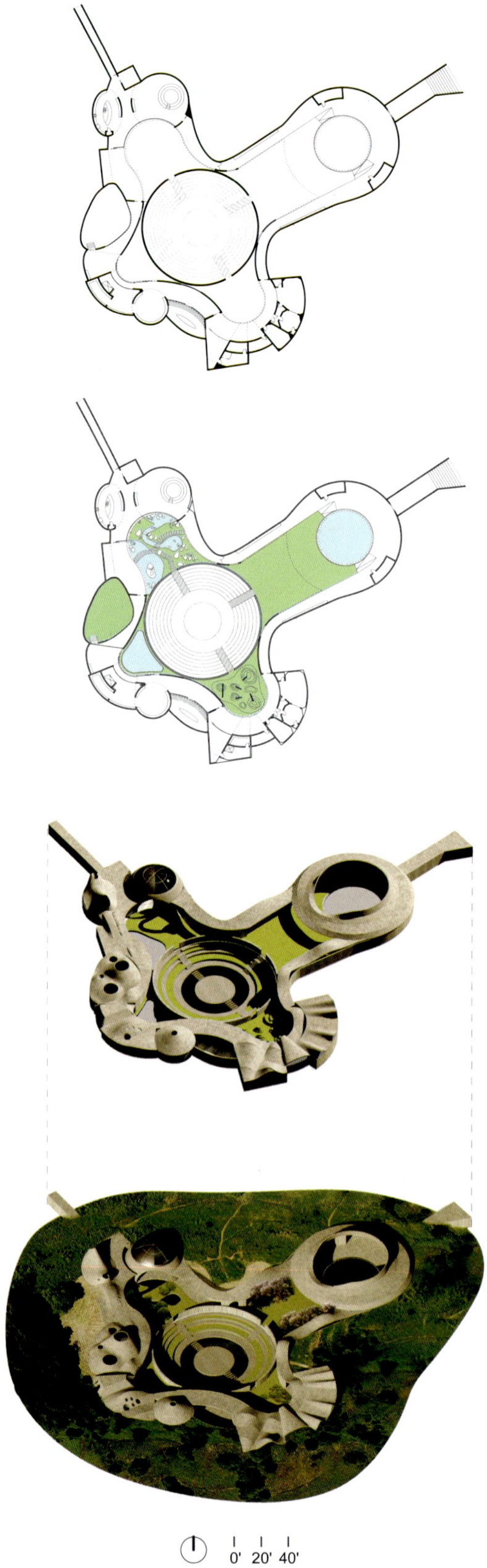

Each interior space is analyzed according to the acoustic notions of reflected, focused, diffused, and resonating sound. By studying how the sounds shape the space, there is a symbiotic relationship between the earth being used as formwork, the static nature of the resulting textured concrete forms and the ethereal nature of the sounds produced in these spiritual spaces.

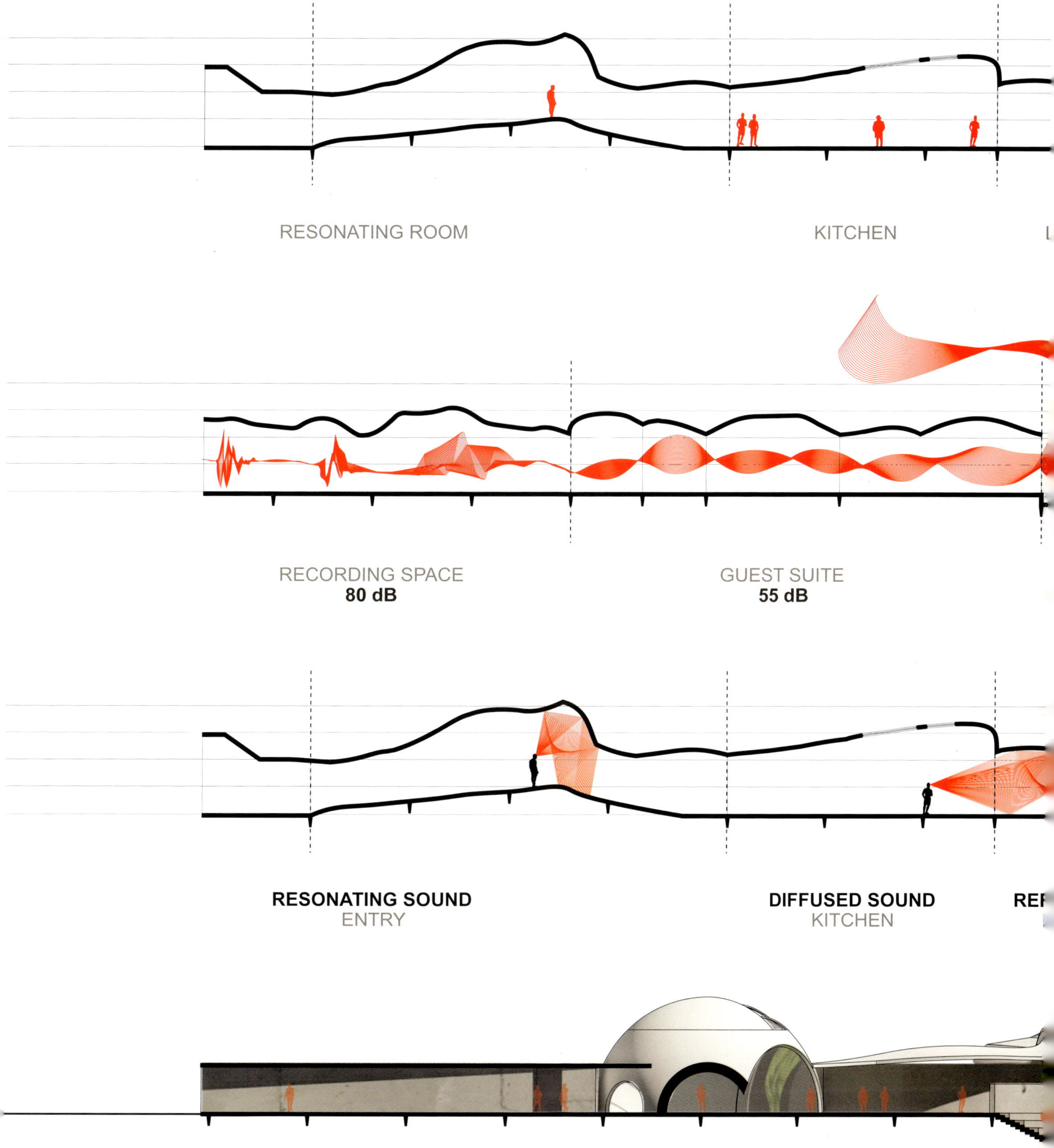

NO BORDERS | PERCEPTION

ARCHITECTURE OF SOUND: AN UNREAL SPACE

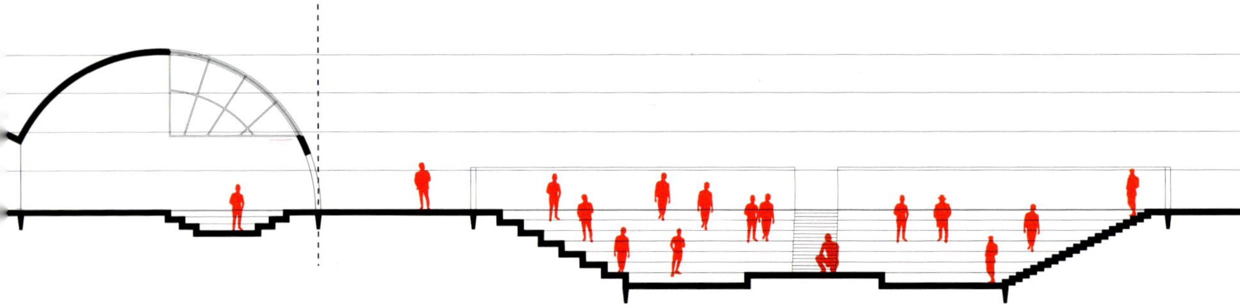

G SPACE

PERFORMANCE SPACE

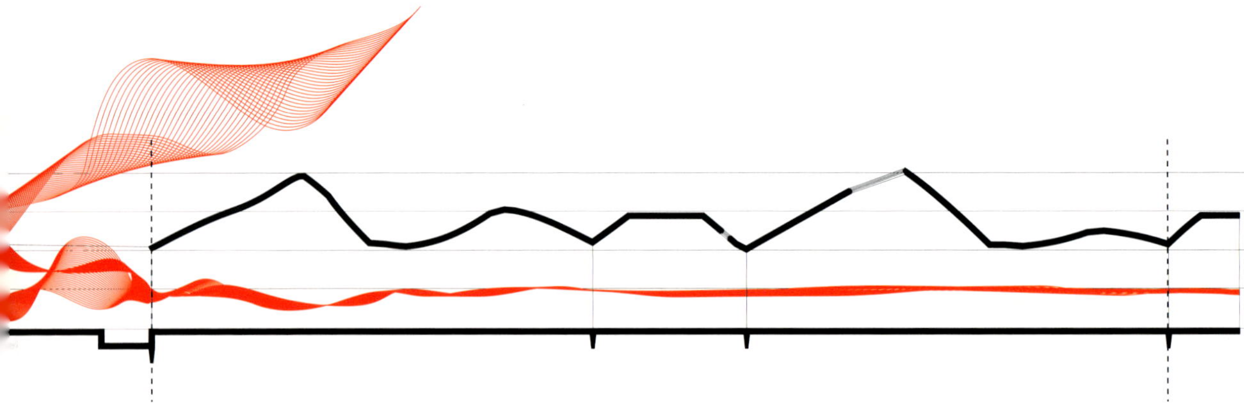

RTYARD
10 dB

MASTER SUITE
20 dB

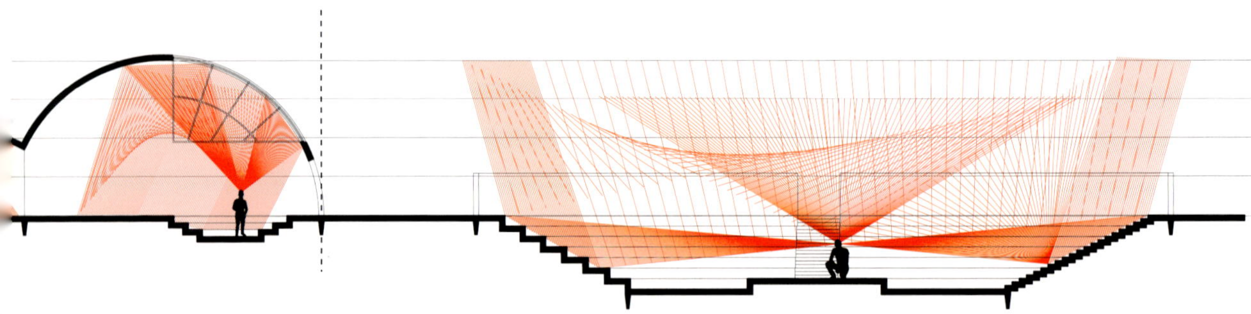

TED SOUND
G SPACE

FOCUSED SOUND
PERFORMANCE SPACE

Two species, traveling along the same path, are facing crises—one environmental and one political—but the Sanctuary offers a safe haven for both. The climate crisis has endangered the monarch butterfly population. From Michocan, Mexico the monarch's make a yearly migration to the Great Lakes in the Northeastern US, however, in 2020, 26% of the migratory monarchs died due to the effects of climate change and deforestation.[155] As the temperatures continue to increase the forests, which are populated with milkweed that is needed to feed the monarchs, will require a new location that preserves the habitats of the butterflies. Through ecological research, an island was identified that could house the new migratory safe haven for the monarchs. With a newly funded research center for studying the changing ecology of climate change on endangered species, the researchers and butterflies would coinhabit the island as a testament to the interdependence on one another. Tourists could also come to view the forest in a controlled protected living habitat. The sparsely populated island would continue to keep its residential

base, but expand upon this rural setting with the construction of housing for the researchers, an aviary, conservatory, and a research center for the butterflies and climate change. As a further acknowledgment of the concept of No Borders, the researchers would be pooled from the immigrants of DACA (Deferred Action for Childhood Arrivals), which protects immigrant youths, who came to the US as children, from deportation. This population are victims of the social construct of Borders and how appropriate to hire these immigrants to take care of the butterflies who freely migrate across the borders of Mexico and the US. The relationship between immigrants and immigration is not lost on the Sanctuary, which becomes a commentary on the social constructs of politics creating borders instead of having humans and animals living in harmony without being restricted by the arbitrary boundaries of nation states. If humans are to continue as a species, then the Sanctuary will encourage this process of living together between humans and the animal kingdom.

+

MIGRATION

MONARCHS

The first modern passports were issued by the newly formed League of Nations in 1920 as a result of World War I. This meant that prior to 1920 people were allowed to travel anywhere in the world without fear of being stopped at a border. This free migration from country to country is inconceivable today as restrictive borders have now become normalized and populist politics has ensured a sense of nationalism. Tribalism has always existed throughout human history. Today, the destabilization of the political world on a global stage has resulted in the increase of restrictive immigration policies including the building of more walls on borders by sovereign countries, the refusal to accept refugees, and increasing the policing of borders, etc. By hiring Dreamers in the US and providing a home for the monarch butterflies, the Sanctuary tries to reinstate the belief in No Borders.

IMMIGRATION

DREAMERS

Every 20 minutes a living plant, or animal has become extinct because of the actions of humans. The most recognized butterfly, the monarch, is no exception as the degradation of their historic habitats are being threatened by climate change and people's disruption of the natural world. The proposal of the Sanctuary to create a safe place to land for this fragile species has become increasingly critical. As the temperatures rises globally, so too have the monarch's habitats changed with the loss of fauna and milkweed. The annual migration of the monarchs from Mexico to the US border by Canada has become treacherous as their population decreased by 25% in one year. Through relocation to an island closer to the Canadian border and planting milkweed, the potential for the monarch population to survive is dramatically increased.

SANCTUARY

HAVEN

The current issues of immigration and migration can come together and represent a sacred place for both protection and inspiration in the Sanctuary. Humans and animals are under siege by restrictive immigration policies and the animals and plants are becoming endangered by the loss of habitat. In response to climate change, the Sanctuary is conceived wholistically as an ecological place to live, a research center and an aviary where the monarch butterflies, humans, and wildlife can learn to come together as a prototype model for the future. The design of the Sanctuary acknowledges that local site conditions are changing globally and the fauna and migration of animals will substantially be altered because of human intervention. The Dreamers working to protect and interact with the migrating monarch butterflies is poetic in its symbolism.

MONARCH BUTTERFLIES

2002
80% OF MIGRATING MONARCHS KILLED AT A RESERVE IN MEXICO IN CLIMATE RELATED EVENT.

2006
HABITAT OF MONARCH BUTTERFLIES CONTINUES TO BE PUSHED UPWARDS AS A RESULT OF GLOBAL WARMING.

2012
FILM: *FLIGHT OF THE BUTTERFLIES* CAPTURES THE MIGRATION OF MONARCH BUTTERFLIES IN A DOCUMENTARY.

CLIMATE CHANGE

2001
IPCC REPORTS GLOBAL WARMING HAS REACHED EXTREME LEVELS WITH CERTAIN IMPACT IN THE FUTURE

2006
AN INCONVENIENT TRUTH: A FILM RAISES GLOBAL AWARENESS OF GLOBAL WARMING.

2011
GLOBAL AVERAGE TEMPERATURE DROPS, SKEPTICS ARISE.

DREAMERS

2001
UNDOCUMENTED YOUTH IDENTIFY THEMSELVES AS DREAMERS.

2012
1,684 DACA DREAMER
1.7 MILLION ESTIMATE
DREAMERS ELIGIB
FOR DACA.

POLITICS

2001
DREAM ACT IS INTRODUCED TO CONGRESS.

2007
DREAM ACT FAILS TO PASS.

2012
DACA IS ESTABLISHE
THROUGH EXECUTIV
ORDER UNDER OBAM

2022

2015

ꟷITE HOUSE ISSUES ꟷONAL STRATEGY TO ꟷTE THE HEALTH OF ꟷCH BUTTERFLIES ꟷHER POLLINATORS.

2020

26% PLUNGE IN MIGRATORY: MONARCHS DUE TO CLIMATE CHANGE + DEFORESTATION.

2030

CLIMATE BEGINS TO STABILIZE AND THE MONARCH BUTTERFLIES START TO REBOUND.

2022

2015

ꟷGREEMENT IS ꟷED. 2°C UPPER ꟷESTABLISHED.

2018

CO_2 EMISSIONS GREW AT A RAPID 10% SINCE 2010.

CONTINUED COVID-19 PANDEMIC REDUCES DEMAND FOR FOSSIL FUELS.

2030

100% OF NEW ENERGY PRODUCED IN DEVELOPED COUNTRIES ARE CLEAN AND RENEWABLE.

2022

2020

643,560 CURRENT DACA DREAMERS. 1.3 MILLION ESTIMATED DREAMERS ELIGIBLE FOR DACA.

2025

1.9 MILLION DREAMERS ARE ACCEPTED BY THE DREAM ACT.

2030

1.9 MILLION DREAMERS BECOME US CITIZENS.

2022

2017

TRUMP ANNOUNCES A PLAN TO ABOLISH DACA.

DACA IS REINSTATED THROUGH EXECUTIVE ORDER UNDER BIDEN.

2025

THE DREAM ACT PASSES THROUGH CONGRESS.

2030

THE FIRST WAVE OF DREAMERS OBTAIN PERMANENT RESIDENCY.

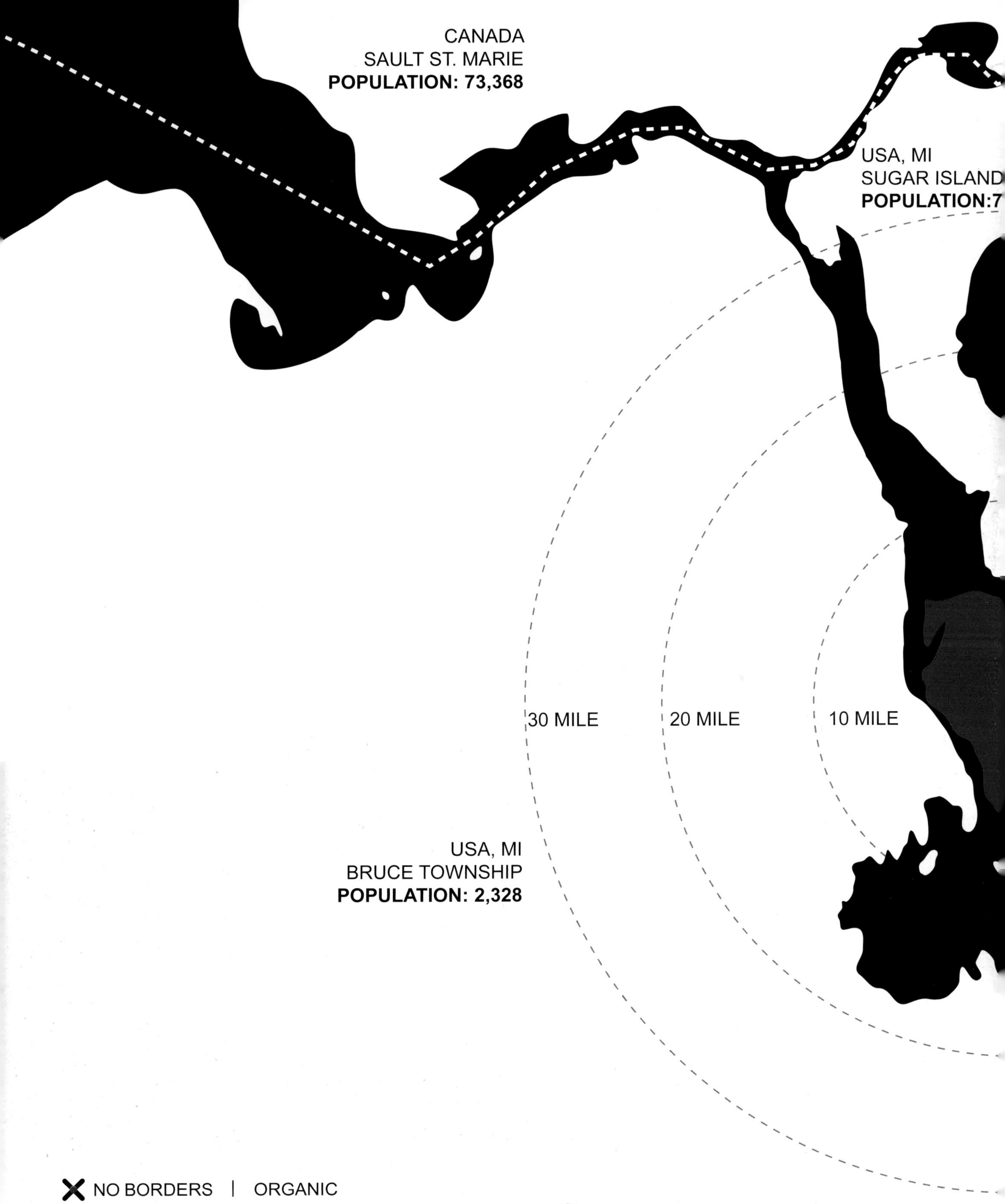

✕ NO BORDERS | ORGANIC

BORDERS: USA | CANADA[158]

ISLAND CONSERVATORY

CANADA
SYLVAN VALLEY
POPULATION: 1,609
CANADA
ST. JOSEPH ISLAND
POPULATION: 2,031

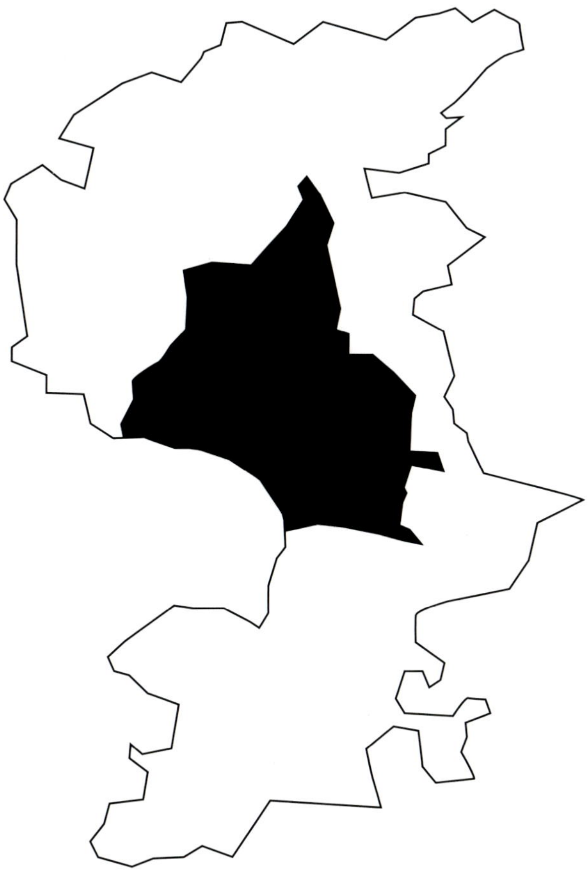

15 MI²

MEXICO, MICHOACAN
WINTER MIGRATION RESERVE

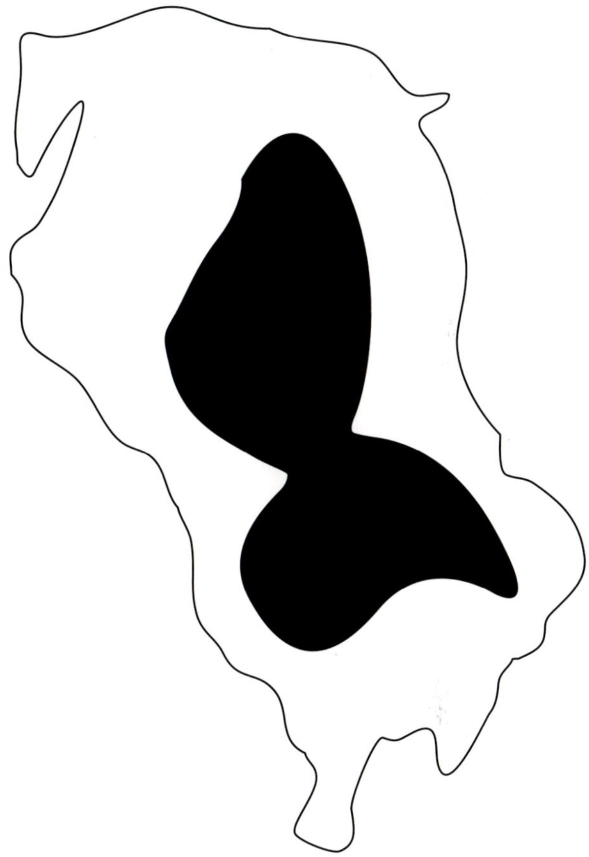

18 MI²

USA, NEEBISH ISLAND
FALL MIGRATION RESERVE

NO BORDERS | ORGANIC

MIGRATION: MEXICO TO USA

0 Mile 1 Mile

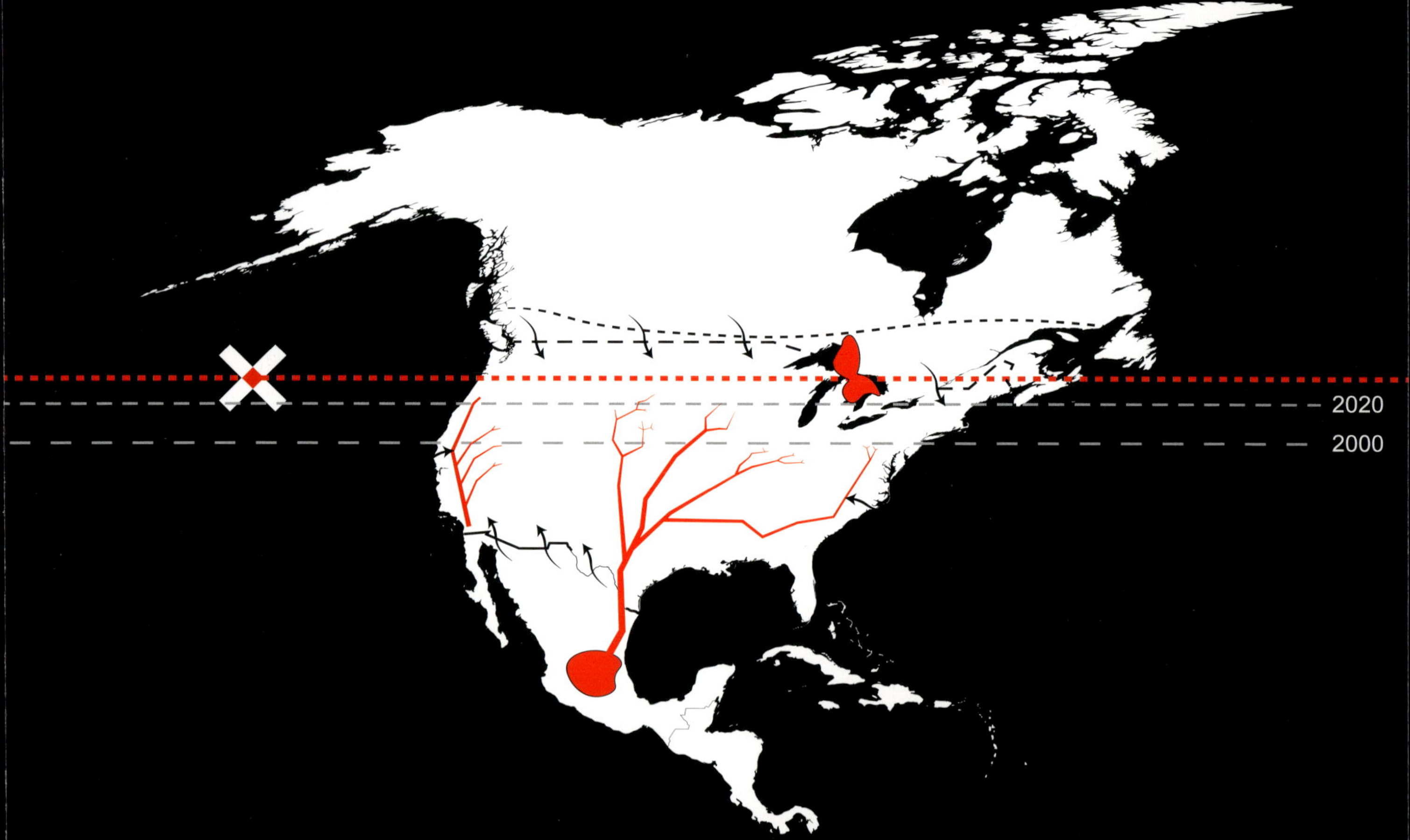

FINAL DESTINATION FOR THE MIGRATION NORTH

2020: AS A RESULT, THE SITE WILL MOVE AN ADDITIONAL 175 MILES.

2000: IF NOTHING IS DONE, 2040 MAY SEE A +1.5 °C TEMPERATURE ANOMALY.

IMMIGRATION   MONARCH MIGRATION   BORDER   MILKWEED LIMIT   0'   500 MILES

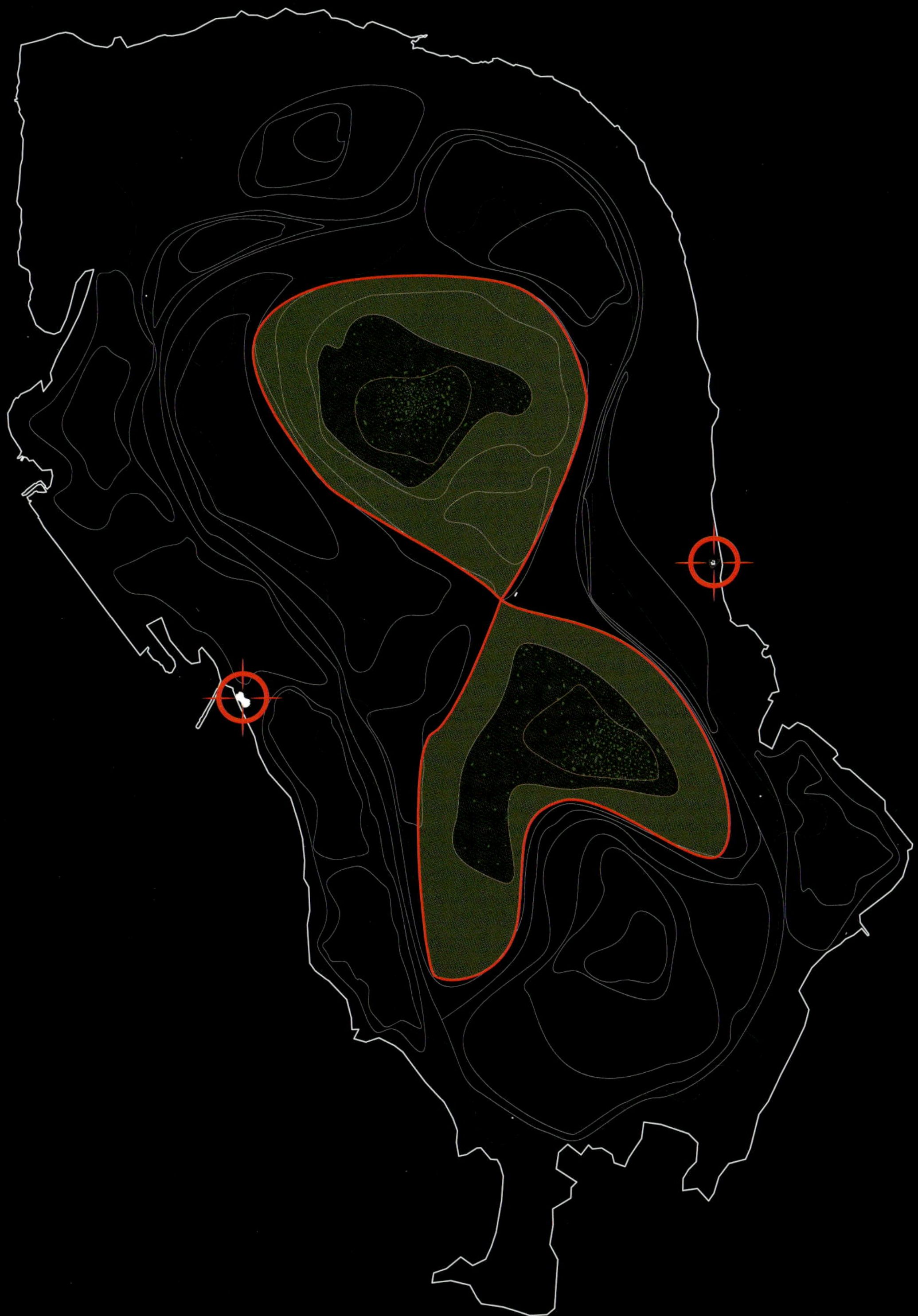

NO BORDERS | ORGANIC

IMMIGRATION: HOUSING THE DREAMERS

CORE

BUFFER ZONE

1 MONARCH RESERVE
2 AVIARY
3 CONSERVATORY
4 DREAMER HOUSING
5 HARBOR CAMPING
6 FERRY DOCK
7 ARCHIPELAGO TOWNSHIP
8 ISLAND HOUSING
9 MONARCH HOTEL

HOUSE WINGS

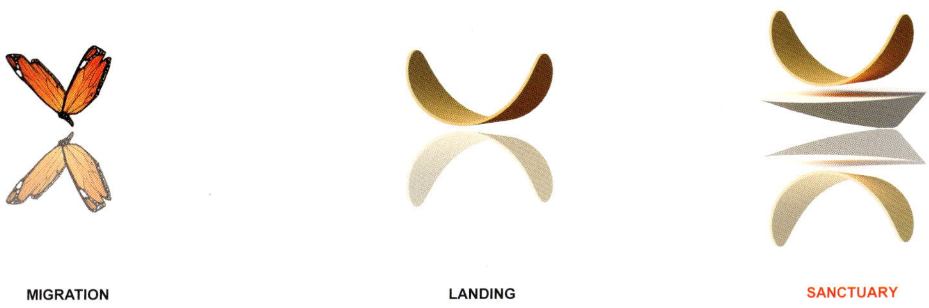

AVIARY

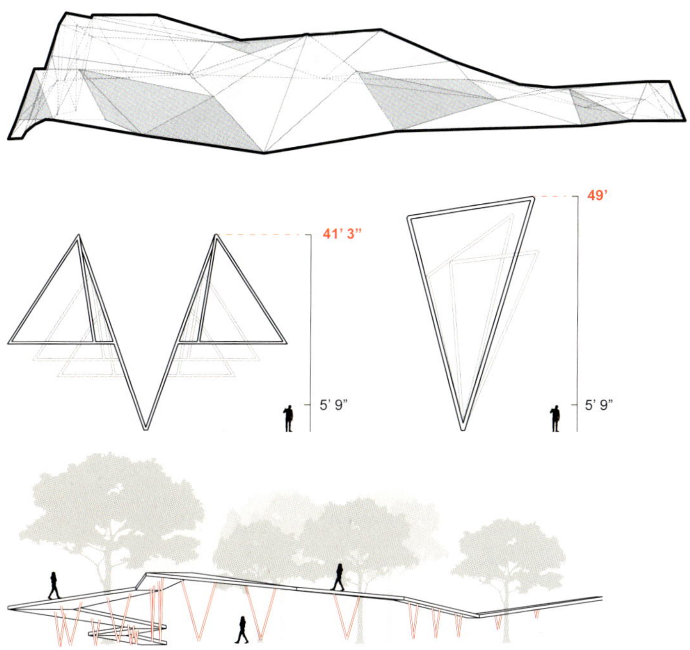

CONSERVATORY

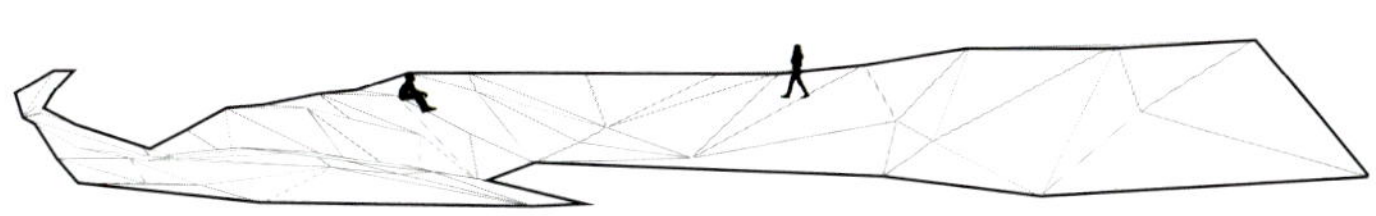

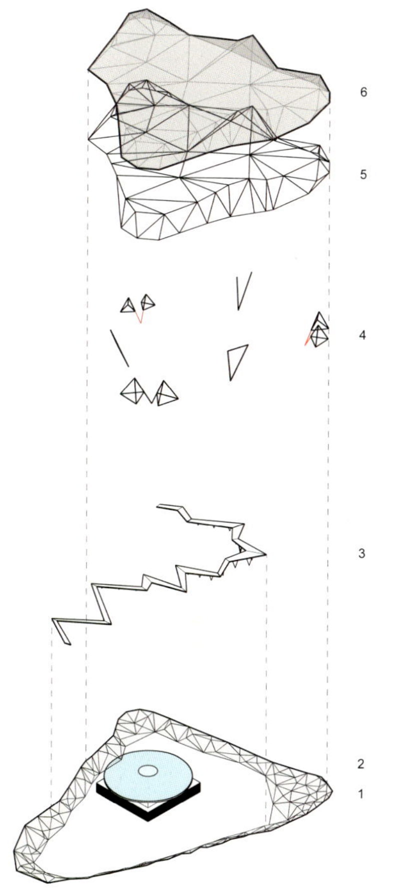

NO BORDERS | ORGANIC

AVIARY: MONARCHS HAVEN

1 PERFORATED MESH 2 RESEARCH LAB 3 PATH
4 STRUCTURAL TUBES 5 TENSION CABLES 6 ALUMINUM NET

The aviary cantilevered over the water is composed of stretched metal fabric held aloft by steel tensegrity bars enclosing a smorgasbord of plants and trees conducive to the appetites of butterflies. Fabric draped over the structure and strategic heat sources are used to keep the plants growing year-round with suspended walkways allowing tourists to mingle in the garden.

The twenty-first century is marked by the further exploration of inner and outer space. Advances in studying the brain and venturing into planetary travel has extended the reach of No Borders both on a human scale learning about the body, to a macro scale of understanding the mysteries of the universe. IOS (inner + outer space) located in Florida is centrally placed between NASA and Disneyworld. These two influences created a combination of a technological approach and one of entertainment when building an interactive, multimedia museum showcasing the potential of humans. IOS is designed as an island in the everglades reachable by an air boat from a pier that is easily accessed within a short car ride from Disneyworld, or NASA. Team Lab studio exhibitions and Metaverse experiences take the participants into other virtual and augmented realities that allow people to explore the inner and outer spaces that are unique to the twenty-first century. The architecture is

INNER + OUTER SPACE (IOS)

highlighted by the integration of the closed circulation of the fuselage taken from the technology of rockets and airplanes and introduced in IOS as a system that propels visitors to each level in a circular path. The pneumatic skin of the building uses air to regulate the interiors allowing a variety of multi-media projections and expansive spatial experiences to capture the concept of No Borders by altering the perception of the human mind. The dialogue between the isolation of the island and the tension of the inner and the outer fully immersive experience of the museum sets up a true conversation of what it is like to live in the contemporary world and questions which reality humans will want to pursue in the future. The historical progression of understanding space has been the pursuit of architects in the real world. The IOS museum exhibits the new technological perception of space in the Metaverse that will push the boundaries of the real into the new possibilities of the virtual world.

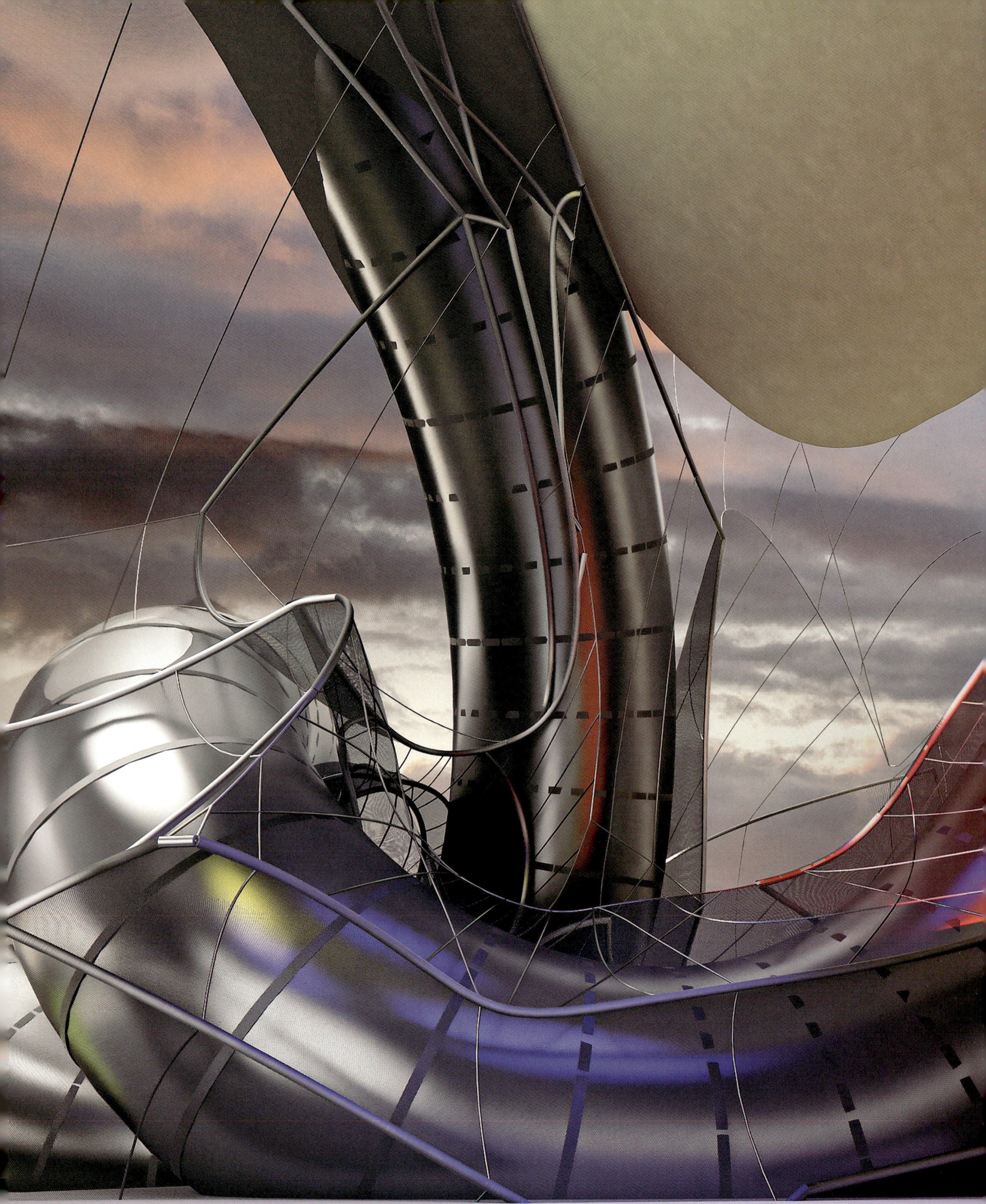

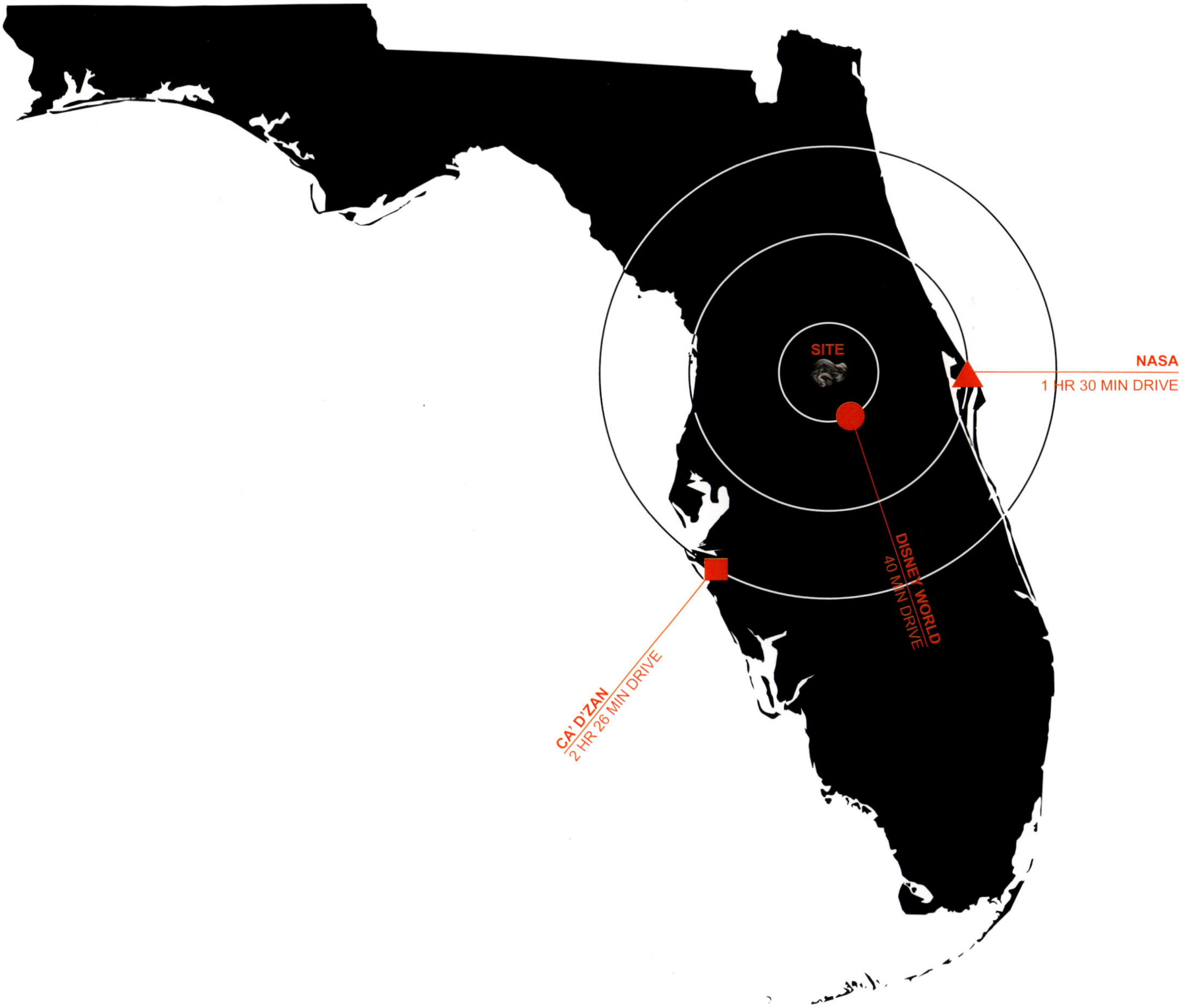

NO BORDERS | IOS

NASA + DISNEYWORLD: FLORIDA EVERGLADES

Located in Florida, a short distance from NASA and Disneyworld, the IOS museum is positioned on an island surrounded by the grasses and alligators of the Everglades. The reflections and misting on the surface of the water contribute to the blurring of reality and suspending belief of borders into another world.

STRINGERS

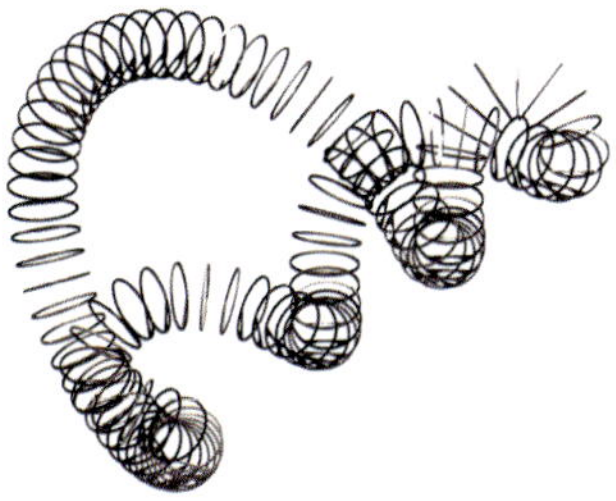

FRAMES

ALUMINUM CLADDING

FUSELAGE

ALUMINUM MONOCOQUE STRUCTURE: A CIRCULATION POD

PLASTIC SHEETING

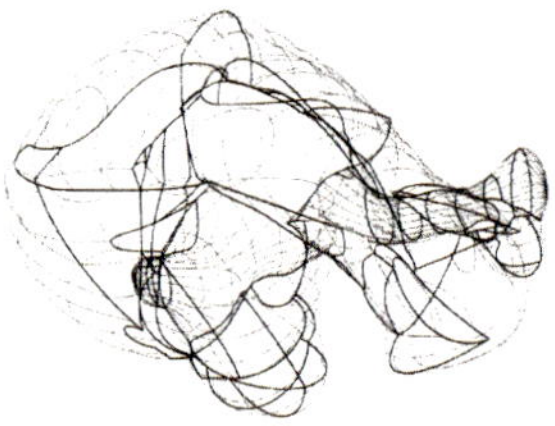

CABLES

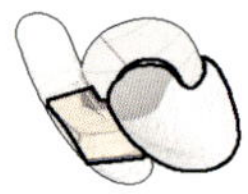

FERROCEMENT

PNEUMATIC

NYLON INFLATABLE: AIR STRUCTURE

The structure of the IOS is divided into two concepts, The fuselage used for circulation is made of stringers, frames, and has aluminum cladding, while the Pneumatic air structure is bound by cables and is reinforced with ferrocement under the land surface. The intertwined systems create a dynamic curiosity in experiencing the innovative structural dynamics of air and contained space.

NO BORDERS | IOS

THE ISLAND: EXPLORING THE REAL + THE VIRTUAL

The Everglades is a unique environment that is teeming with life on a horizontal plane. By elevating the IOS on an island the building immediately takes on the aura of an object, a castle out of reach, that holds the treasures of a forgotten or dream-like world. The extraordinary structural systems beckon the traveler to experience this new reality.

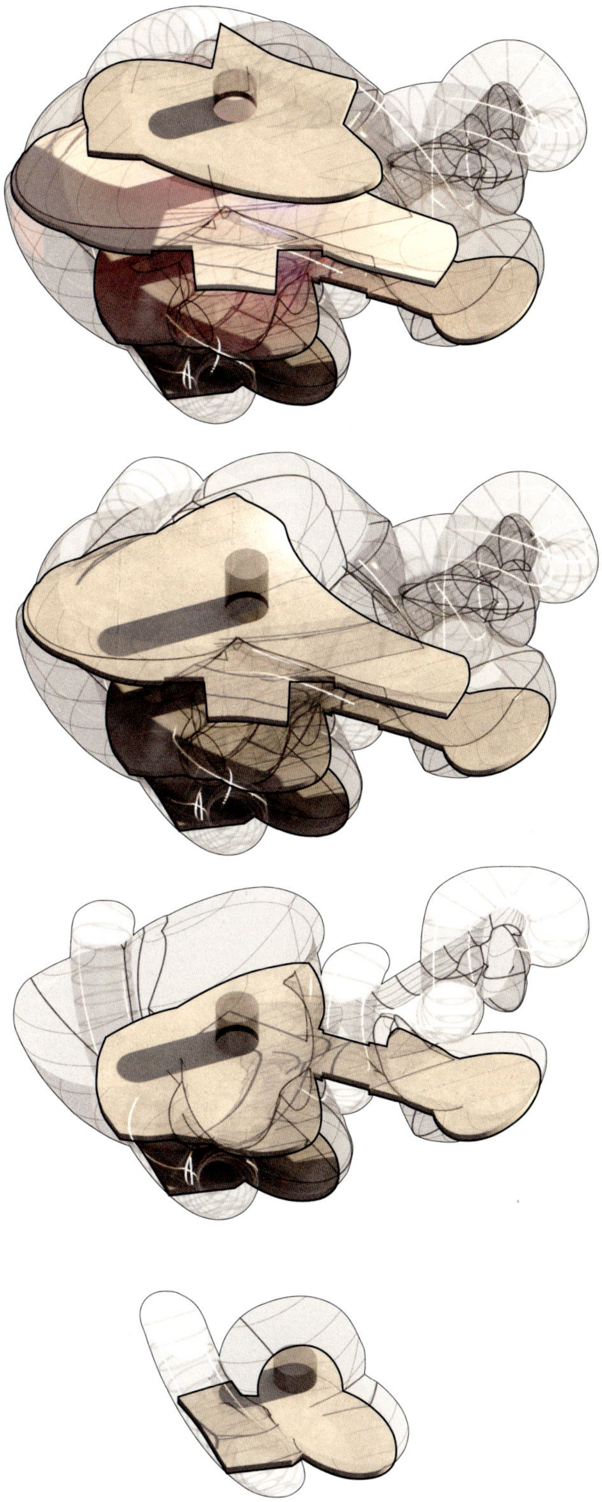

NO BORDERS | IOS

MUSEUM: A SPATIAL EXPLORATION

0' 5' 10' 20'

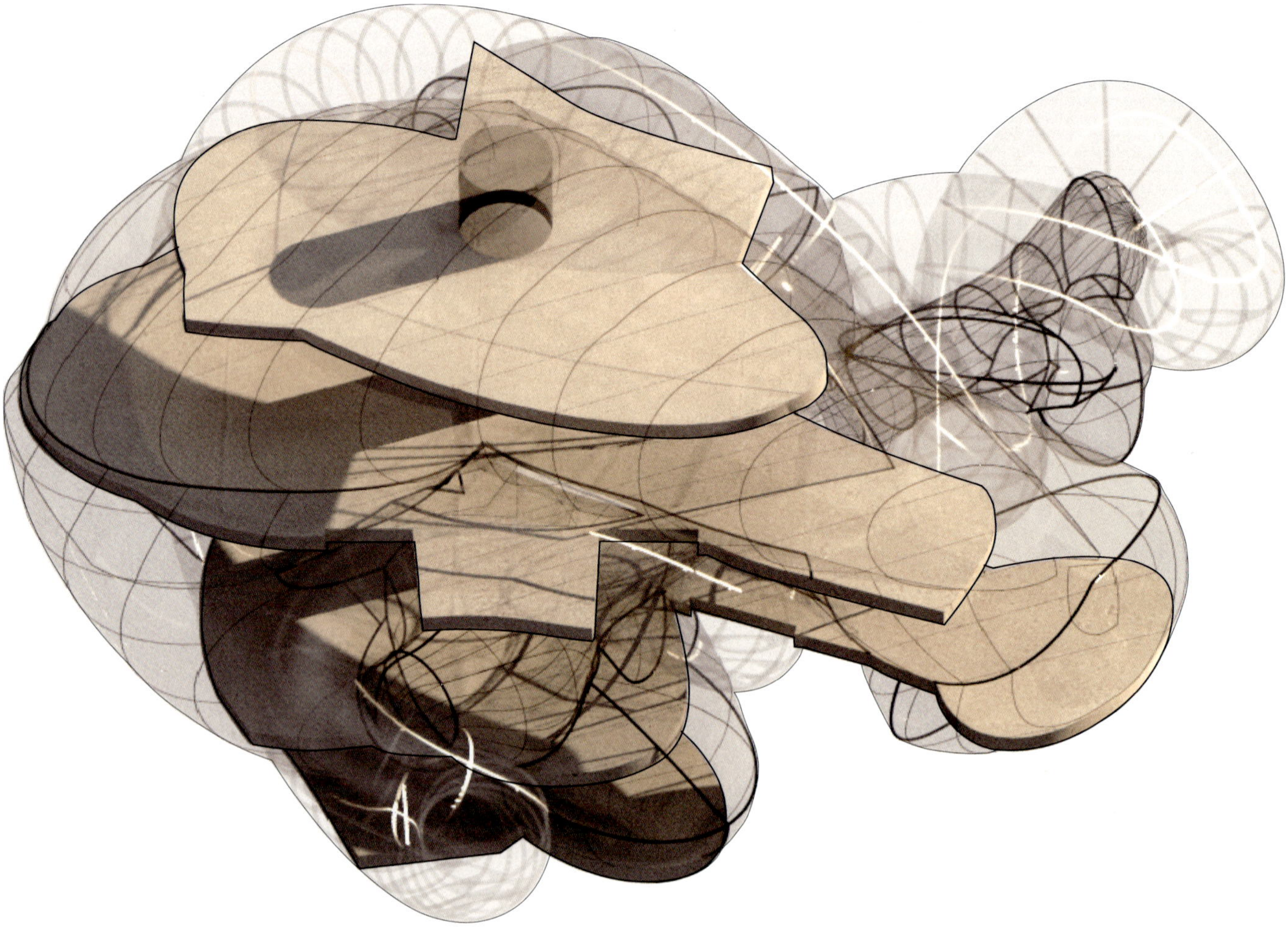

IOS FLOOR LEVELS

F1: RESTROOMS / MECHANICAL ROOM / STORAGE
F2: CONFERENCE, GIFTSHOP, ENTRANCE, RECEPTION
F3: INTERACTIVE DIGITAL GALLERIES, OFFICES
F4: THREE-DIMENSIONAL EXHIBITION / STORAGE

The four levels of the IOS present different functions of the museum. The lower level below grade contains the auditorium, restrooms, and the mechanical equipment. The second level contains the entry, public elevators and stairs, offices, and shop. The third and fourth levels are the primary floors for the immersive, multimedia permanent and temporary exhibitions.

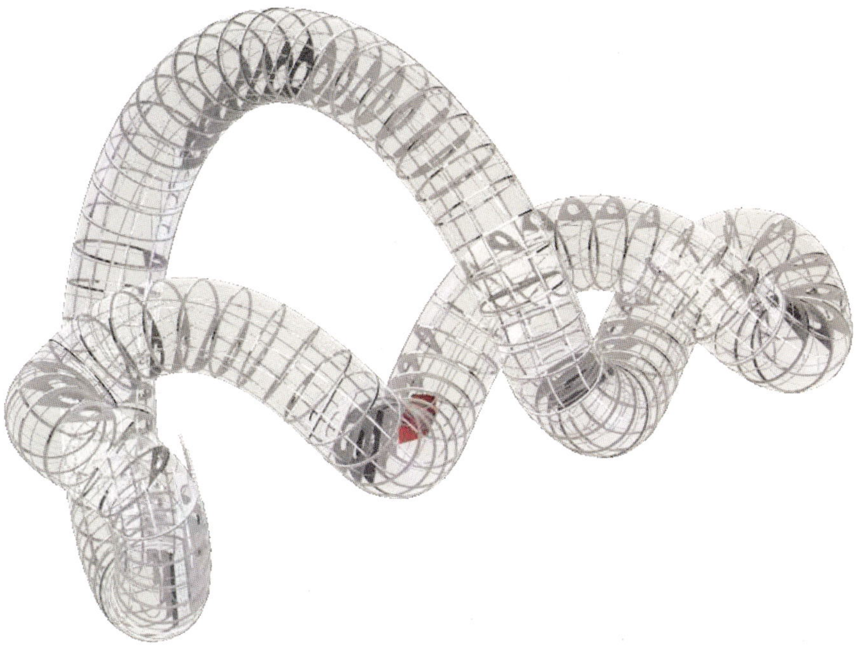

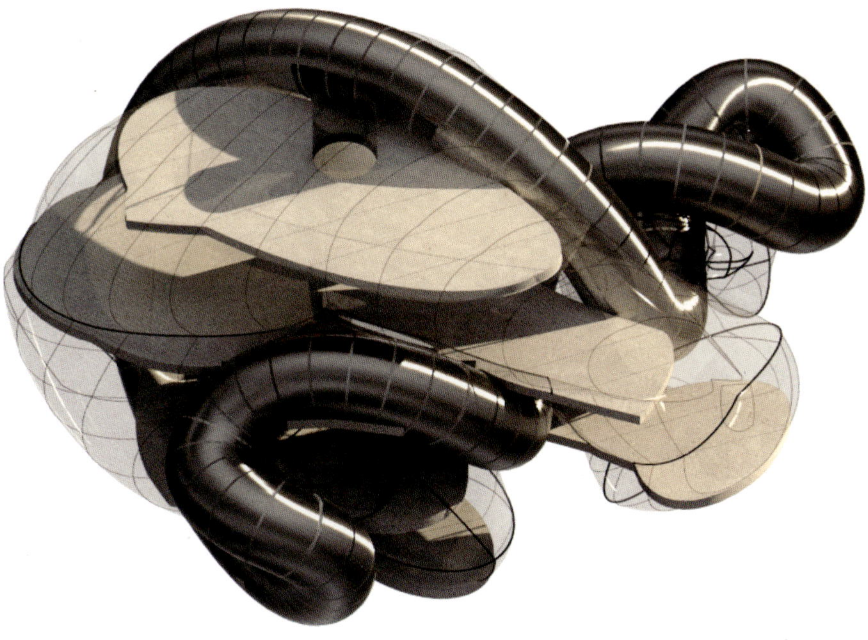

NO BORDERS | IOS

IMMERSIVE EXHIBITION: AN EXPERIENTIAL SPACE

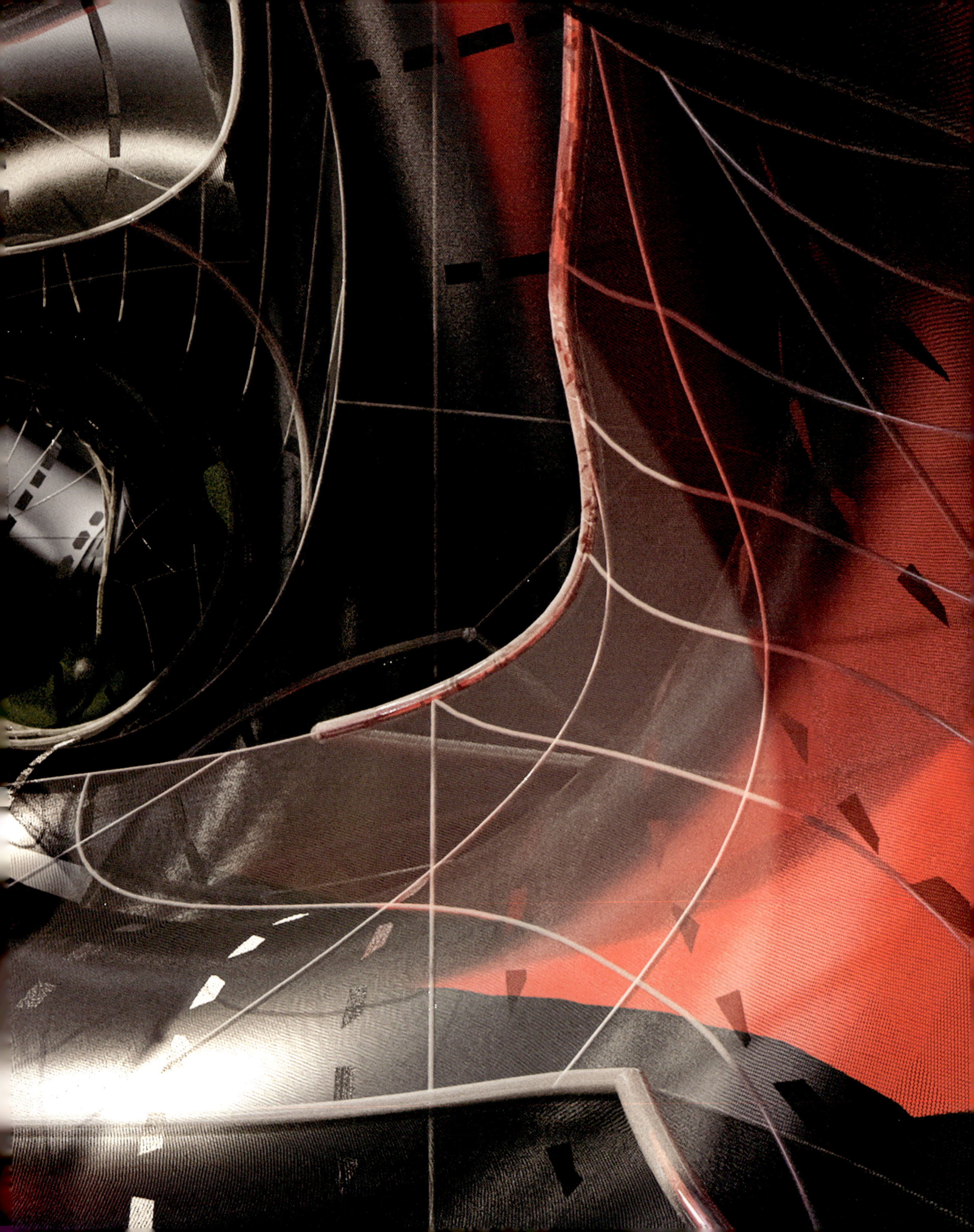

As commercial flights for civilian space travel increase over the next decade, the race to Mars and to revisit the moon has once again captured the imagination of the public.[159] The Last Iceberg is conceived as a Spaceport that is composed of a public / private consortium of entrepreneurial space technology companies, including SPACE X, Virgin Galactic, and NASA that plan to construct a tourist / training / research center for private citizens to learn about space travel and to become prepared for traveling to Mars. The facility will house a training center for up to one thousand people taking courses, enrolling in physical workouts, researching growing hydroponic plants, developing strategies for building on Mars, and learning how to live in a hostile environment. The site is built upon a glacial iceberg that is located north of Greenland in a region that experiences the aurora borealis. The unique construction of building into the ice is used as formwork for the building skin

A SPACE EXPLORATION CONSORTIUM

and takes advantage of the controlled atmosphere using modular building systems. The power for the facility uses a geosynchronous reflecting shield that points to the solar cells of the building. This concept allows for an extended period of time for the solar rays to be collected to power the station. The attraction of the public's interest in space travel has gone viral on social media and The Last Iceberg plans to capture this momentum in accelerating a knowledgeable group that has the capacity to fill the spaceships that will be needed to inhabit other planets. This need for survival is heightened as the earth is subjected to the effects of climate change and becomes ravaged by the lack of water, increasing sea rise, and catastrophic natural disasters.[160] The name of The Last Iceberg was conceived as a response to the conditions of global warming causing the polar ice caps to melt. It is imagined that humans making a last-ditch effort to save humanity, decide it is time to leave earth permanently as The Last Iceberg melts behind a ship going off into space.

"YOU DEVELOP AN INSTANT GLOBAL CONSCIOUSNESS, A PEOPLE ORIENTATION, AN INTENSE DISSATISFACTION WITH THE STATE OF THE WORLD, AND A COMPULSION TO DO SOMETHING ABOUT IT. FROM OUT THERE ON THE MOON, INTERNATIONAL POLITICS LOOK SO PETTY. YOU WANT TO GRAB A POLITICIAN BY THE SCRUFF OF THE NECK AND DRAG HIM A QUARTER OF A MILLION MILES OUT AND SAY, 'LOOK AT THAT, YOU SON OF A BITCH.'"

EDGAR MITCHELL, APOLLO 14 ASTRONAUT[161]

LANDING ON THE MOON

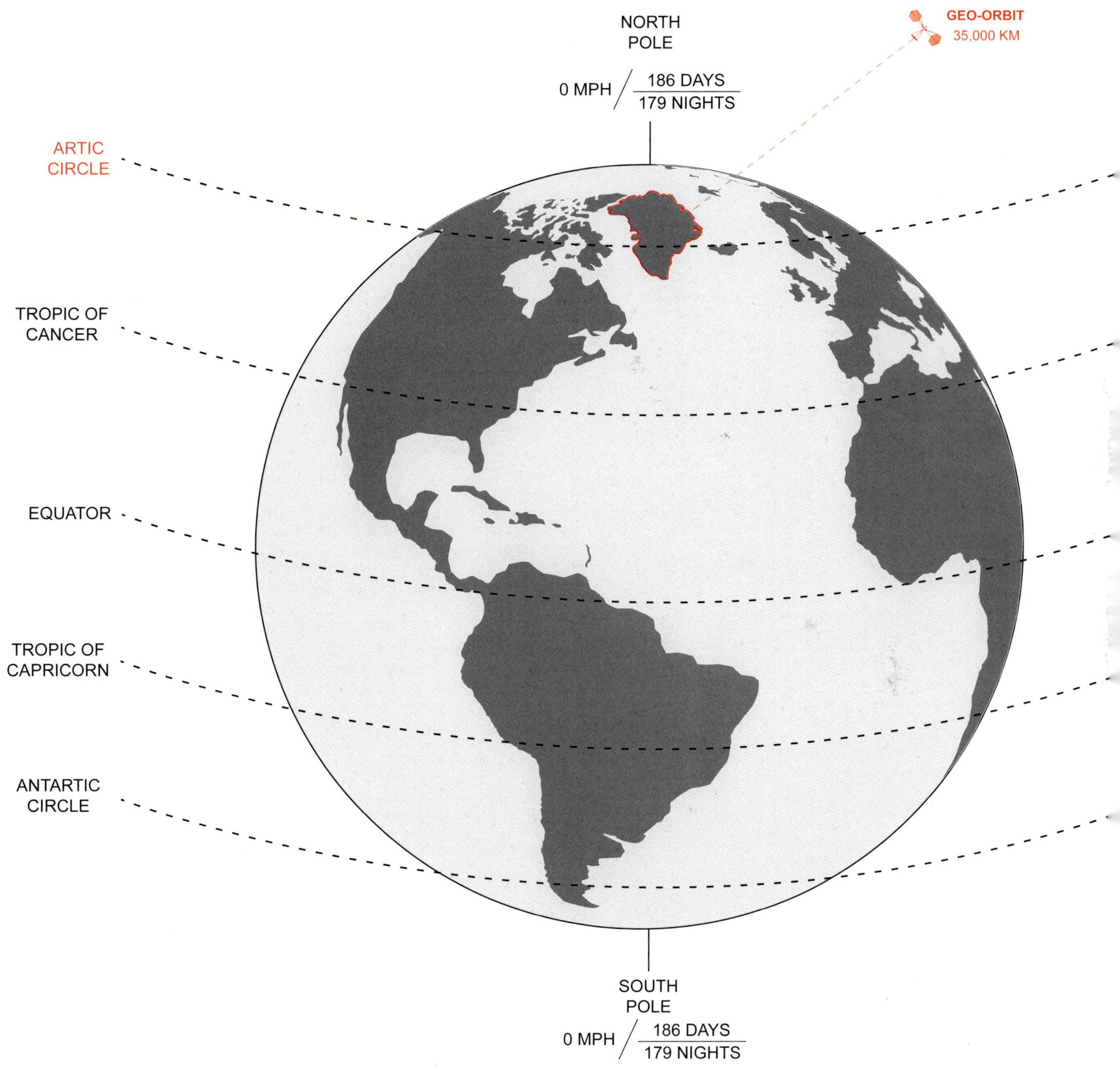

GEOSYNCHRONOUS ORBIT

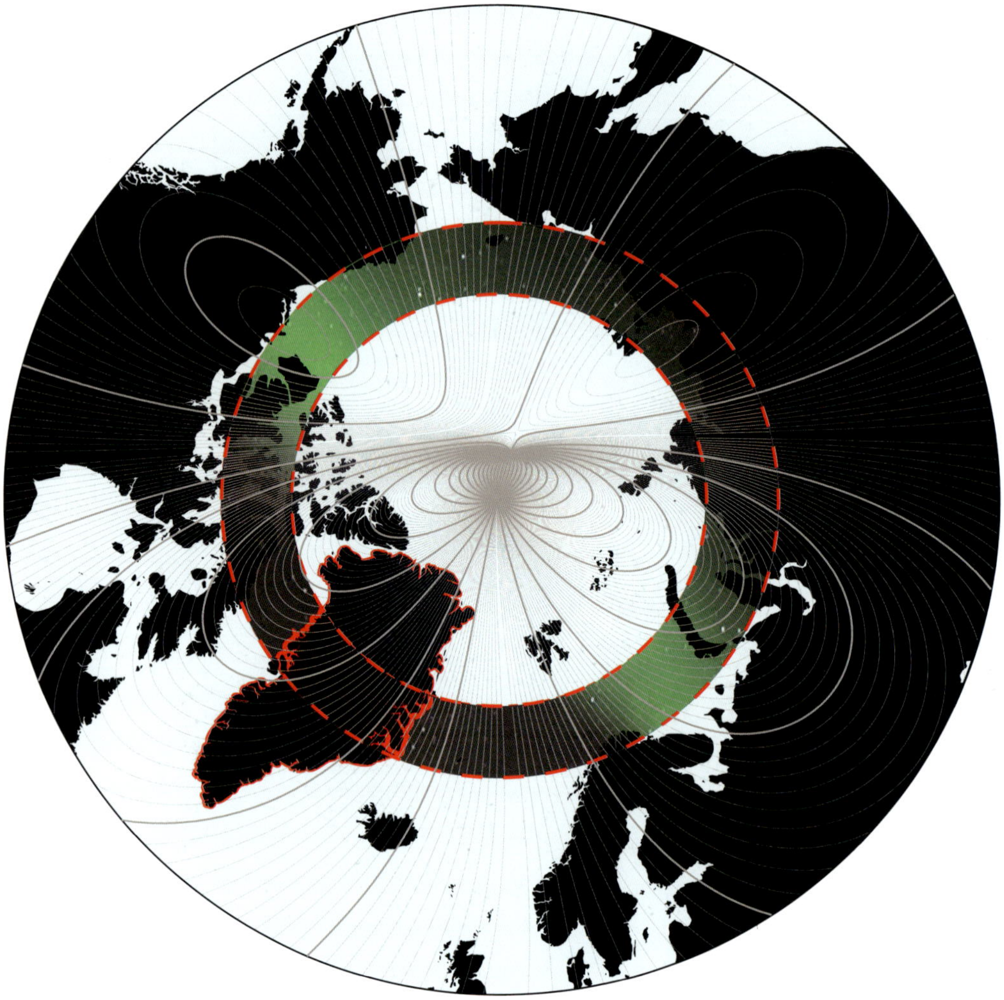

AURORA ZONE

The LAST ICEBERG, located in the Aurora Zone which exhibits extraordinary natural light shows in the sky, is enhanced by using a geosynchronous orbit satellite with a reflective shield. It is possible to increase the sunlight reaching the station by reflecting the light from the satellite during the 179 nights of continuous darkness at the north pole.

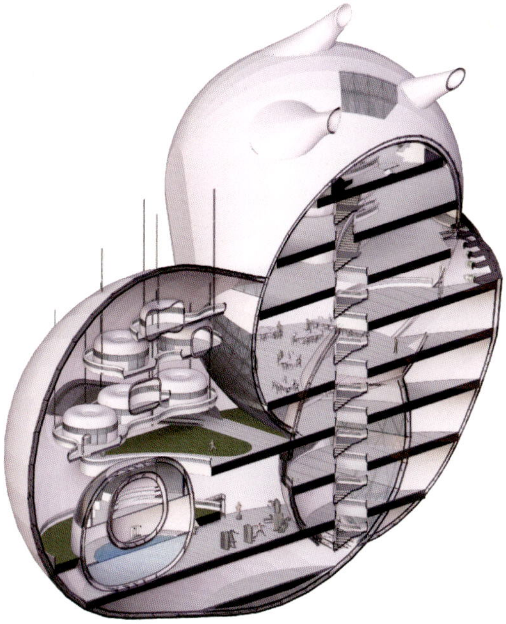

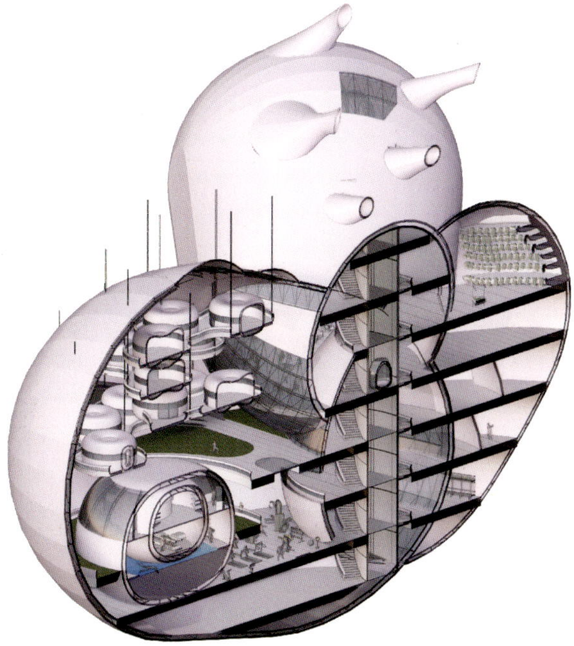

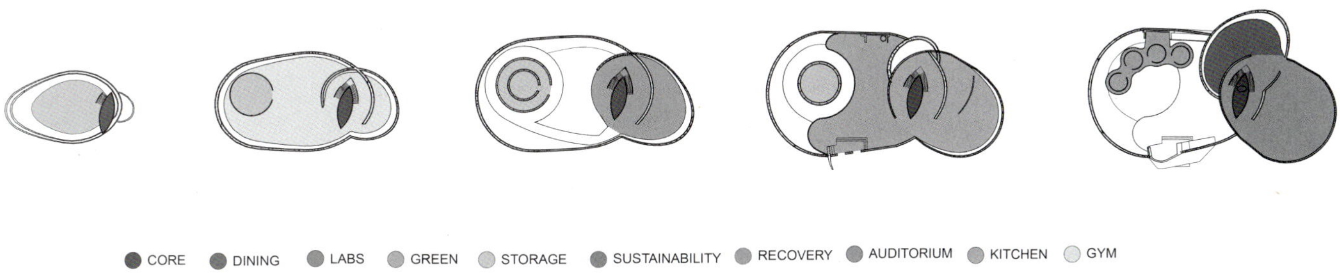

XTREME LIVING: RESEARCH BASE

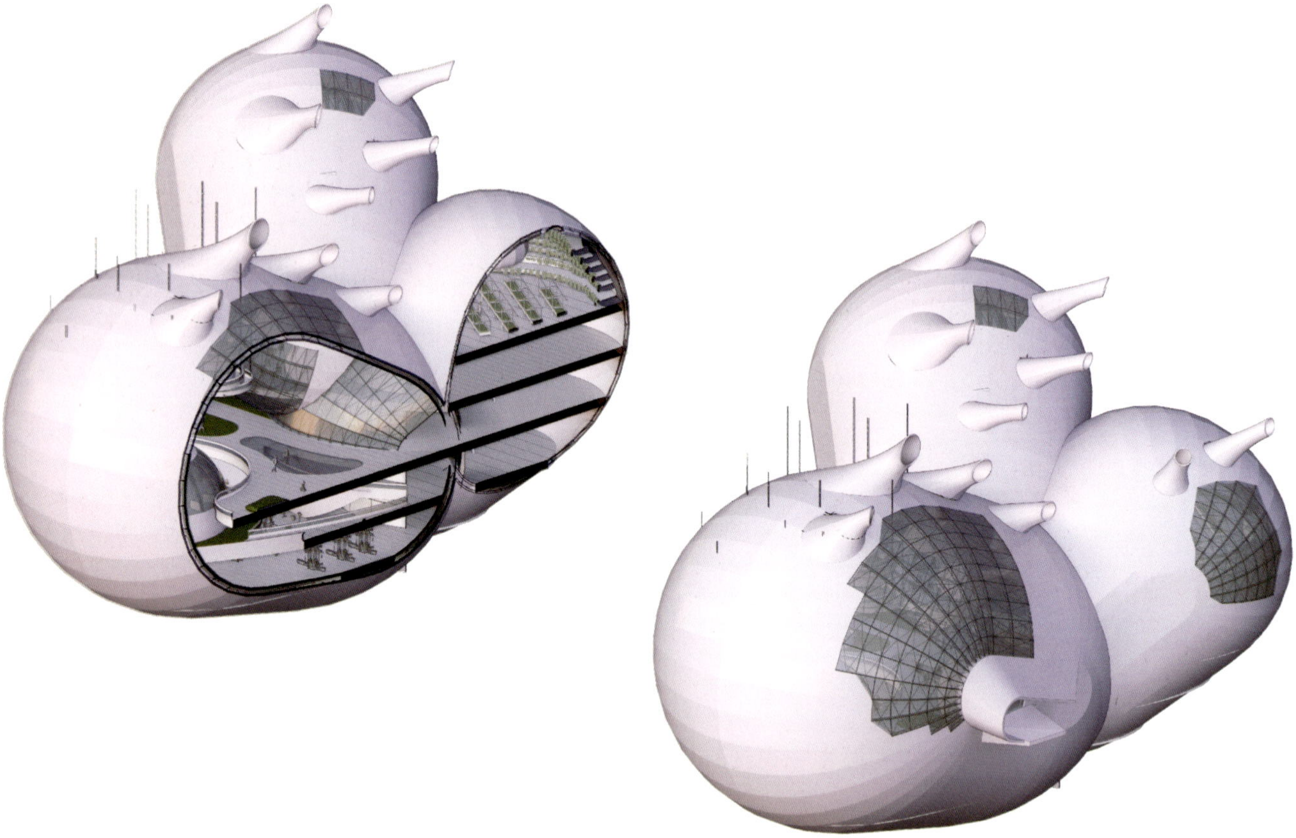

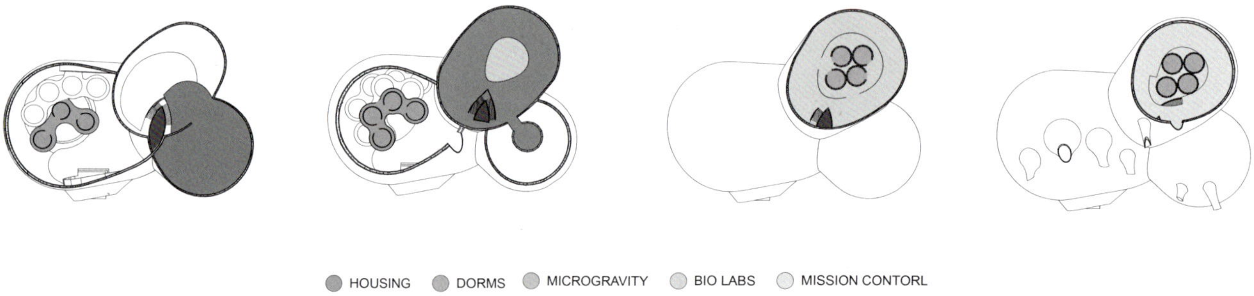

The sectional quality of the research base allows for the layering of zones for research, housing, entertainment, and training. The intersecting circular forms of the exterior allow for the efficient use of material and space and modular construction. It is from mission control that the coordination of all activities and monitoring of exterior temperatures, pollution, sea rise, etc., is monitored.

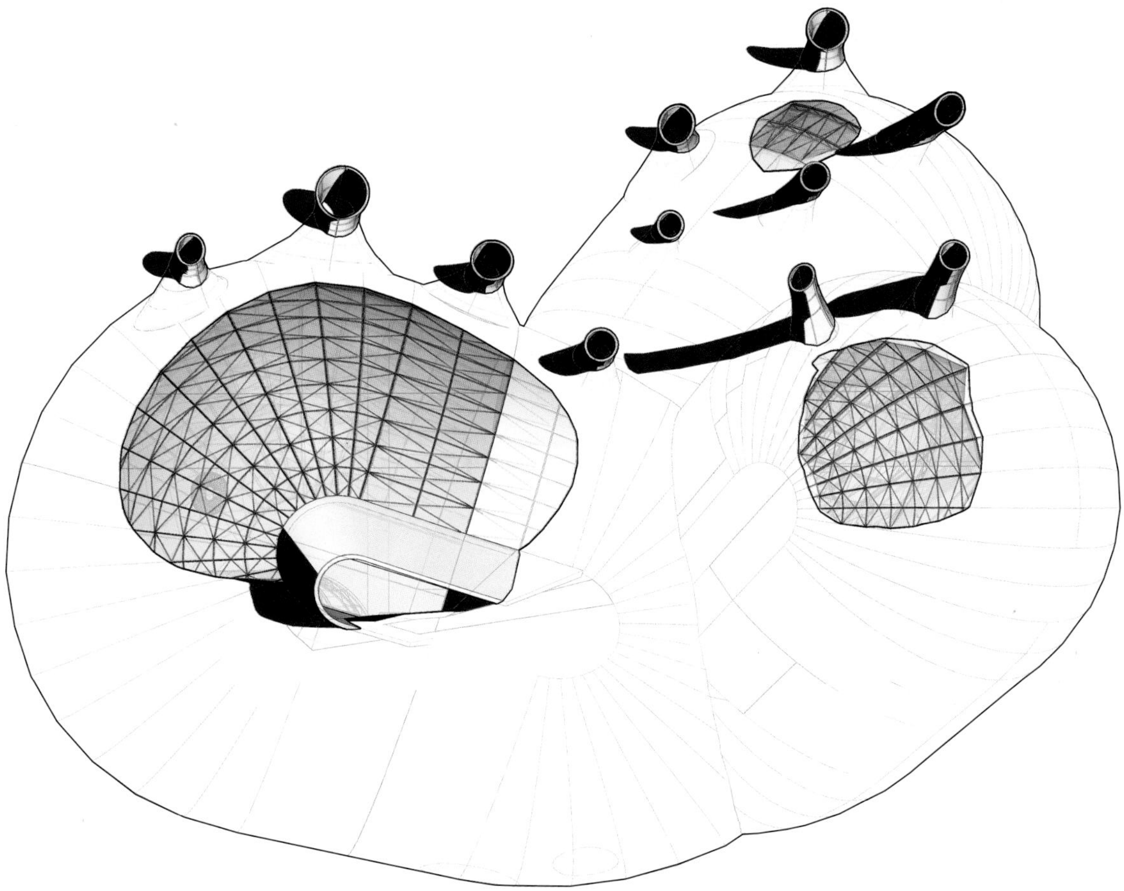

NO BORDERS | SPACE

MAGNETISM: INTERACTING LIVING PODS

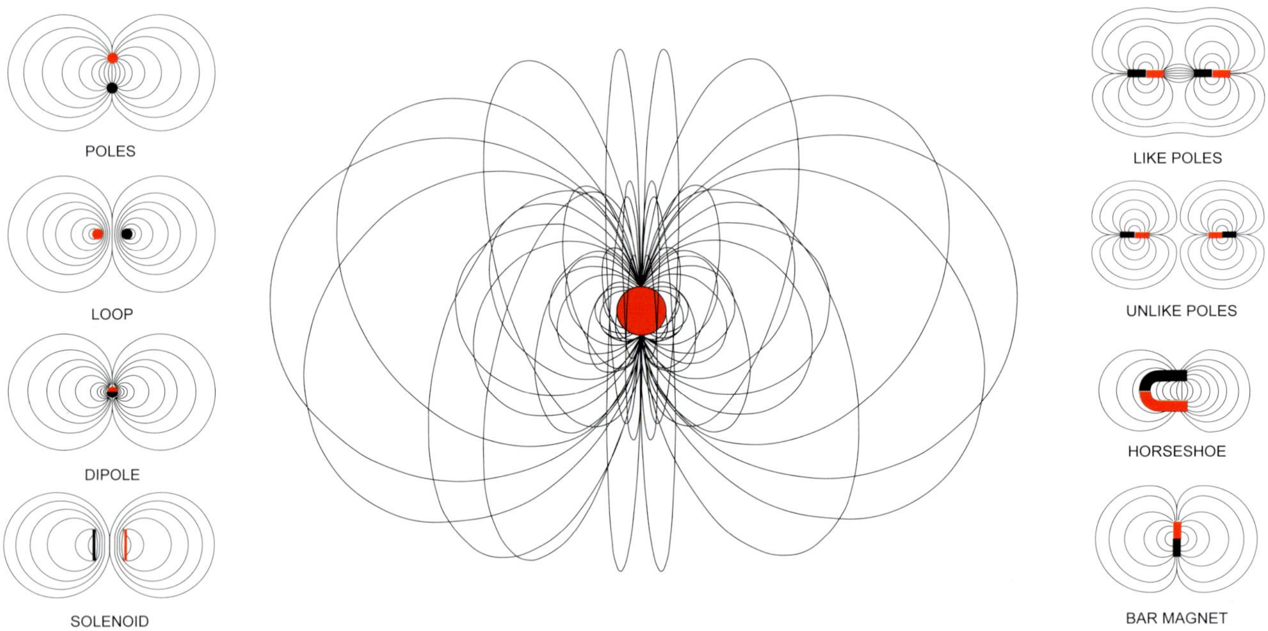

LINES OF FORCE[162]

The initial design of the center was based upon establishing magnetic forces that positioned the primary spaces of the facility. The difference between true north and magnetic north[163] was calculated in locating the actual orientation of the project. Sky scoops have been oriented to the sun positions in this cold region.

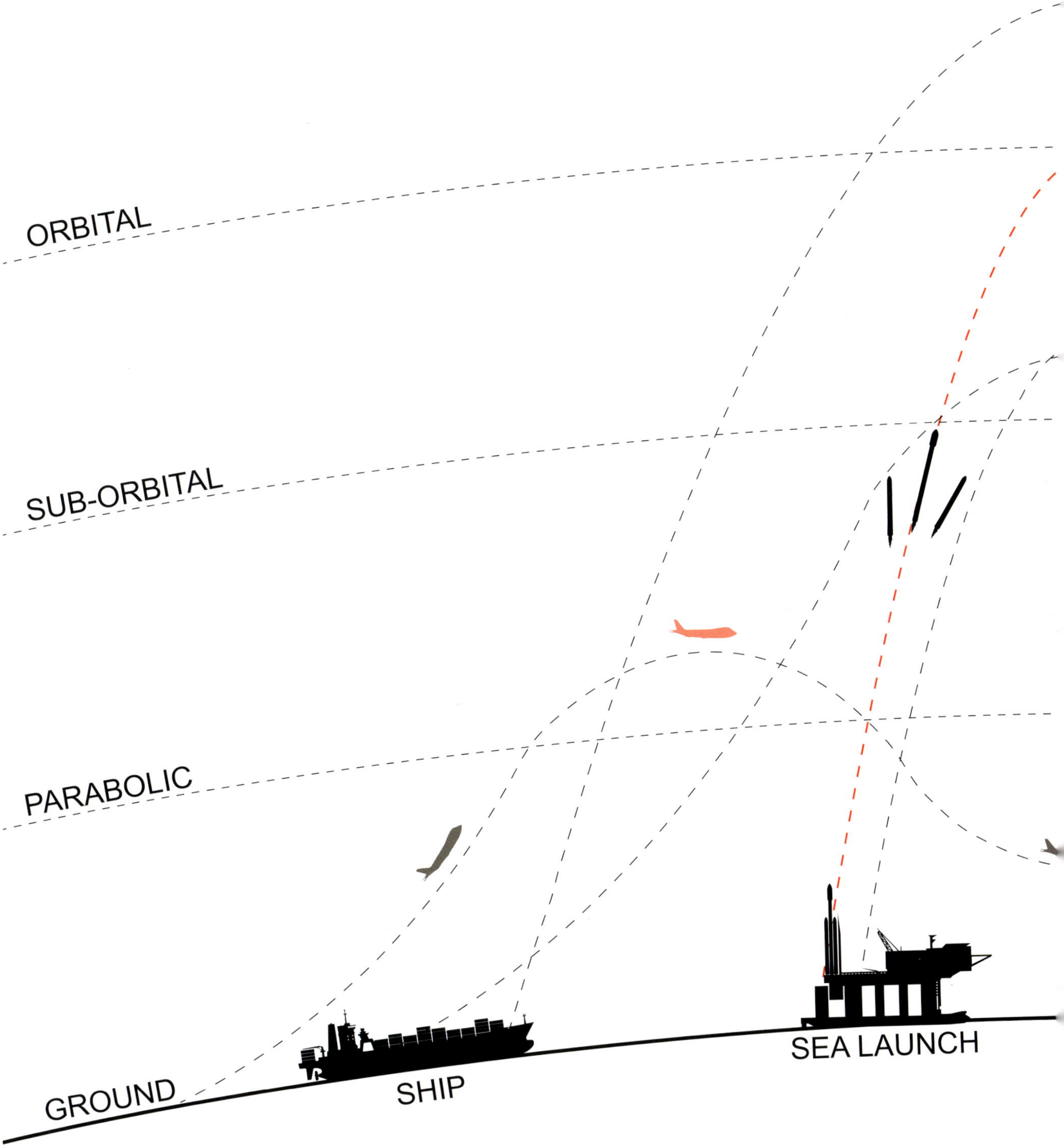

NO BORDERS | SPACE

FLIGHT PATTERNS

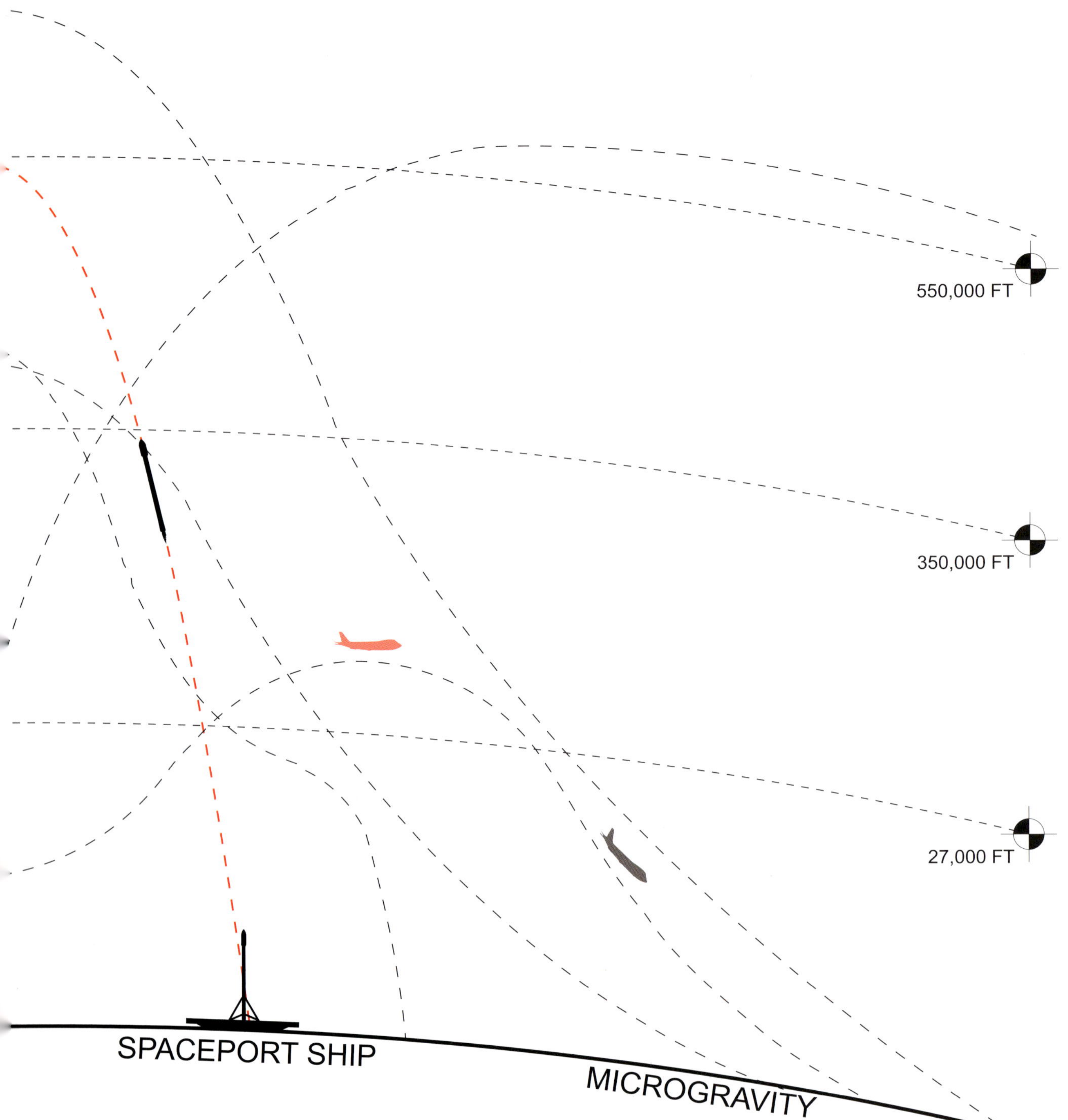

The flight patterns of the earth's atmosphere is mapped for certain types of air travel at the Spaceport. The layers of the atmosphere are critical for those training for space to experience weightlessness. The rockets launched from a platform in the sea returns to the same landing spot to be reused for each flight.

DEFCON 5

2022 1.0°

VERY LARGE

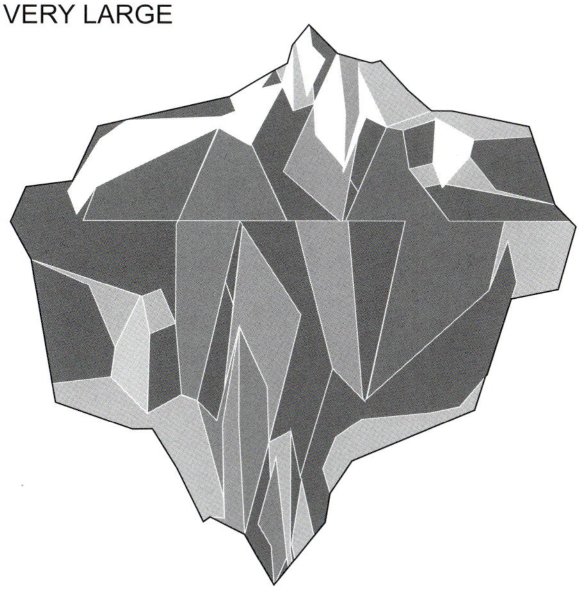

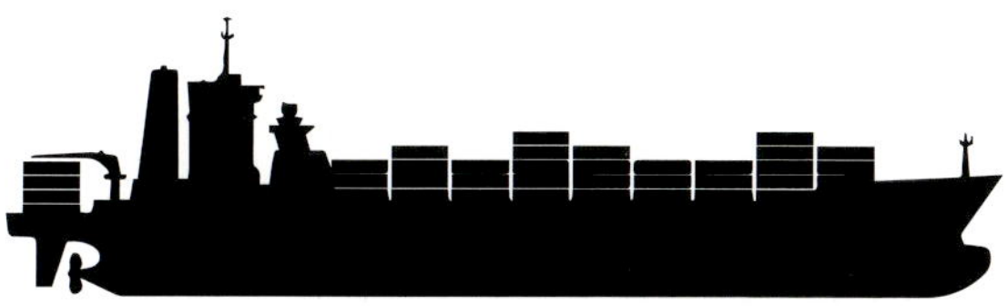

CARGO 16,515 M³
WEIGHT 10,000 T
LENGTH 475 FT

DEFCON 4

2030 1.5°

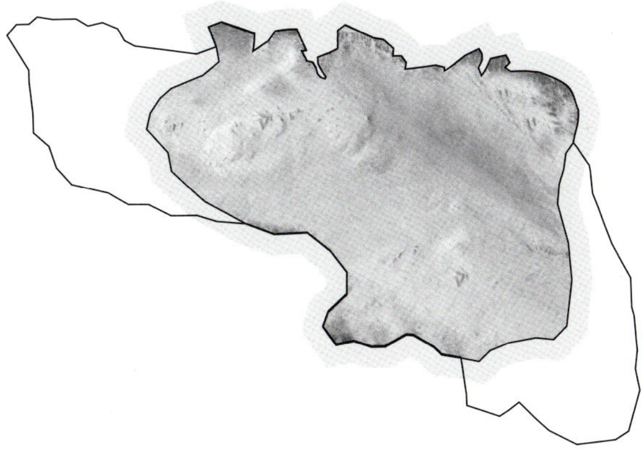

LARGE

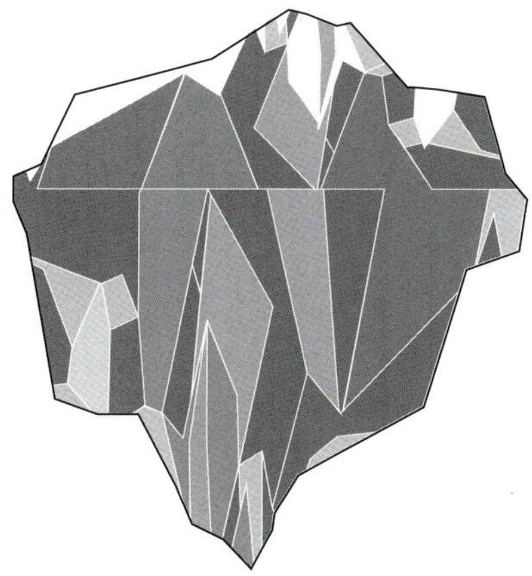

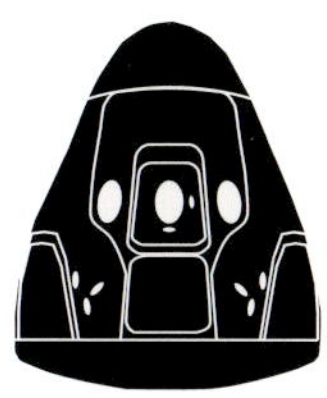

CAPACITY 4
WEIGHT 6,614 LBS
HEIGHT 26.7 FT

DEF

205

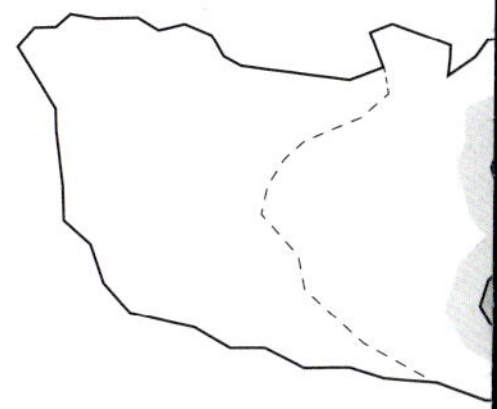

MEDIUM

CAPACI
WEIGH
HEIGH

NO BORDERS | SPACE

DEFCON 1: SURVIVAL OF HUMANITY

PILOT PASSENGER

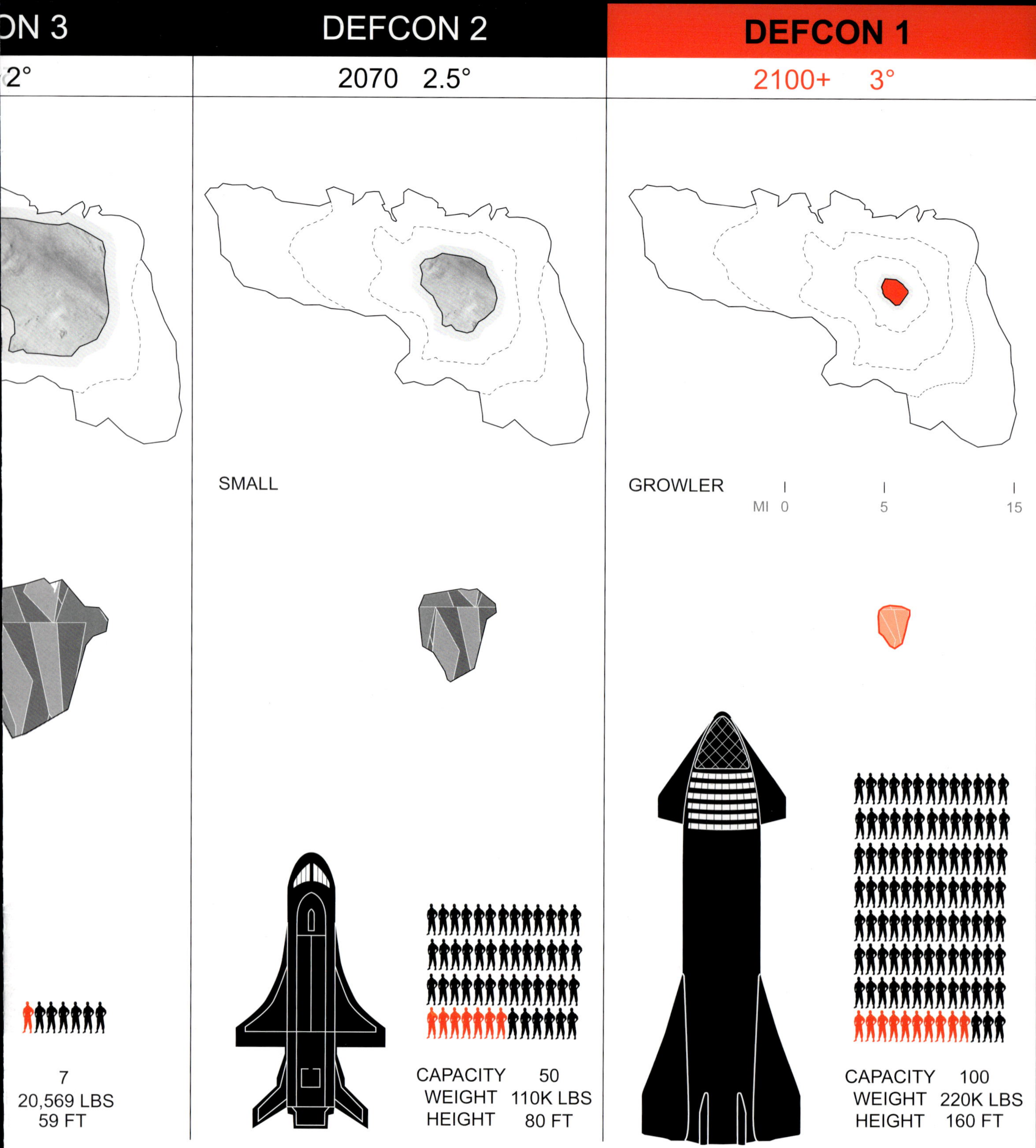

During each stage that the DEFCON alert system for global warming[164] is increased the capacity of the Spaceport increases. As DEFCON 1 is achieved, The Last Iceberg is activated and humans who trained at the facility are now ready to permanently leave the earth and become Martians. This is the final act to save humanity.

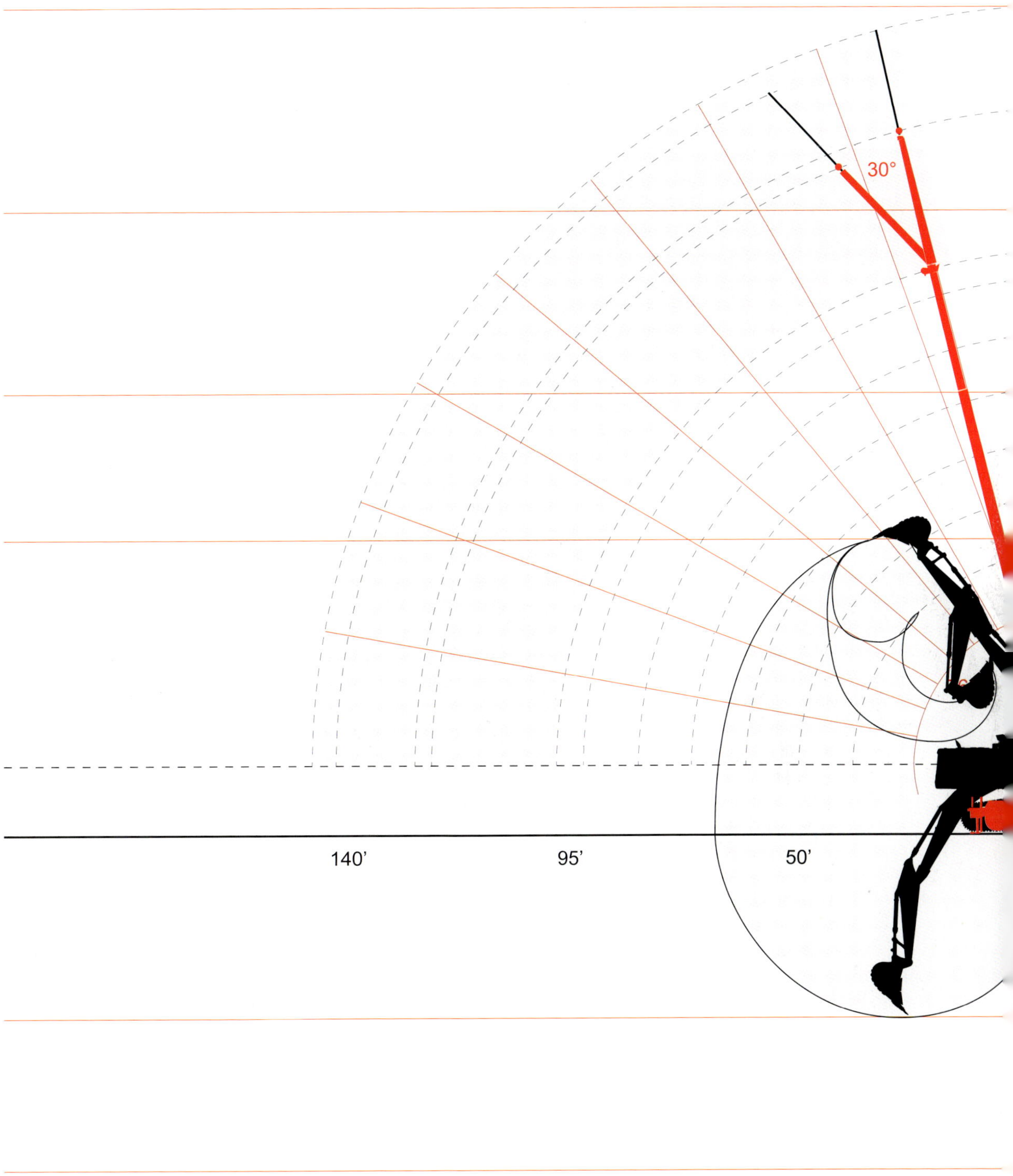

FORMAL CONCEPT: CONSTRUCTION RANGE

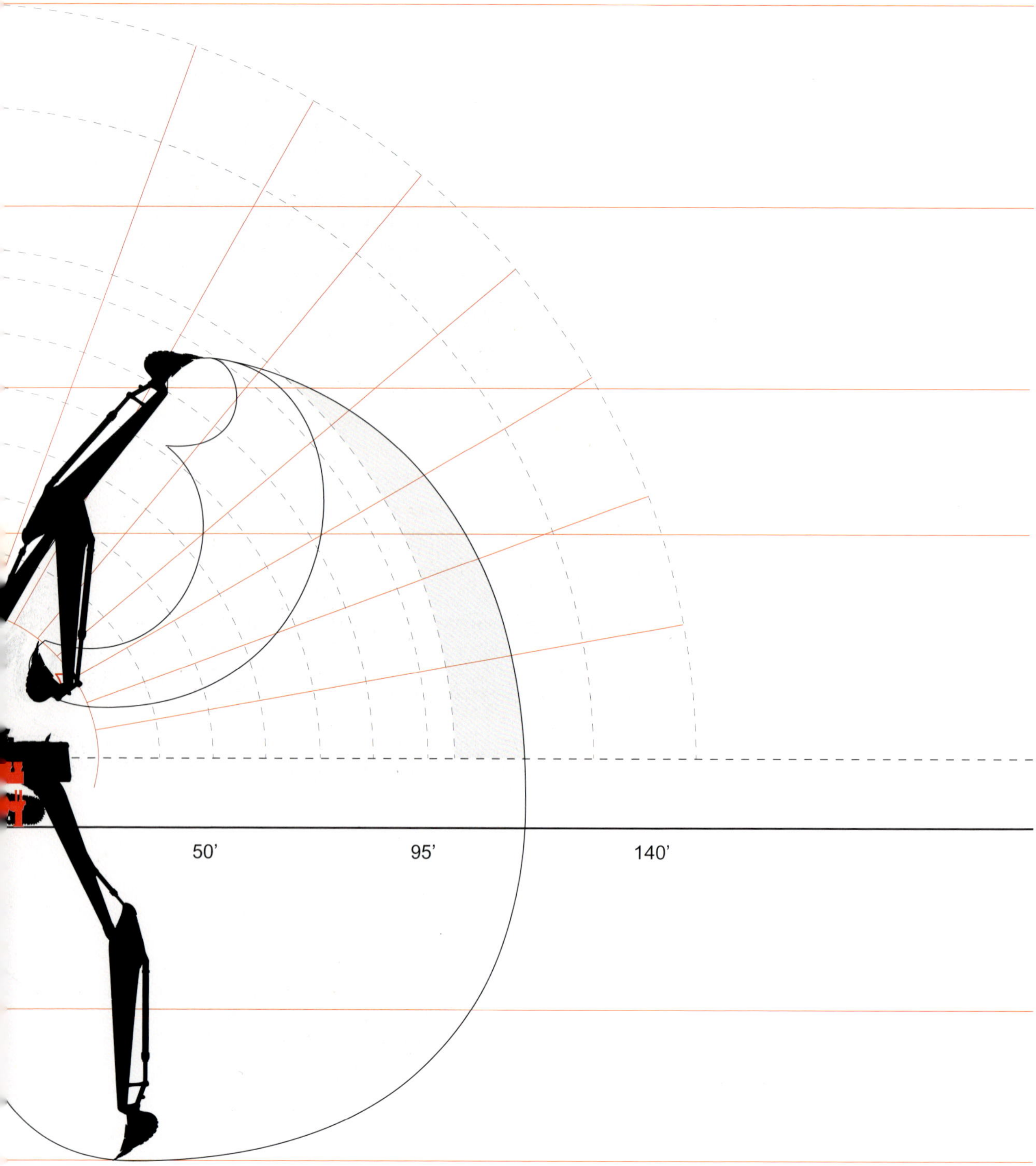

The innovative construction of the Spaceport begins with a modified excavation digger truck with an arm span that can evacuate the snow within the iceberg. This system of burrowing into the snow and removing a cavity for the construction of the center is efficient, has a controlled atmosphere, and extends the building season.

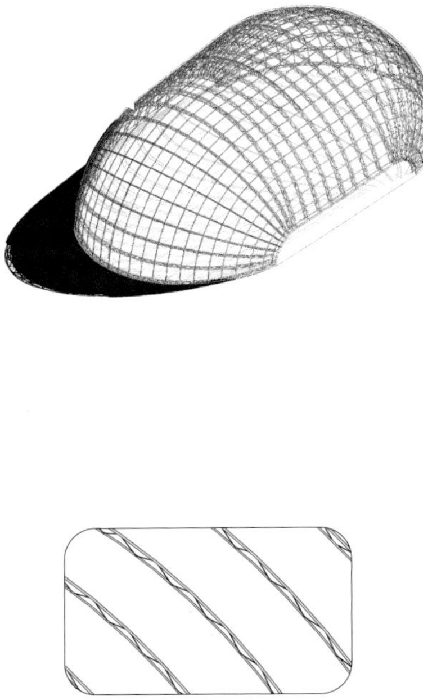

CARBON FIBER STEEL

STRONGER / LIGHTER THAN STEEL
HIGH TENSILE STRENGTH
LOW THERMAL EXPANSION

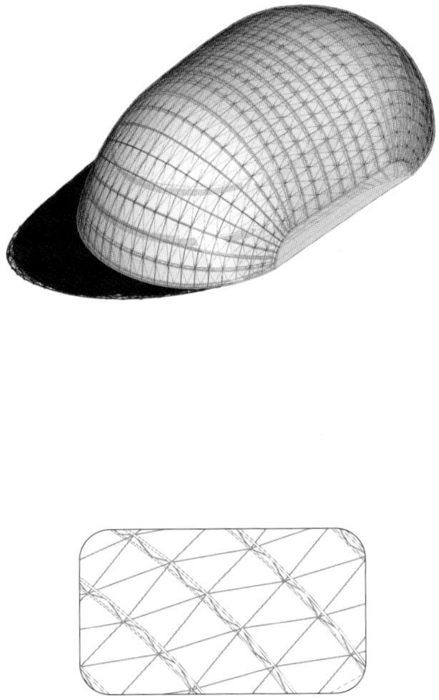

NANO-AEROGEL

LOW DENSITY SOLID
HIGH STRENGTH
R-20 PER INCH

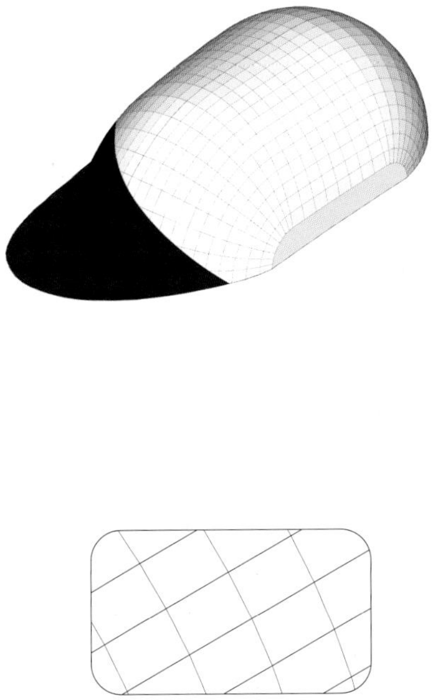

GRP PANEL

GLASS FIBER REINFORCED PLASTICS
POLYISOCYANURATE INSULATION
R-6.5 PER INCH

NO BORDERS | SPACE

ENVELOPE: MATERIAL + LIGHT SCOOPS

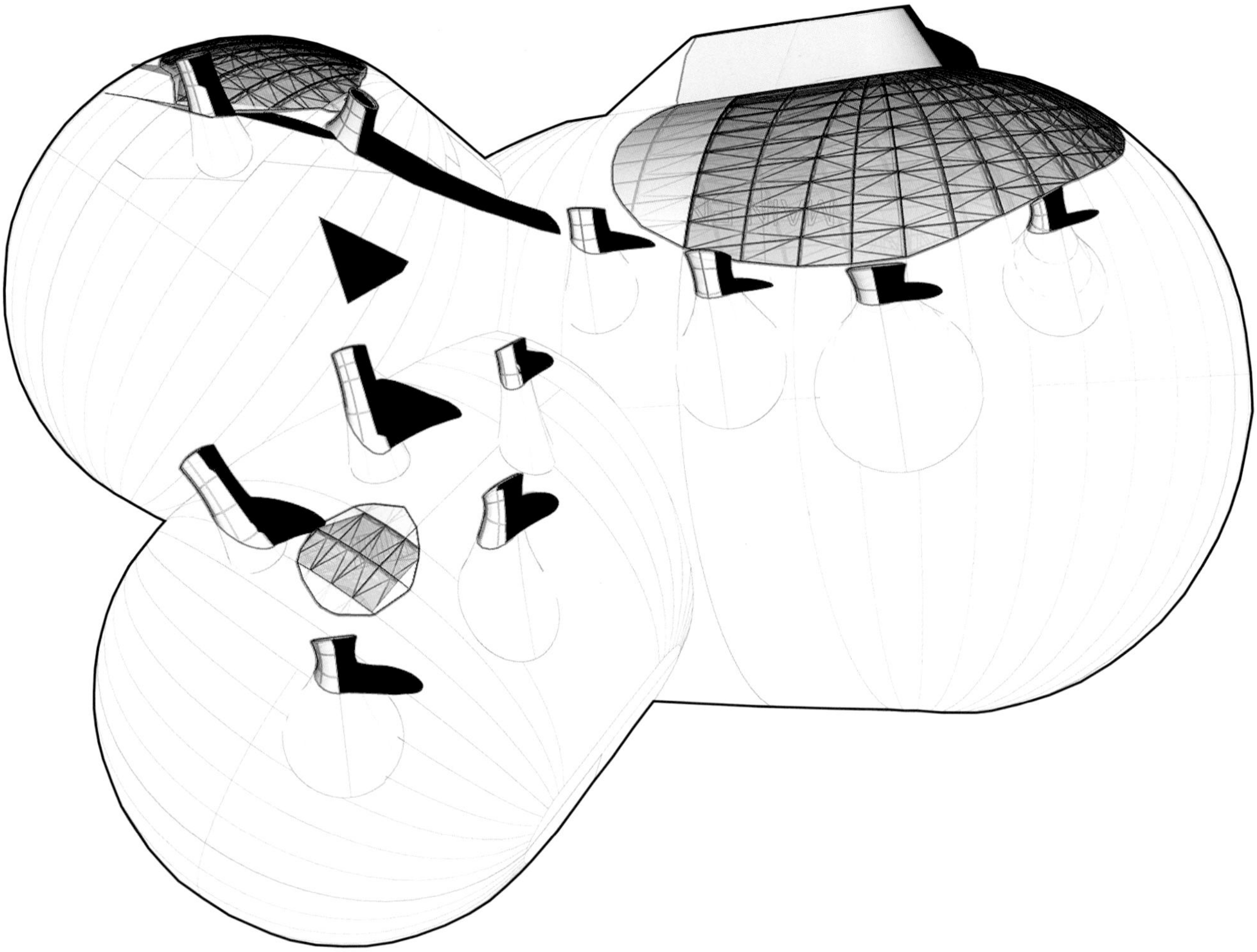

The material used for the construction of the Spaceport draws upon the latest technologies being developed. The lightweight, high strength carbon fiber is used for the structure and walls with GRP panels as sheathing. The nano-gels used in the light scoops capture natural light deep into the interior, while providing high strength and insulation.

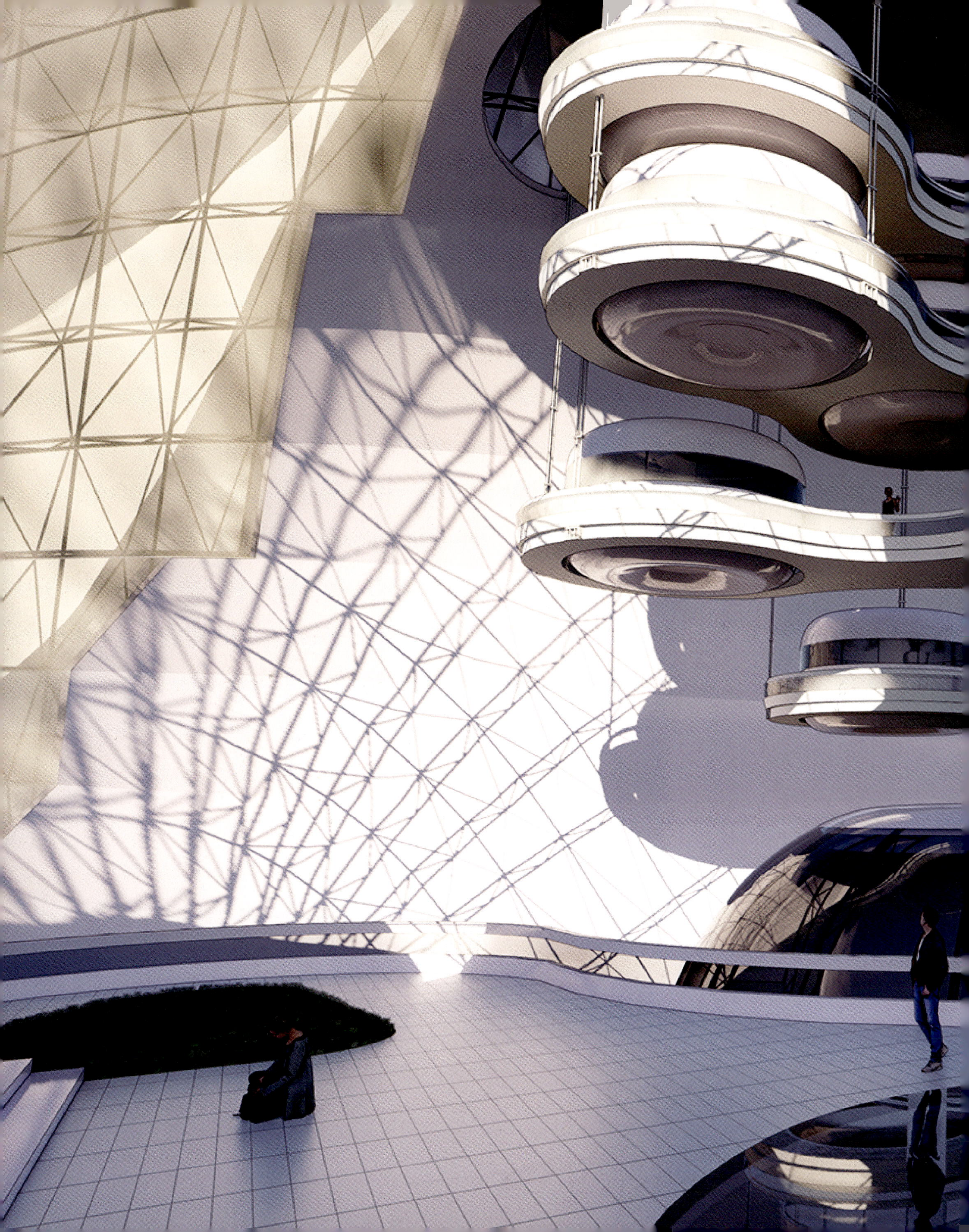

The Last Iceberg is a publicly funded research center composed of a consortium of universities and public institutions. One important focus is to create hydroponic gardens that will become a prototype for living on other planets. The research center will use hydrogen as a source of water for the plants,[165] with solar radiation harvested from the geosynchronous orbit of the satellites above the site which are used for power[166] and diffusing the light rays for growing the plants.[167] The researchers are housed on site in living pods for sleeping and bathing under the extreme conditions of the glacier. The diets are balanced for healthy living, with the trainees undergoing physical and mental training for living on Mars. The calibrated hydroponic gardens are curated for the needs of communities that are being planned for the initial trips to Mars. Courses are given for preparing for travel on spaceships planned for takeoff at the Spaceport. The harsh living conditions on Mars dictate a regimen and discipline from all the trainees coming to the facility. Innovative construction techniques are also taught, from rudimentary levels to more advanced classes, in order for the scouting expeditions to set up camp for the future trips to Mars. Because of the accelerating conditions on Earth due to climate change and the apparent lack of preparations for the worsening weather conditions, lack of water, and sea rise[168] it has become of utmost concern at the Spaceport because the timeschedule is being moved up for those at the facility. The icebergs have already started melting at an alarming rate as DEFCON 1 will be reached shortly.[169] The pleas of scientists, activists, and those wanting the earth to continue to suppport life are so discouraged that humanity has now turned its attention to The Last Iceberg to carry human life to another planet and to try to ensure that humans will continue to exist.

HYDROPONICS: A REAL FOOD SOURCE

NO BORDERS | SPACE

XTREME SELF-CONTAINED ENVIRONMENT

The self-contained pods within the Spaceport allow for a variety of microstructures to exist in innovative ways. Year-round gardens with hydroponic farms and housing pods for the staff, researchers, and tourists create interesting alternative environments that can adapt to changing architectural conditions for those learning to live in extreme environments.

URBAN

GAM

E ON
SUBURBAN

As one looks at the urban realm today the desire to go ever more vertical on small lots has become the new norm. An alternative to this disruption to existing neighborhoods and to the issues of overpopulation,[170] rising property values,[171] and orientation and access to the sun at both the street level and in the air[172] is to understand the possibility of building on top of existing buildings. A Portal to the urban future proposes building a mixed-use structure that anticipates keeping the existing fabric of NYC, but reinventing the technological innovations of making the complex net zero, connecting to mass transit and existing gardens, and establishing a new typology of courtyards in the air that allows the building to use the natural paths of the sun, wind, and rain. These designs will reconnect the city dweller with nature. The retractable roofs, the exterior gardens in the air, and the central atrium space that is connected to elevators that lead to the street level and to the

subway system below all contribute to making the new multi-block development an enhanced midrise with high density instead of having to go vertical and disassemble the existing fabric of the neighborhood. Architects by nature are optimists in trying to predict the future with programs and services. This project takes the position of architects participating in Game On, a visionary look at how the urban realm can change over time in four levels, yet keep the existing texture of the neighborhood and create innovative programs to respond to the outside forces shaping the city. By integrating the economics, demographics, and natural disasters, etc., the game of projecting a future becomes speculative and lends itself to Game On. These possible scenarios become a blueprint to imagine how communities might reinvent themselves. The city, in a constant state of change and a best case scenario becomes a laboratory, an experiment in how humans might live in the future.

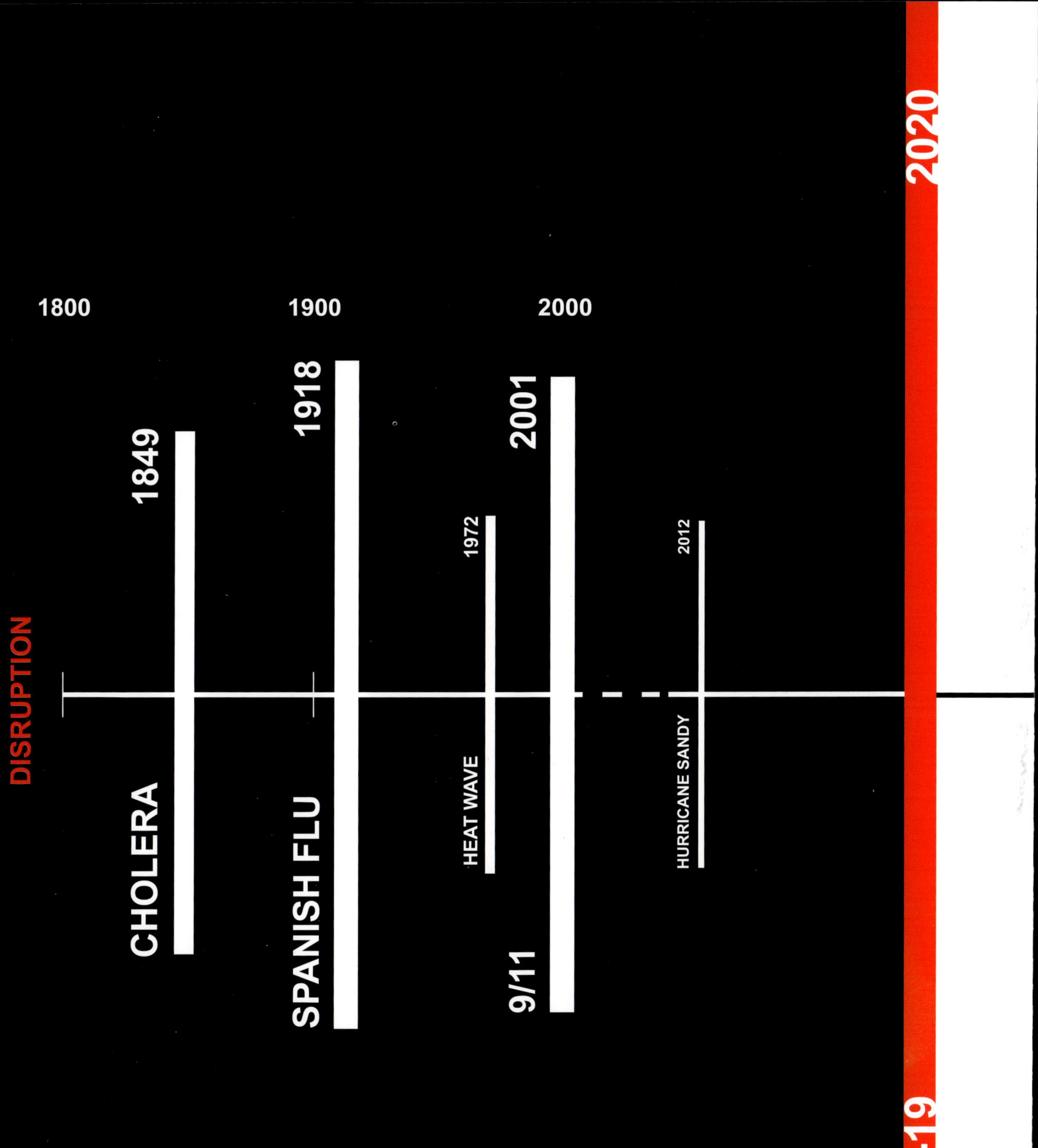

DISRUPTION IN FOUR LEVELS

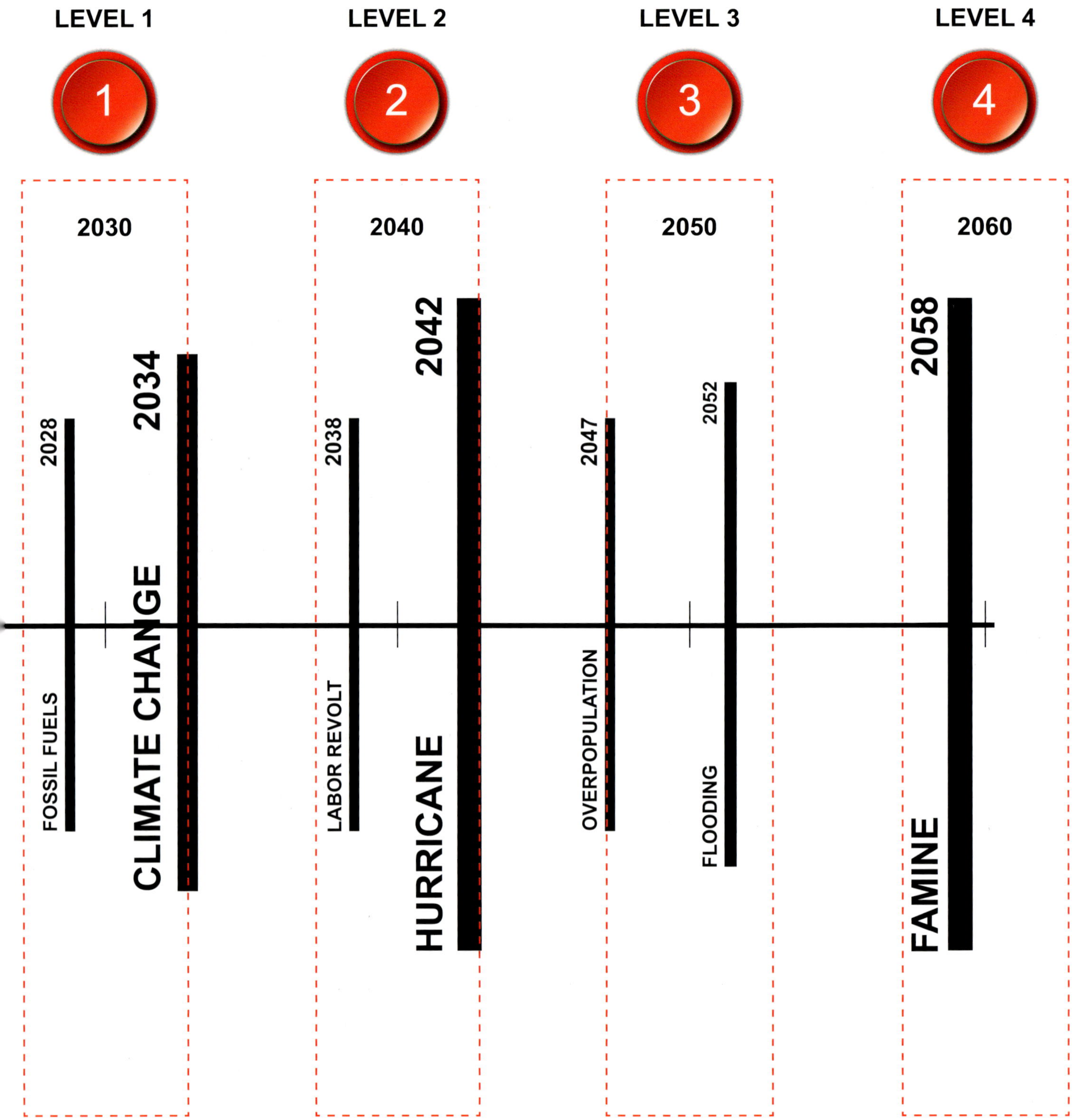

The disruption in the evolution of cities is well documented historically.[173] *However, in the future, so many problems can be classified as being an unpredictable "black swan". The four levels of portals in Game On involve projecting future possible disruptions in the next forty years and allowing the new policies and methods of building to guide the innovative building systems.*

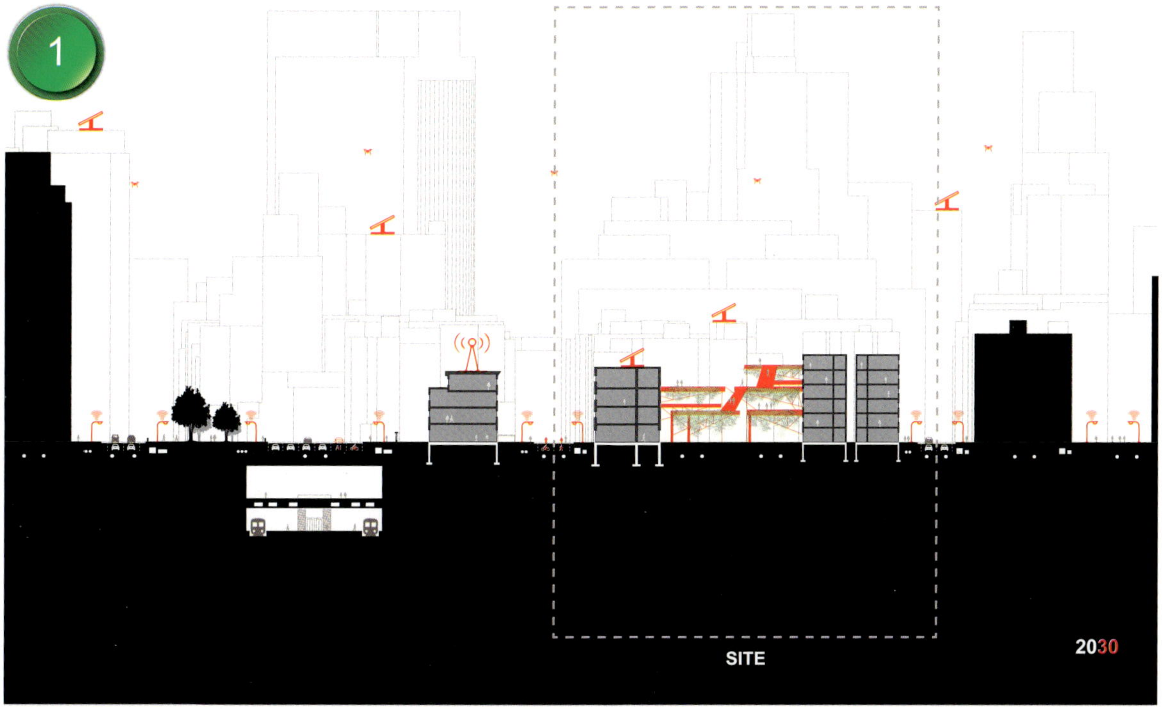
1
SITE
2030

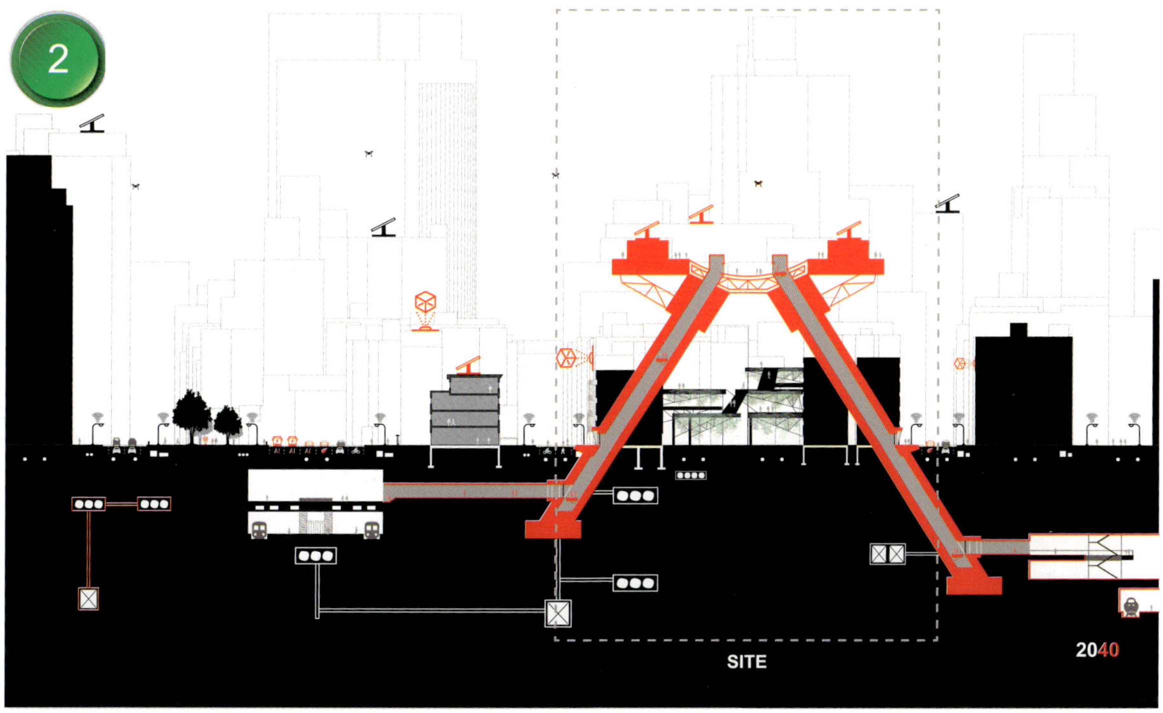
2
SITE
2040

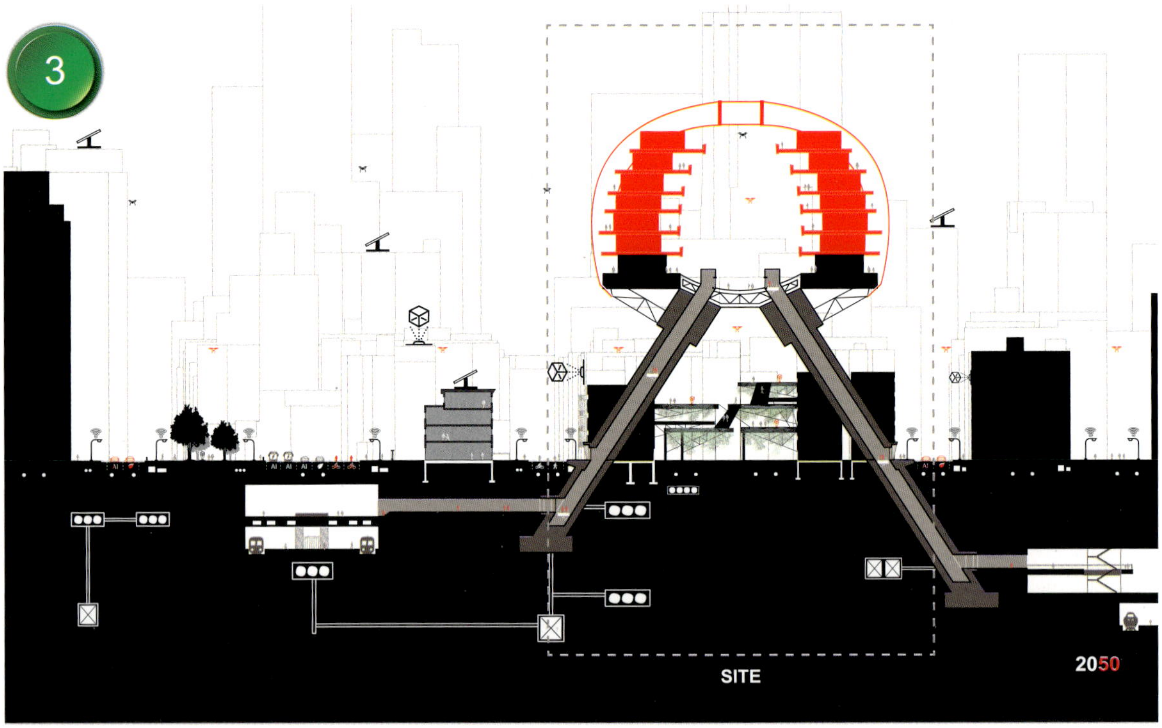

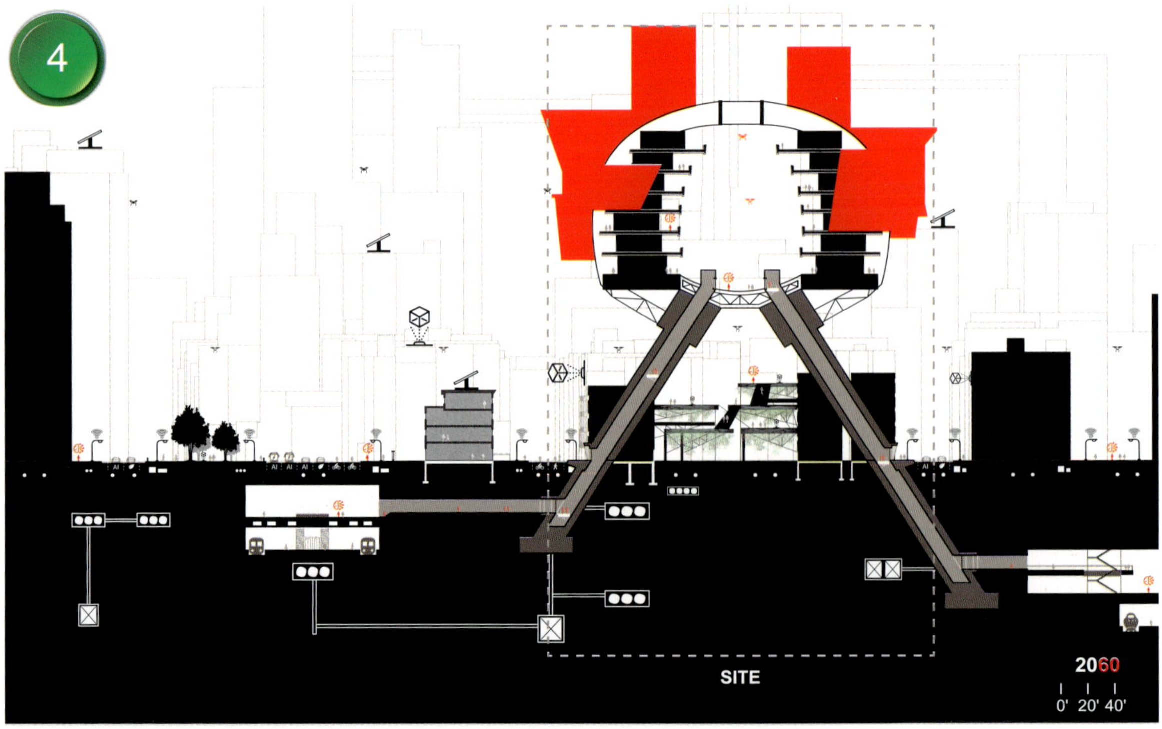

The sectional timeline of the four levels of the portals in Game On is quite dramatic in the attempt to connect to the underground mass transit system and to build on top of an existing city block. The acupuncture of the steel legs piercing the existing buildings is used as support for a platform for construction in the future phases and for all infrastructure pipes and circulation systems.

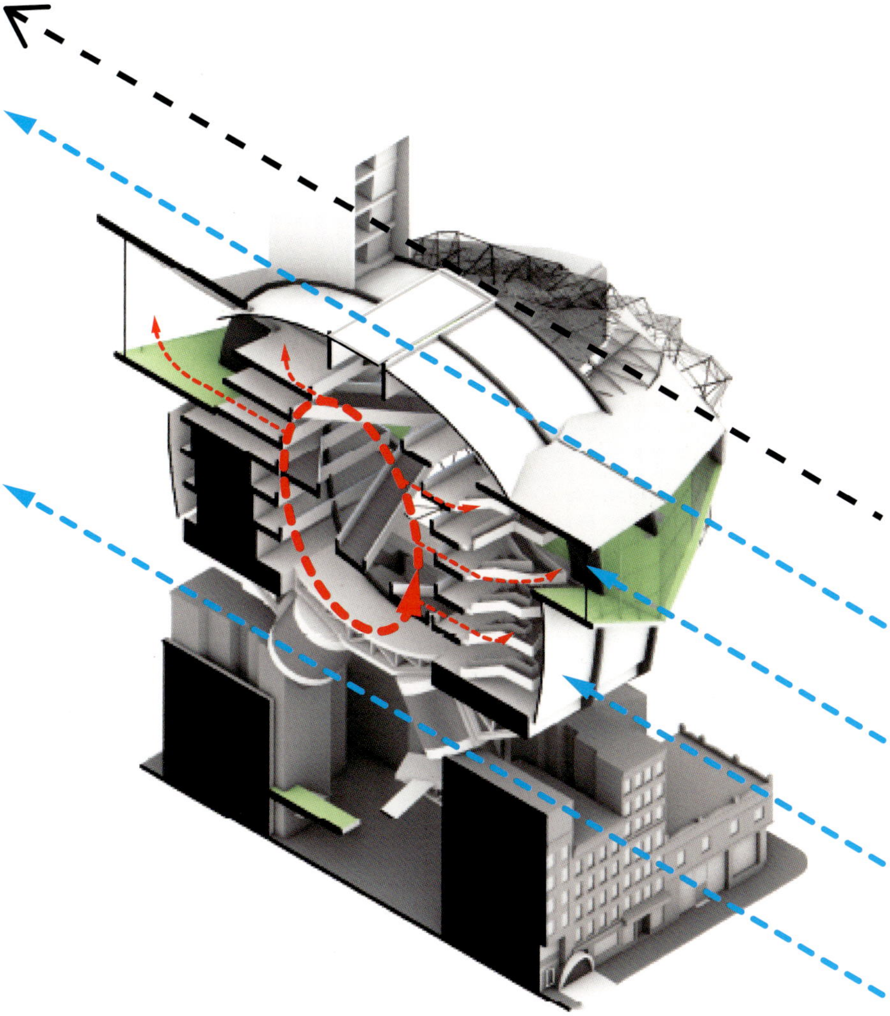

CLOSED

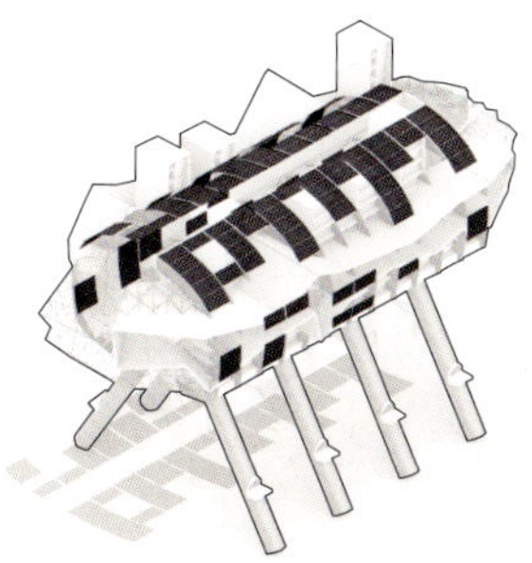

SOLAR COLLECTION
40,000 SQ FT

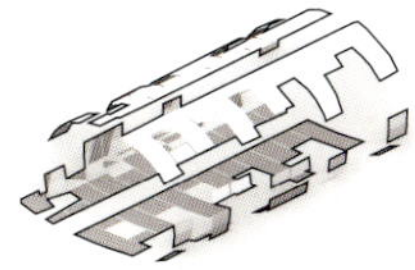

126% OF **ENERGY** NEEDS
PROVIDED BY SOLAR

1,215 kWh PER **PANEL** ON
SMART SLIDING TRACK

GAME ON | URBAN

AN **OPEN AIR** ATRIUM

WARM AIR   COOL AIR   PREVAILING WIND

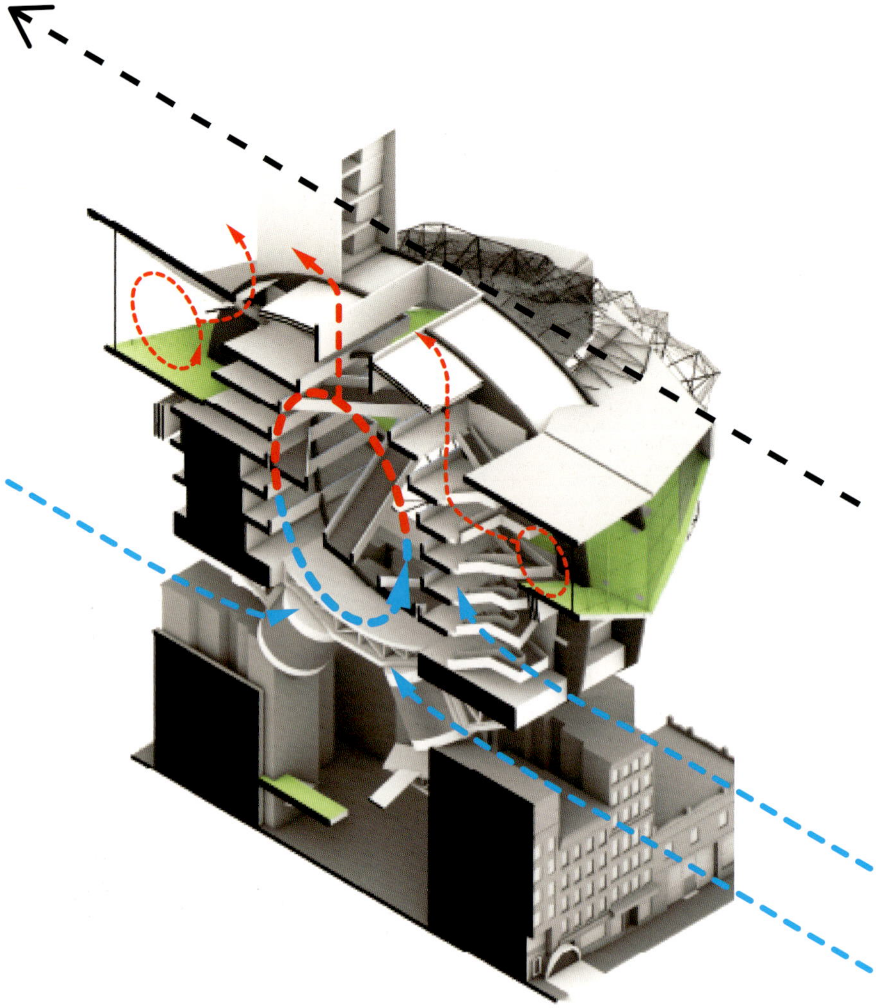

OPEN

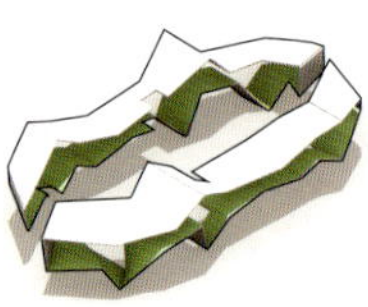

58,000 SQ FT OF PUBLIC GARDENS
FOR LEISURE AND DAYLIGHT

103% OF **FRUIT** AND **VEGETABLE** NEEDS
PROVIDED BY **HYDROPONICS**

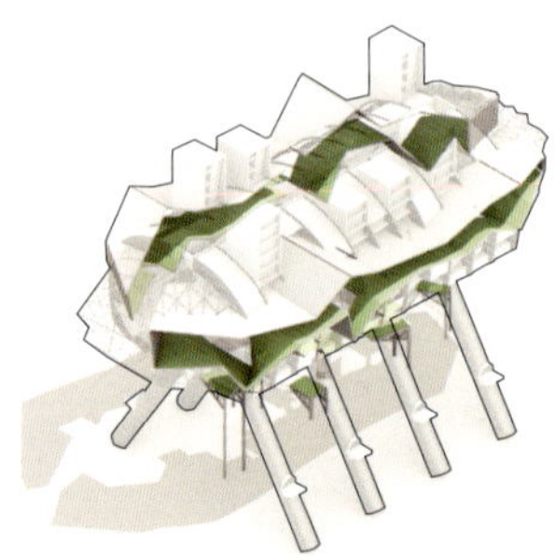

GREEN SPACE
70,000 SQ FT

The central atrium space is designed as an open (summer, fall, and spring) and closed (winter) system. The retractable ceiling is used to circulate air for cooling breezes in the warmer months with negative pressures created at the roof level for containing the heat in the winter months when the roof is closed. These passive systems allow the building to self-regulate the air temperature.

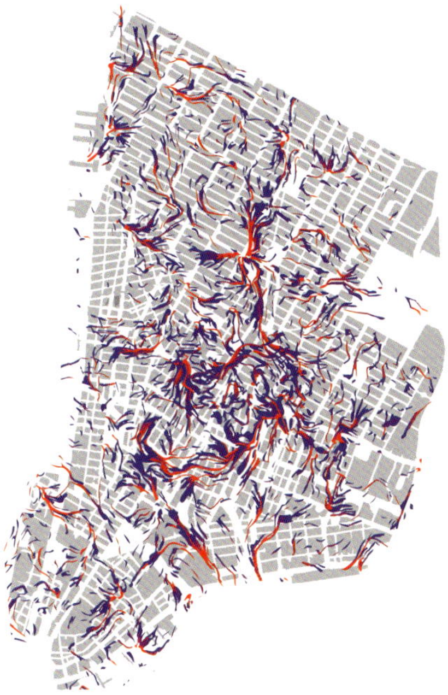

MOST ACTIVITY LEAST ACTIVITY

HUMAN ACTIVITY BEFORE COVID

MOST ACTIVITY LEAST ACTIVITY

HUMAN ACTIVITY AFTER COVID

> 90% OCCUPANCY > 50% OCCUPANCY > 10% OCCUPANCY

BUSINESSES BEFORE COVID

> 90% OCCUPANCY > 50% OCCUPANCY > 10% OCCUPANCY

BUSINESSES AFTER COVID

AN URBAN **CONTEXT**

During the pandemic NYC, like other cities around the world, experienced a retraction of human activity that allowed nature to return with animals reappearing in the city[174] and less pollution in the atmosphere[175] among other documented changes. The level of business activity in the city also diminished, which affected the residential flight to the suburbs during the pandemic.[176]

RESIDENTIAL PASSPORT TO THE METAVERSE

PERSONALIZE AN AVATAR: 2022

NOMAD RASCAL GUARDIAN FASHIONISTA WARRIOR

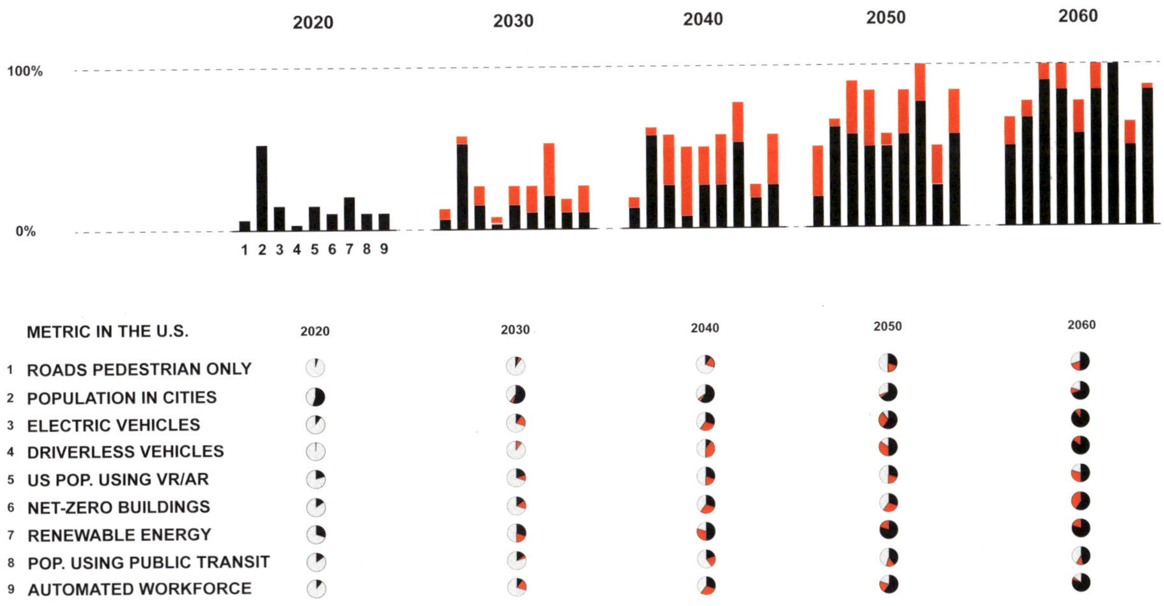

THE EVOLUTION OF TECHNOLOGY: 2020-2060[177]

■ EXISTING ■ PROJECTED

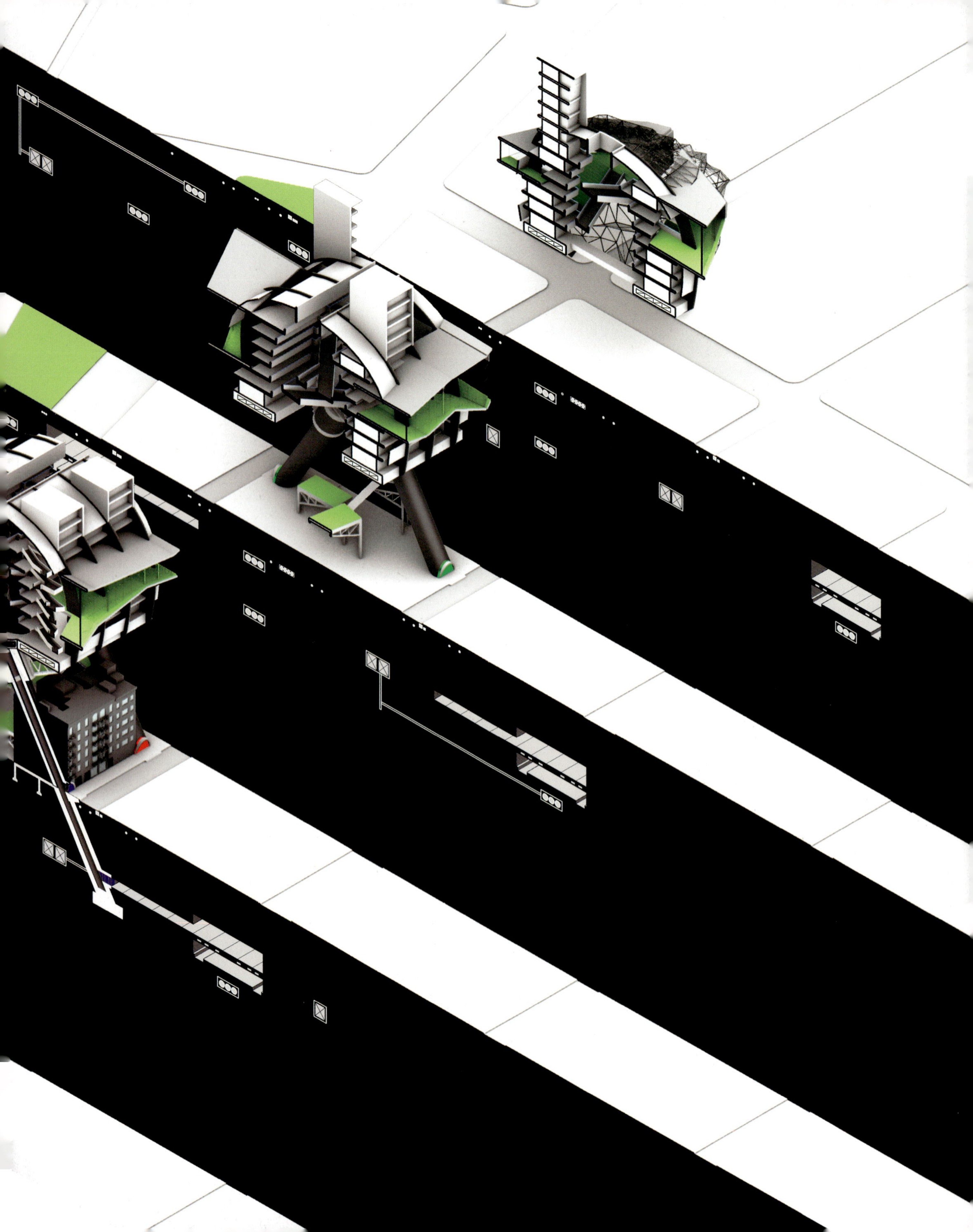

PROGRAM

216,000 SQ FT

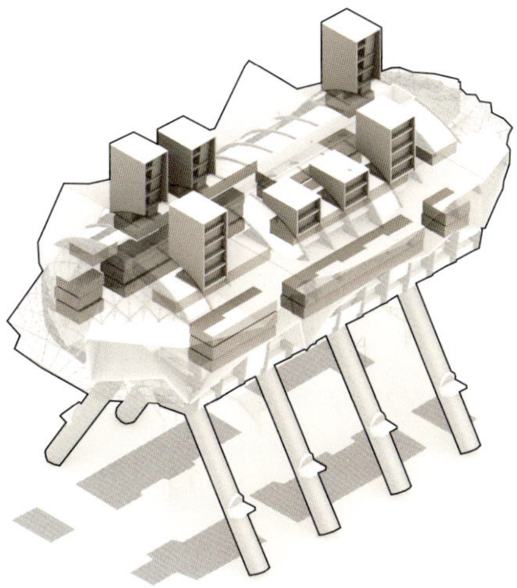

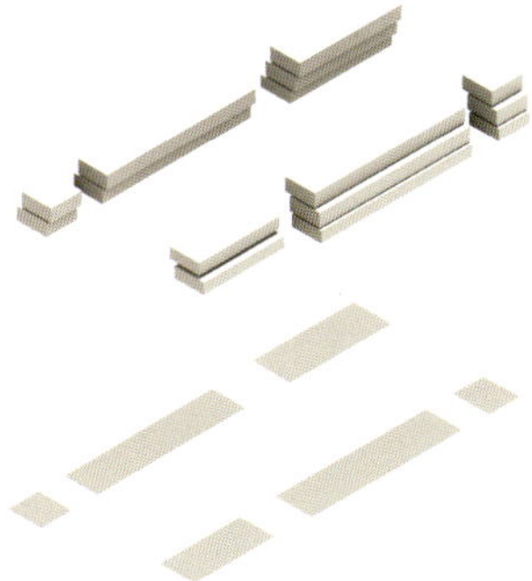

COMMERCIAL

26,000 SQ FT

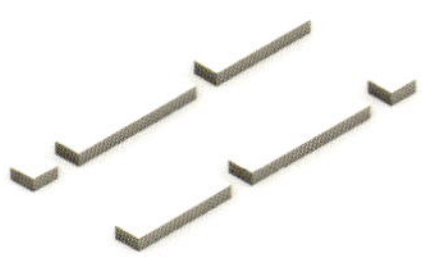

RETAIL - 18 UNITS
1,000 SQ FT

HYDROPONIC MARKET
8,000 SQ FT

RESIDENTIAL

142,000 SQ FT

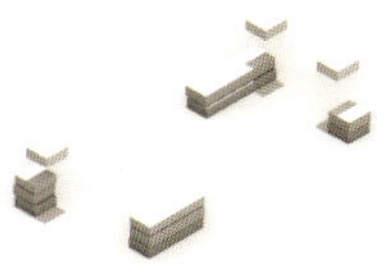

1 BR - 60 UNITS
750 SQ FT

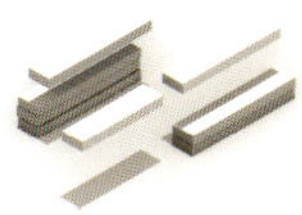

2 BR - 40 UNITS
1,300 SQ FT

PENTHOUSE - 50 UNITS
900 SQ FT

OFFICE

48,000 SQ FT

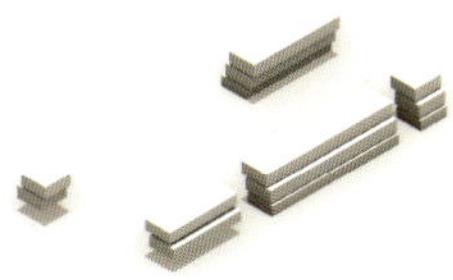

LEASED - 30 UNITS
1,000 SQ FT

CO-WORKING - 4 UNITS
4,000 SQ FT

GAME ON: PROGRAM + STRUCTURE + ENVELOPE

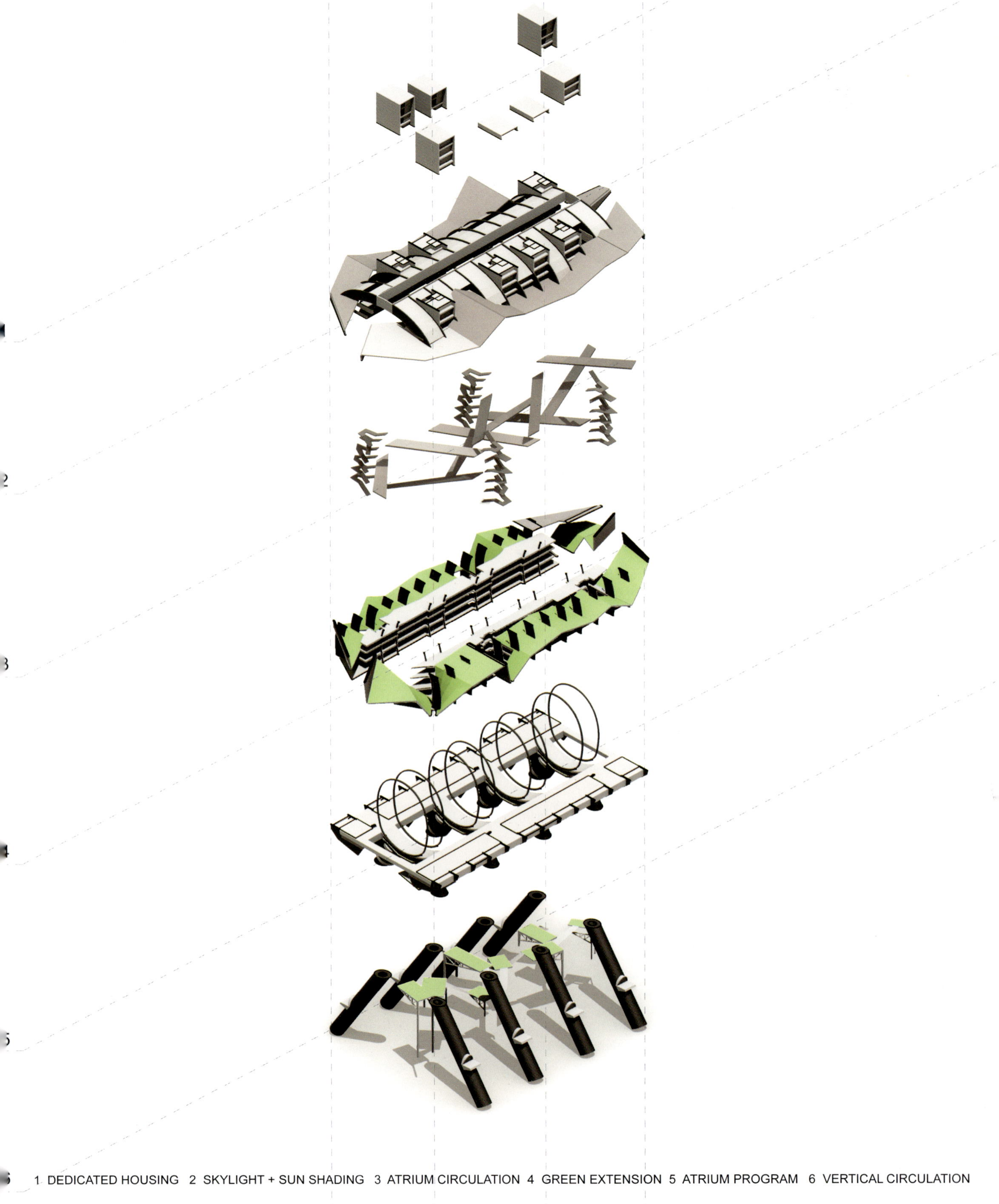

1 DEDICATED HOUSING 2 SKYLIGHT + SUN SHADING 3 ATRIUM CIRCULATION 4 GREEN EXTENSION 5 ATRIUM PROGRAM 6 VERTICAL CIRCULATION

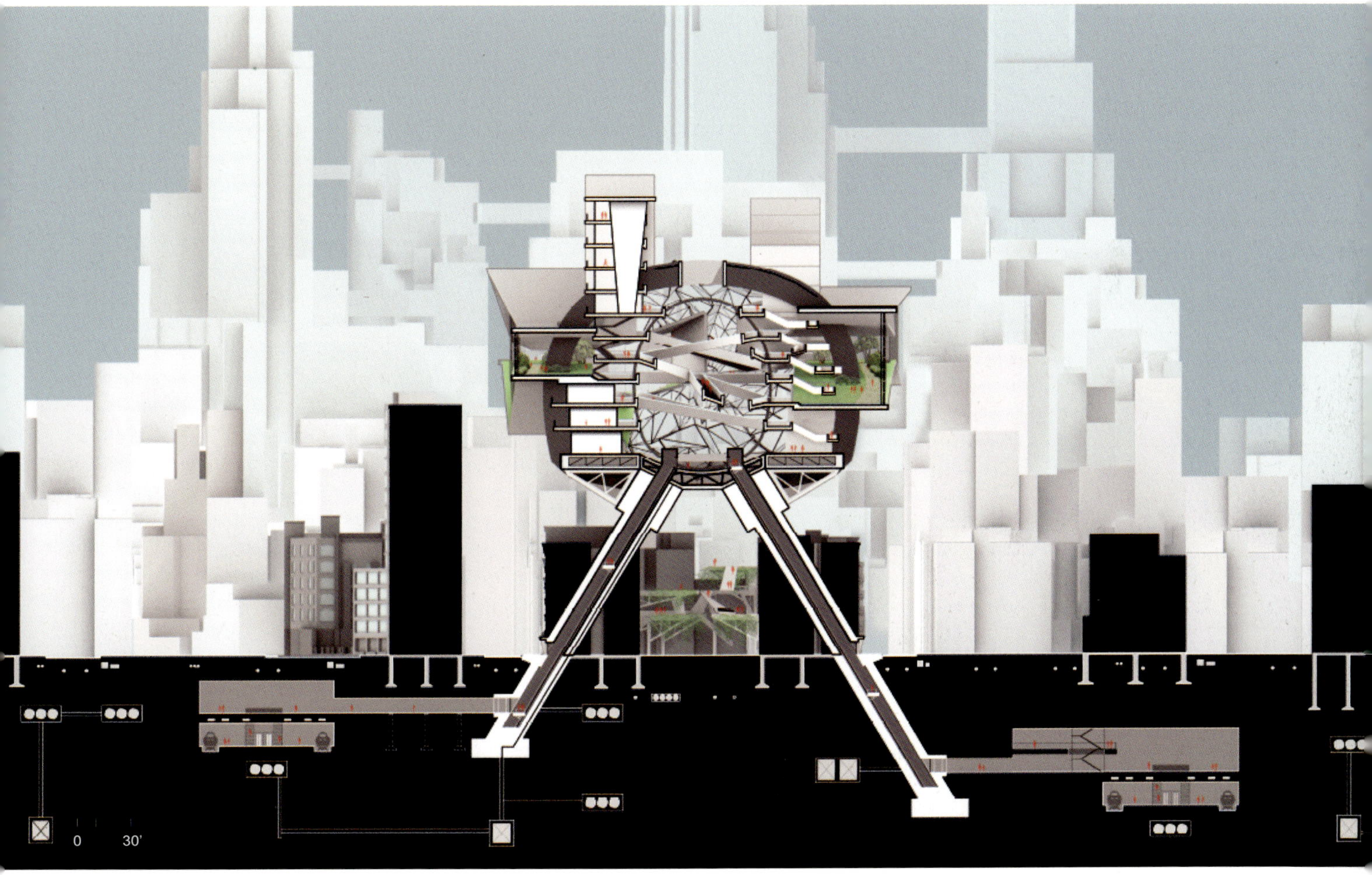

GAME ON: LOOKING THROUGH THE URBAN REALM

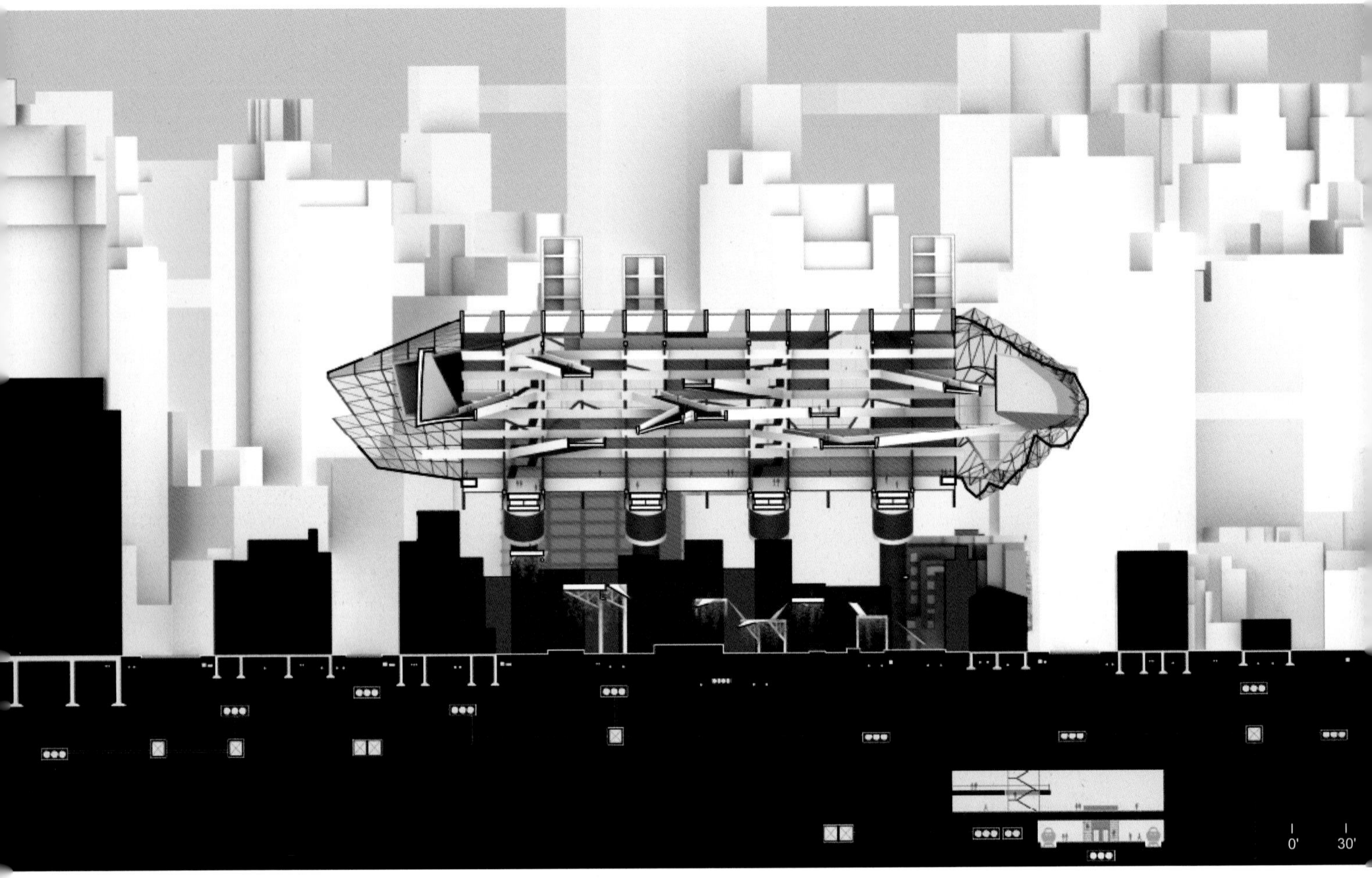

The sections and elevations demonstrate the innovative relationship of the new to the existing buildings by the interaction with the urban fabric. The use of air space above the buildings allows a midrise structure to be built instead of destroying the existing, well established character of the neighborhood. This prototype could serve as a way for cities to move forward.

The nineteenth century city revolved around a central public space that reached only four stories in height and was organized around streets and porches. Cities expanded vertically thanks to elevators and horizontally thanks to the car and mass transit. The problems of density, pollution, equity, safety, the changing relationship of the pedestrian to the street through the new vertical circulation sytems, has produced a new set of challenges to the twentieth century composition of the city. Living in the Clouds reinvents the high rise by placing a neighborhood in the sky with people circulating between buildings in the air. This four-dimensional concept of the city changes the pattern of having to come down to street level, in order to simply go back up an elevator to an adjacent building. Located in Hong Kong, the innovative building construction system in concrete explores creative spatial possibilities that connect the floor levels to the surroundings buildings. The elevators, stairwells,

LIVING INBETWEEN THE CLOUDS

and mechanical shafts serve as major structural supports for the interstitial, hollow concrete building systems that allow a variety of configurations and for expansion as needed. The mixed-use neighborhood uses thin, solid steel columns to support the horizontal plates of concrete with lightweight ash additives to construct the floors. The neighborhood in the air is called, Living with Art in the Clouds and is composed of galleries, residential housing units for visiting artists, workshops with outdoor sculpture gardens, and rental conference / workspaces. The benign weather allows the sliding windows to be opened to the exterior for breezes and to extend the interior gathering areas. The minimal surfaces of the structure creates a sense of transparency that produces a lightness for living in the clouds and appears to dematerialize the large buildings, as if they are floating between the surrounding highrises.

LIVING WITH CLOUDS: A FOUR DIMENSIONAL CITY

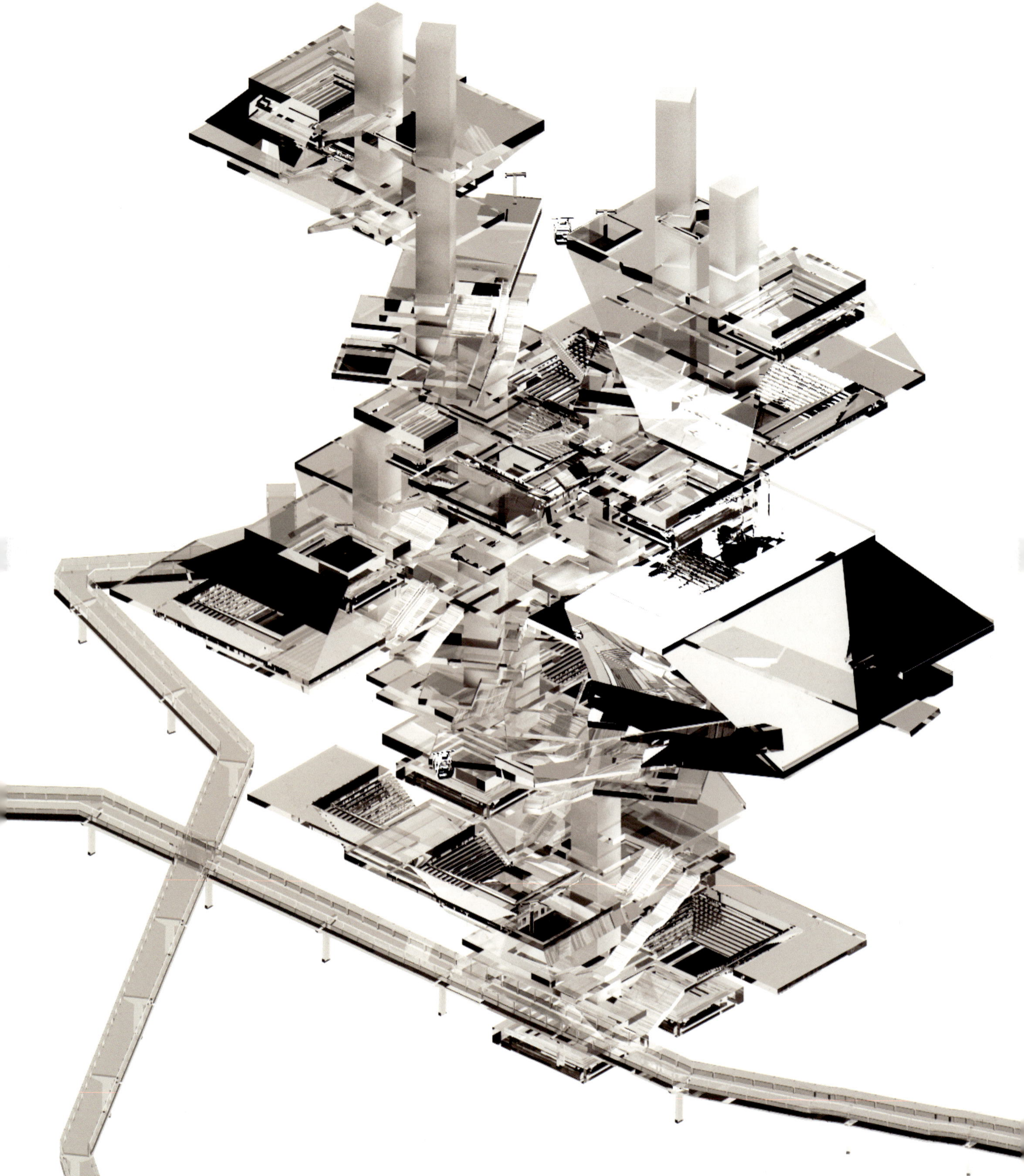

EVOLUTION OF TIME

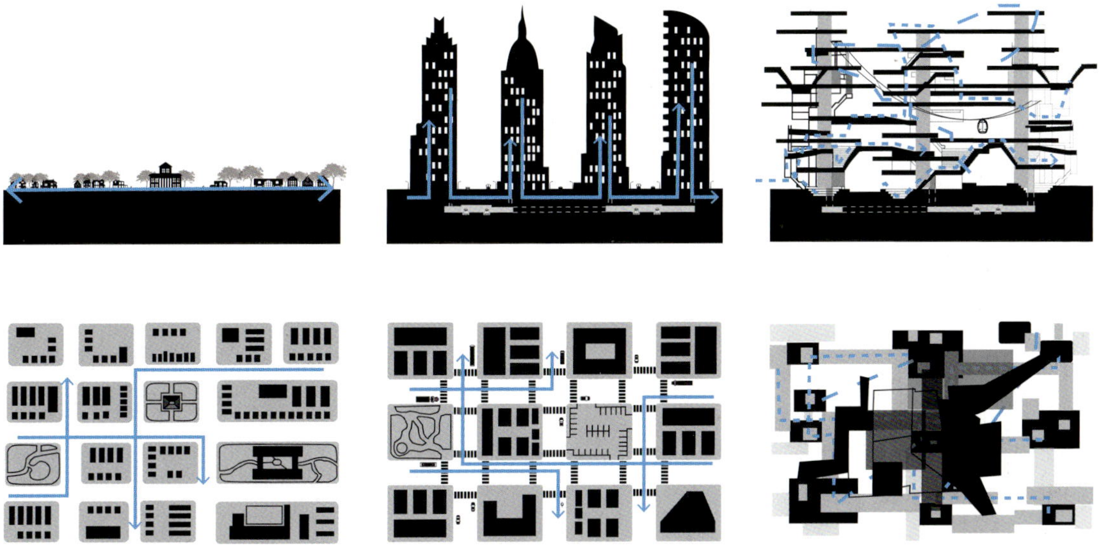

HORIZONTAL GRID
19TH CENTURY

VERTICAL GRID
20TH CENTURY

SPATIAL GRID
21ST CENTURY

HONG KONG

X10 **X100** **X300**

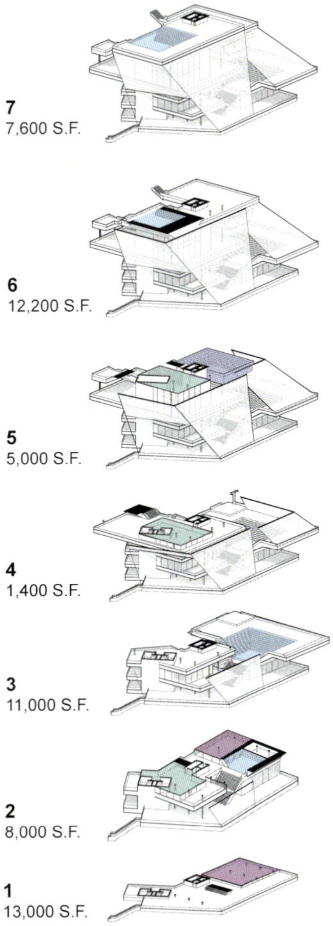

GAME ON | URBAN

SKYSPACE

INTERIOR GALLERIES **1** - 4,172 SF **2** - 1,954 SF
EXTERIOR GALLERIES **2/3** - 3,531 SF **6/7** - 3,531 SF

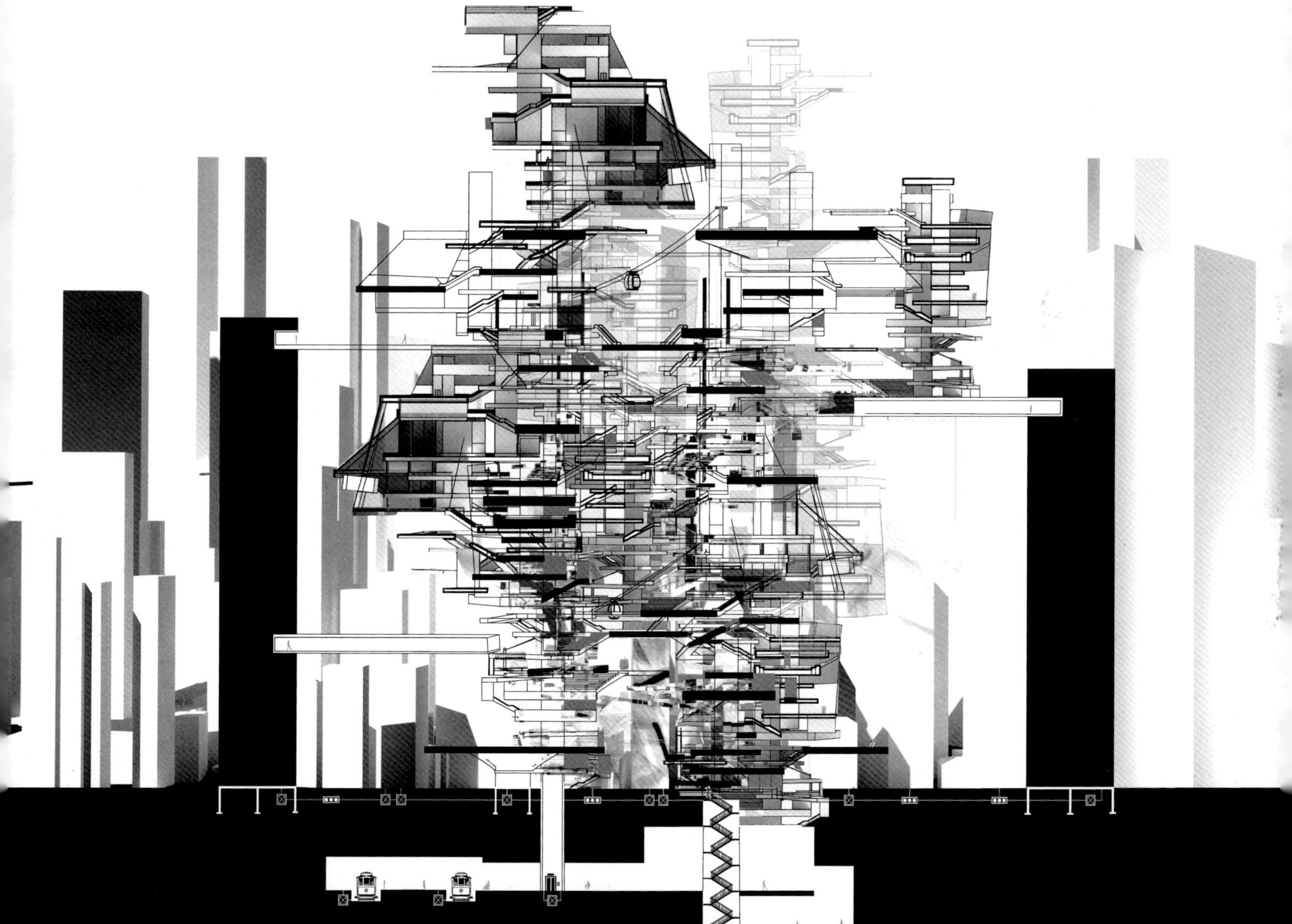

GAME ON | URBAN

A TRANSPARENT CITY: LIVING IN THE CLOUDS

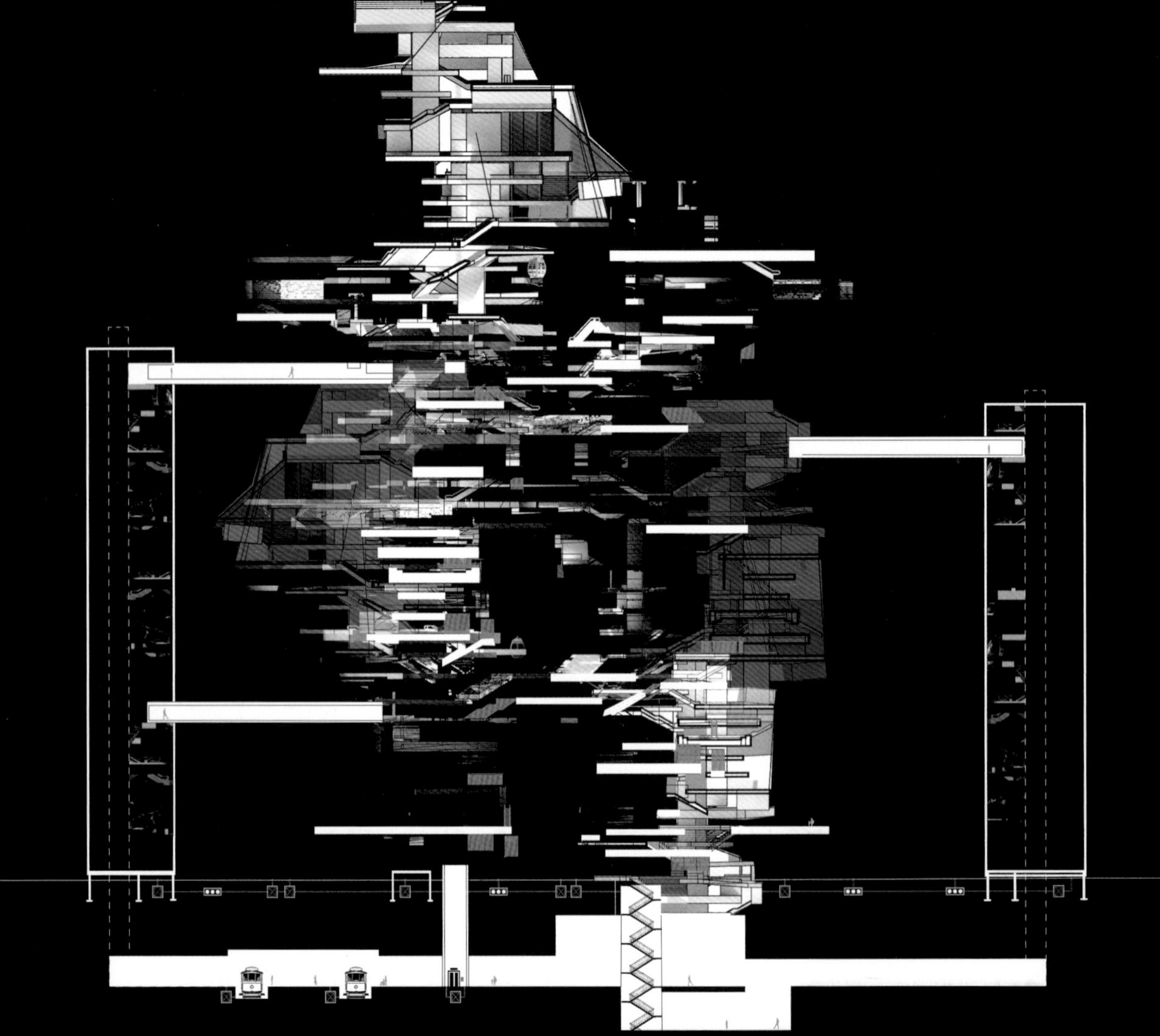

The sectional quality of Living in the Clouds transforms the quality of light during the day and night. The connections to the surrounding structures and the appearance of floating slabs heightens the four-dimensional nature of this twenty-first century typology. Hong Kong, known for its breezeways and bridges between buildings, is a backdrop to this innovative structure.

The first suburb, Levittown in 1951 began as an extension to NYC.[179] The origin of sub- (smaller) and urb- (city) was conceived as a small city outside an urban area.[180] In the twenty-first century the suburb has been reconceived as SUBTOPIA (a small city in utopia) that reunites the neighborhood with nature. Located along a highway in Houston, Texas, the development has proposed a high-speed rail that will dam an existing waterway on the prairie to become the centerpiece of the district. Subtopia creates a community in the four levels of Game On. Technology and entrepreneurial startup companies have developed a living / working environment in a natural setting with schools, daycare, entertainment, and a regional market of organic foods. For the occasional visit to downtown, the hyperrail takes the residents there in five minutes. This ideal community is meant to be a walking city with cars parked by the mass transit with only golf carts allowed into

REIMAGINING THE SUBURBS

the neighborhood, except for the fire truck lanes, EMS, and police access. The net zero community provides as many resources as possible to the residents with the idea that the companies working there give back to the development to promote the philosophy of keeping the workers and staff happy and to maintain high paying jobs. Subtopia has committed to affordable housing to keep people invested in the community's best welfare. The design of the buildings are given to multiple architects and landscape architects, however, the storm drainage system, indigenous plants selection, solar panels necessary to power the community, and a water collection system are all pre-engineered prior to the buildings being built. The streets are made of a hollow concrete, grass-filled water infiltration system that reduces heat and collects water. Subtopia is comprehensive in establishing a way of living that acknowledges climate change, the geology and ecology of the region, and the need for social justice by reimagining the suburbs in the twenty-first century.

SUBTOPIA: LIVING WITH NATURE

SUBURBIA: AN AGE OF DIS

1800 | 1850 | 1900 | 1950

DUTCH HILL
NEW YORK CITY

ANCIENT ALEXANDRIA
ALEXANDRIA, EGYPT

HEIGHTS
HOUSTON, TEXAS

GARDEN CITY
ENGLAND

LINEAR CITY
MADRID, SPAIN

METROPOLIS
NEW YORK CITY

THE RADIANT CITY

BROAD ACRE CITY

EUR
ROME, ITALY

LEVITTOWN
LEVITTOWN, PENNSYLVANIA

AGRICULTURAL CITY

POLITICS
WAR, ECONOMY, SOCIAL INEQUALITY

HEALTH
CONGESTION, AIR QUALITY,

TECHNOLOGY
CARS, TRAINS, INTERSTATE

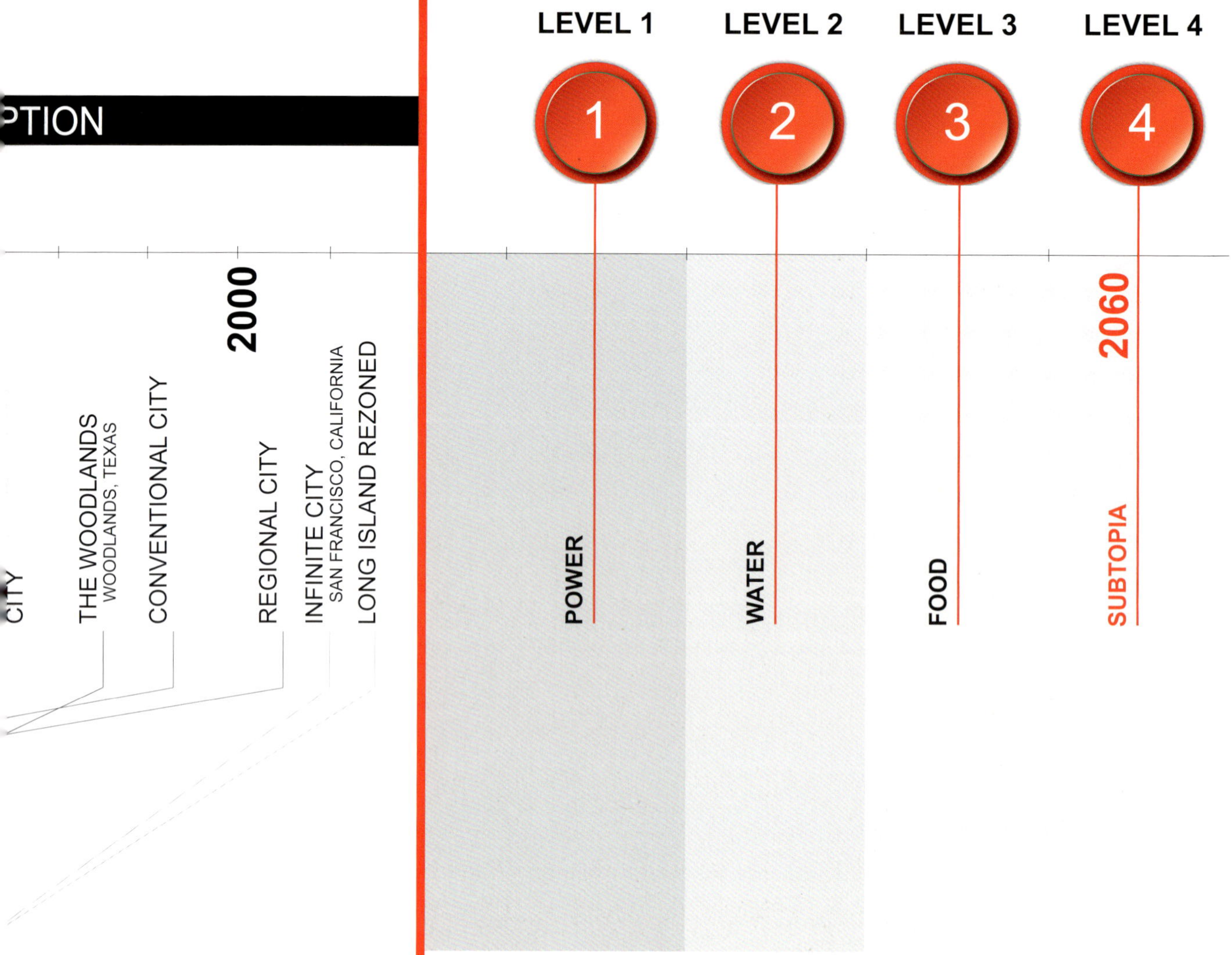

IRONMENT
URAL DISASTERS, CLEAN ENERGY

PANDEMIC

The evolution of the suburbs has taken many turns from the invention of the car and trains to the desire to leave the congestions and poor air quality of the city.[182] *COVID has initiated a new disruption adding to the flight to the suburbs with health risks and the possibilities of remote offices becoming a new reality.*[183]

1

2030

ORGANIZATION

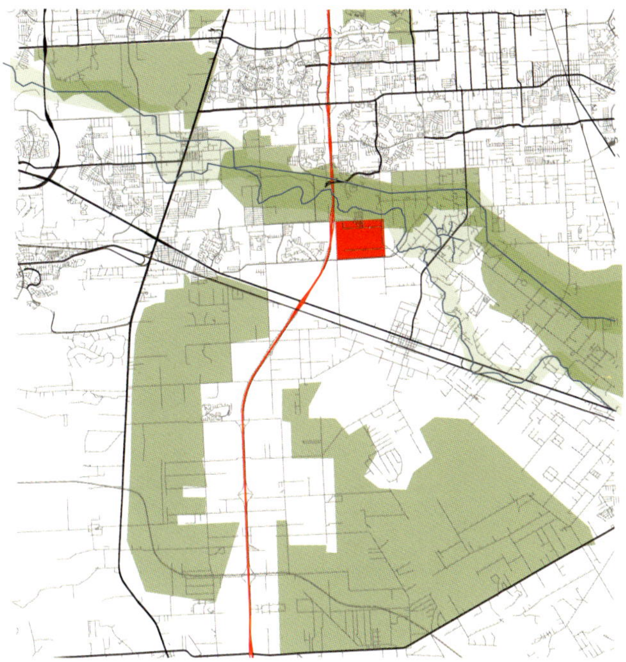

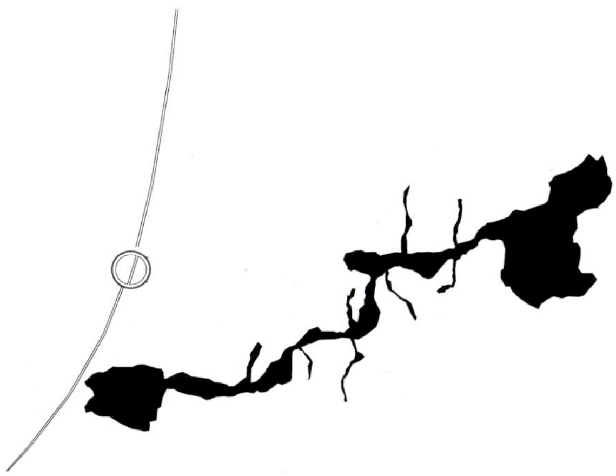

2

2040

GROWTH

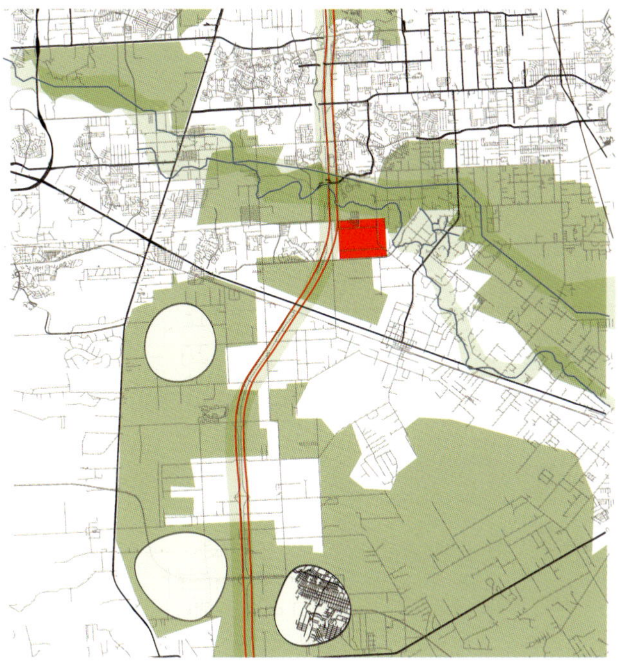

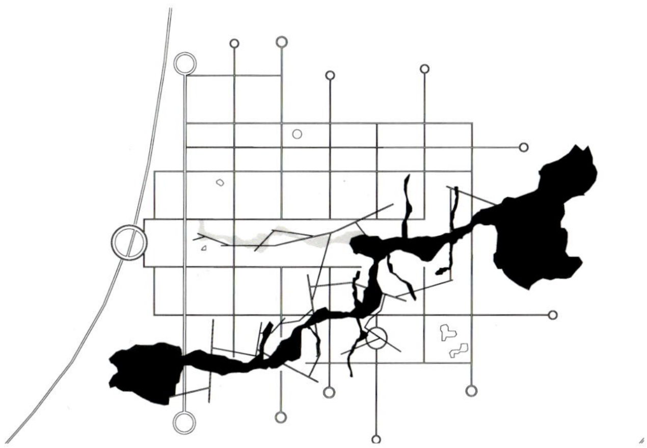

3

2050

REVITALIZE

4

2060

EVOLVE

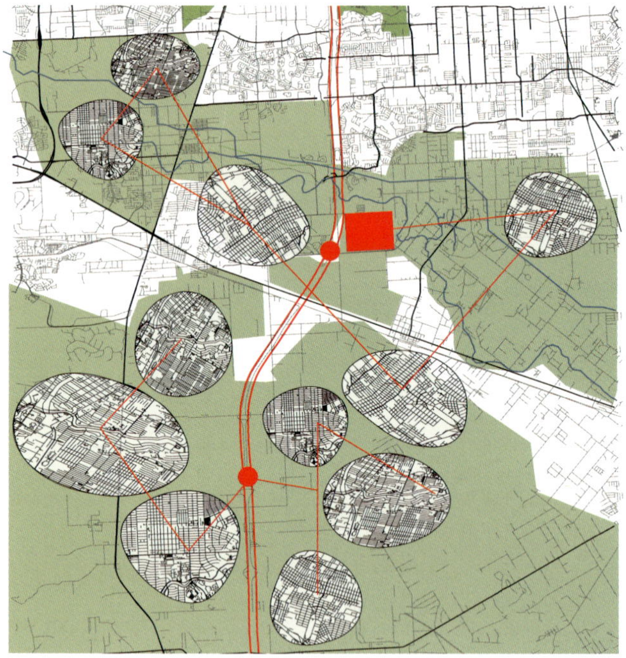

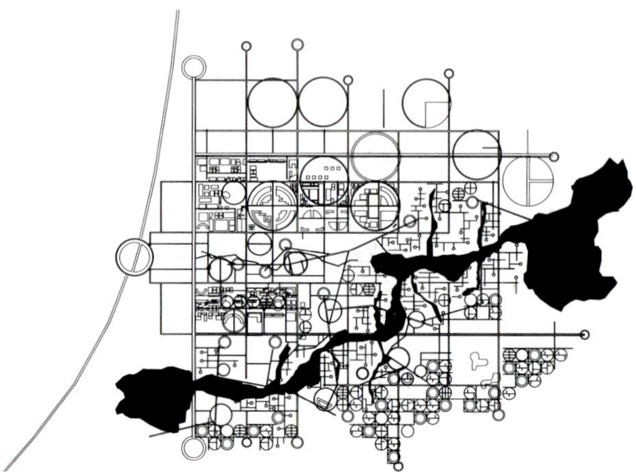

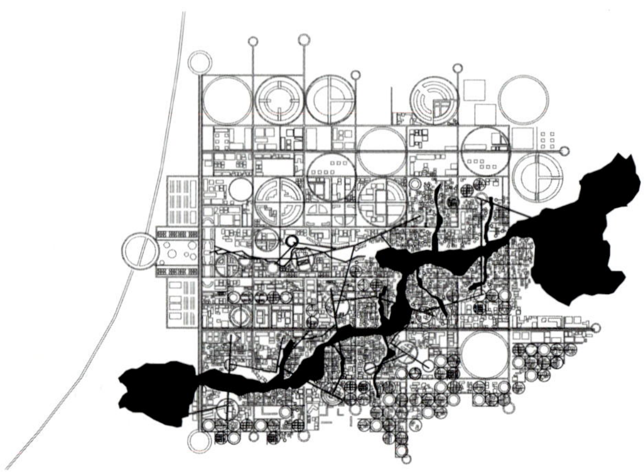

SUBTOPIA will grow organically in the four levels of Game On. The infrastructure of solar power, water collection, and innovative streets with the mass transit to downtown will be built during phase one. The residential, commercial, and school system will evolve in all the levels over time. The controlled footprint of SUBTOPIA fits seamlessly into the prairie's ecology.

SUBURB

SUB·URB | /ˈSƏBƏRB/

SUB: BESIDE, SMALLER, LESS THAN
URB: CITY

AN OUTLYING DISTRICT OF A CITY, OR A SMALLER COMMUNITY ADJACENT TO OR WITHIN COMMUTING DISTANCE OF A CITY. THE SUBURB DEVELOPED HISTORICALLY AS A COMMUNITY DEPENDENT UPON THE CAR.

SUBTOPIA

SUB·TO·PIA | SƏBˈTŌPĒƏ \

SUB: BESIDE, SMALLER, LESS THAN
UTOPIA: NO PLACE, NO WHERE

A DYNAMIC RESIDENTIAL COMMUNITY WITHIN COMMUTING DISTANCE OF A CITY USING MASS TRANSIT TO CONNECT TO DOWNTOWN. A NET ZERO, MIXED-USE DISTRICT THAT IS CONCEIVED AS LIVING WITH NATURE.

SUBTOPIA: INBETWEEN URBAN + RURAL

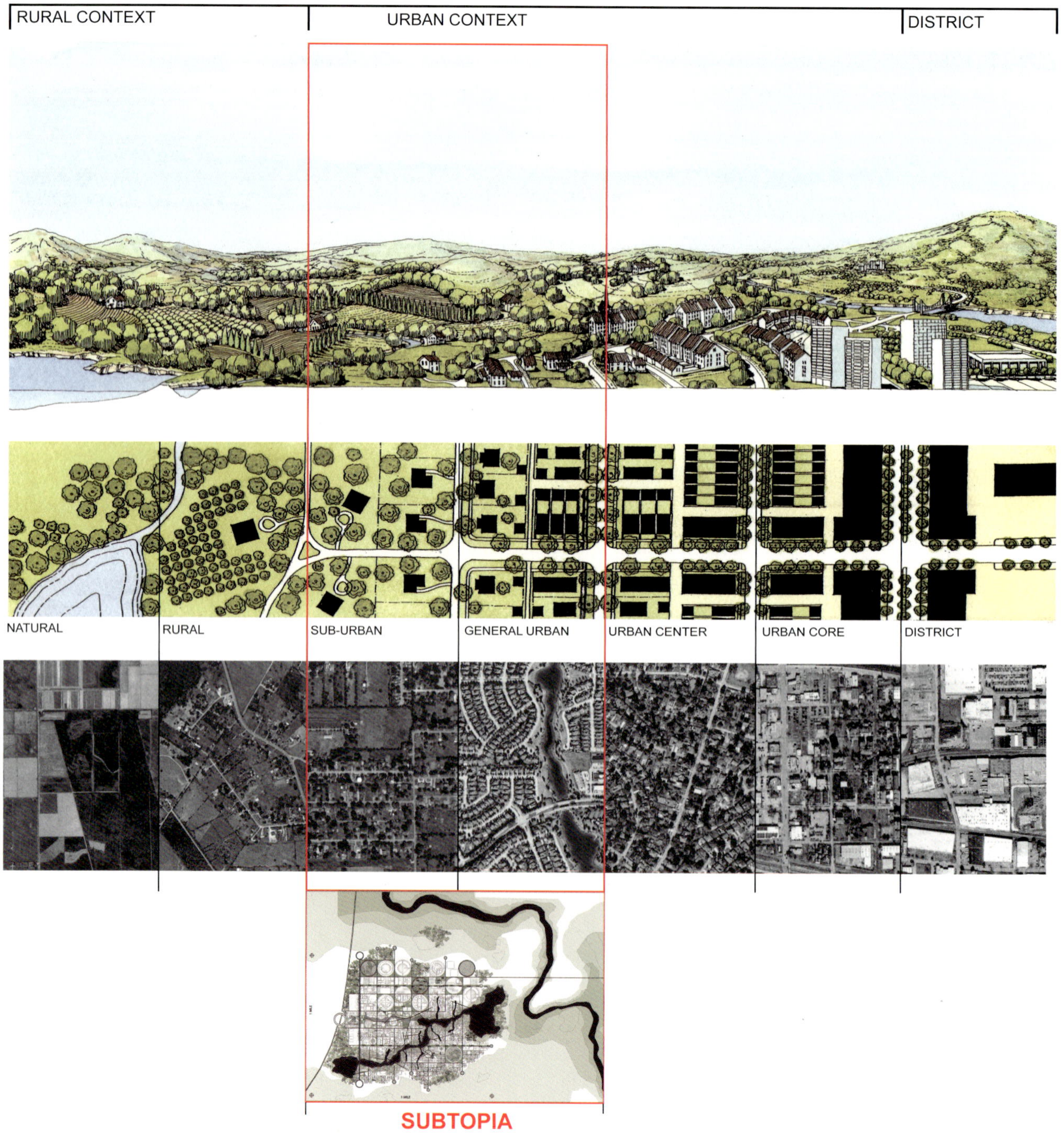

The utopia of SUBTOPIA takes into account the transect of the path from urban to suburban to rural. The hyperloop mass transit allows residents to navigate between the rural and the urban effortlessly with minimal energy use. The concept of the development promotes mixed use, locally grown food, and employs the latest technology to achieve net zero.

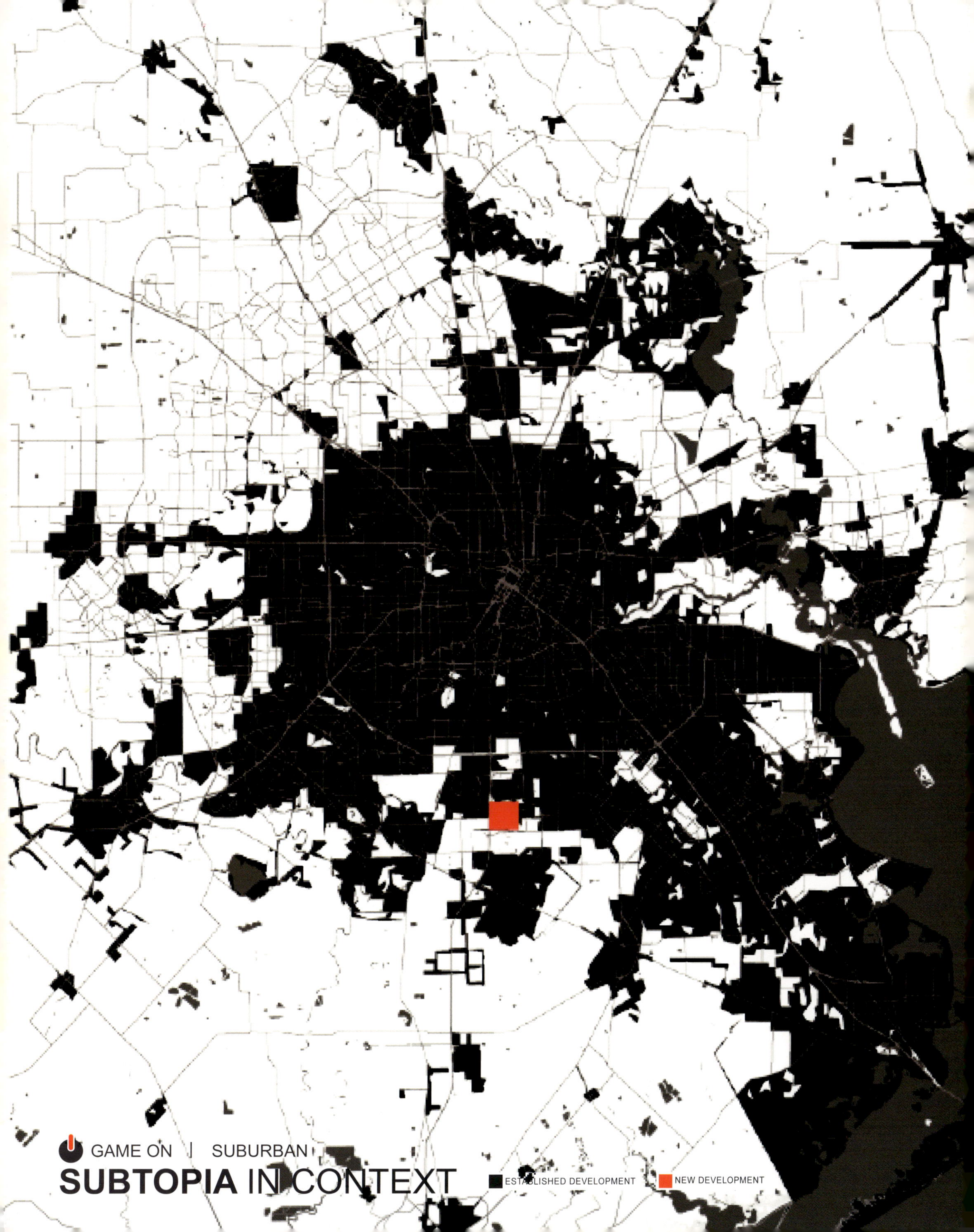
GAME ON | SUBURBAN
SUBTOPIA IN CONTEXT
ESTABLISHED DEVELOPMENT
NEW DEVELOPMENT

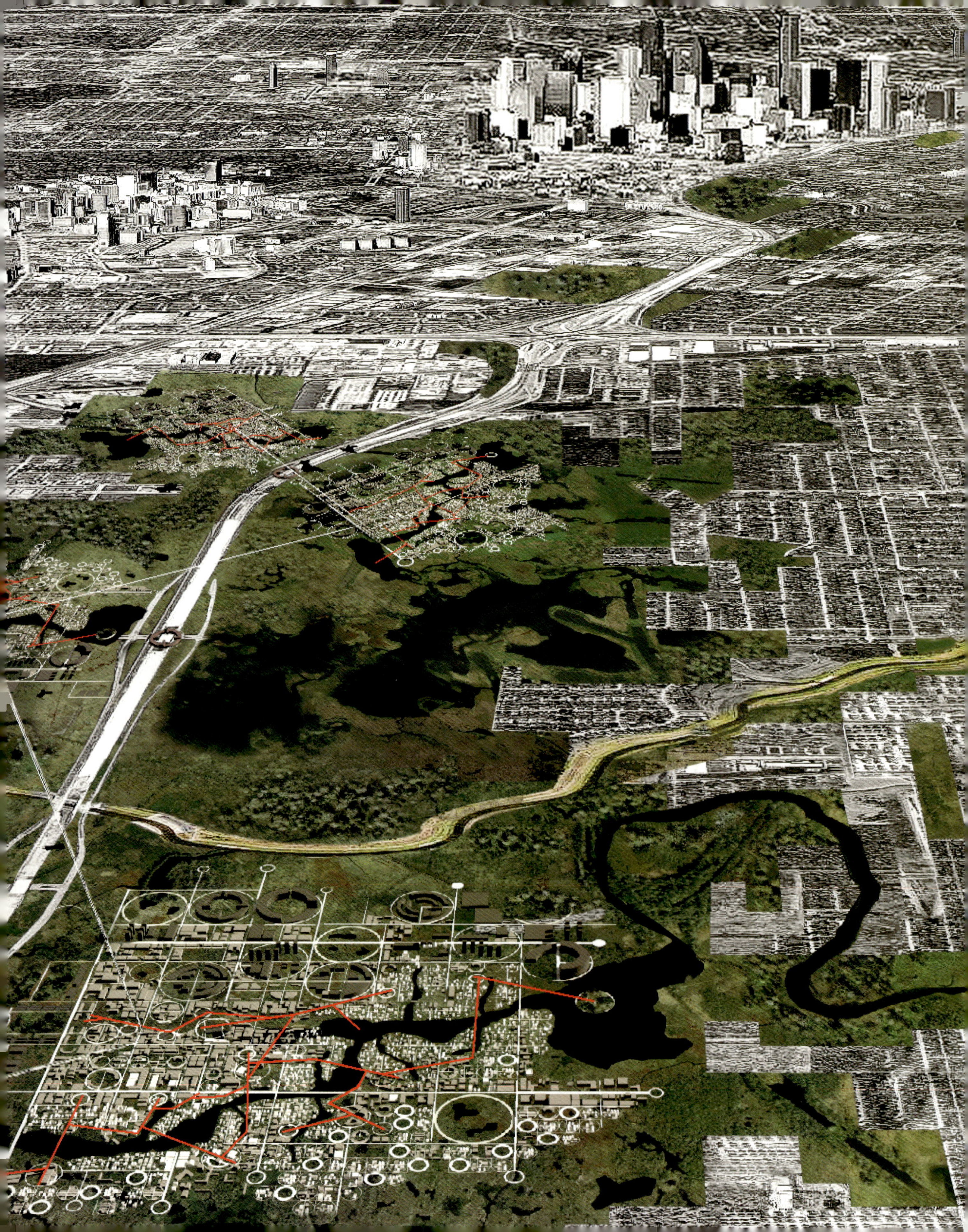

1 MILE
1 MILE

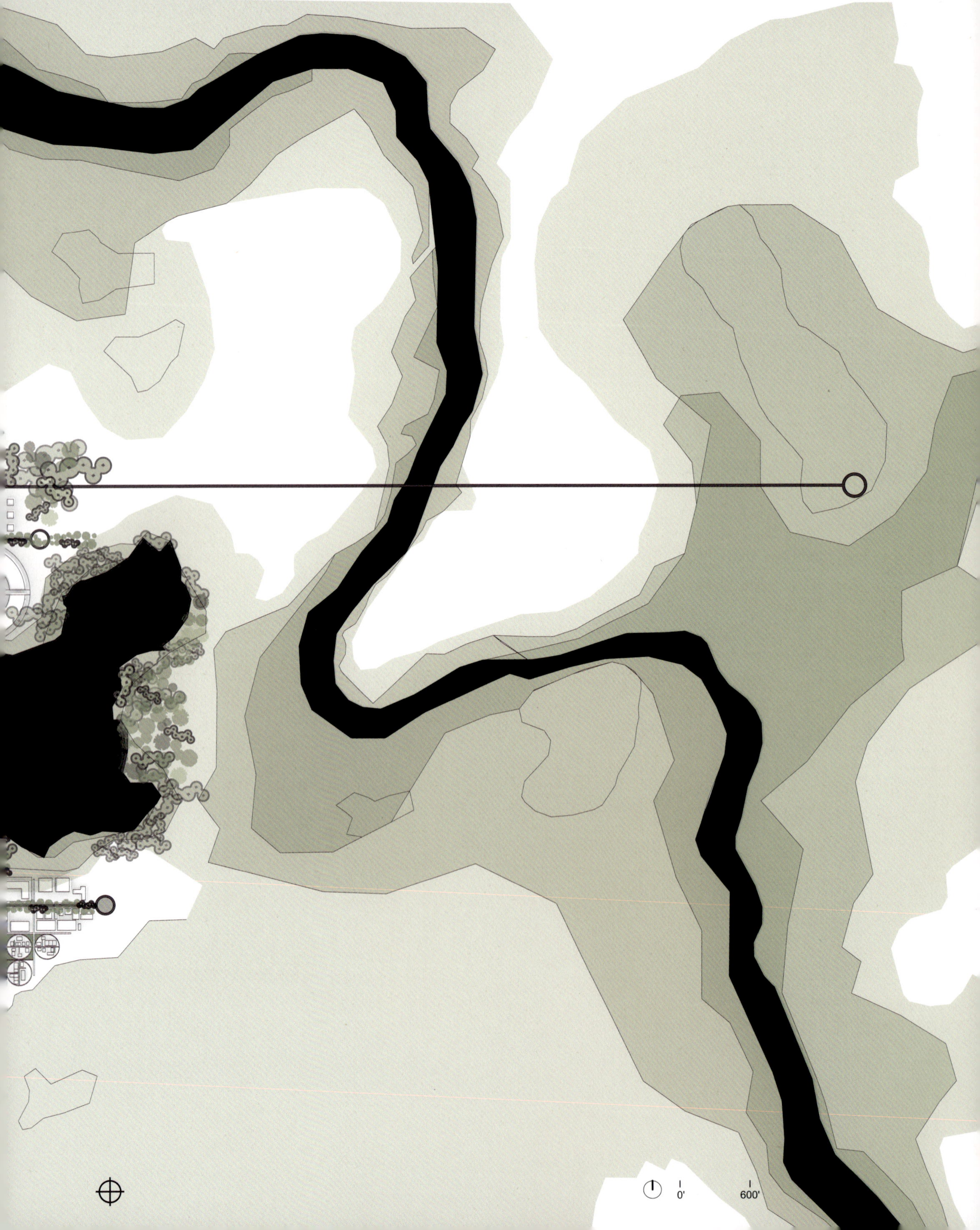
0'
600'

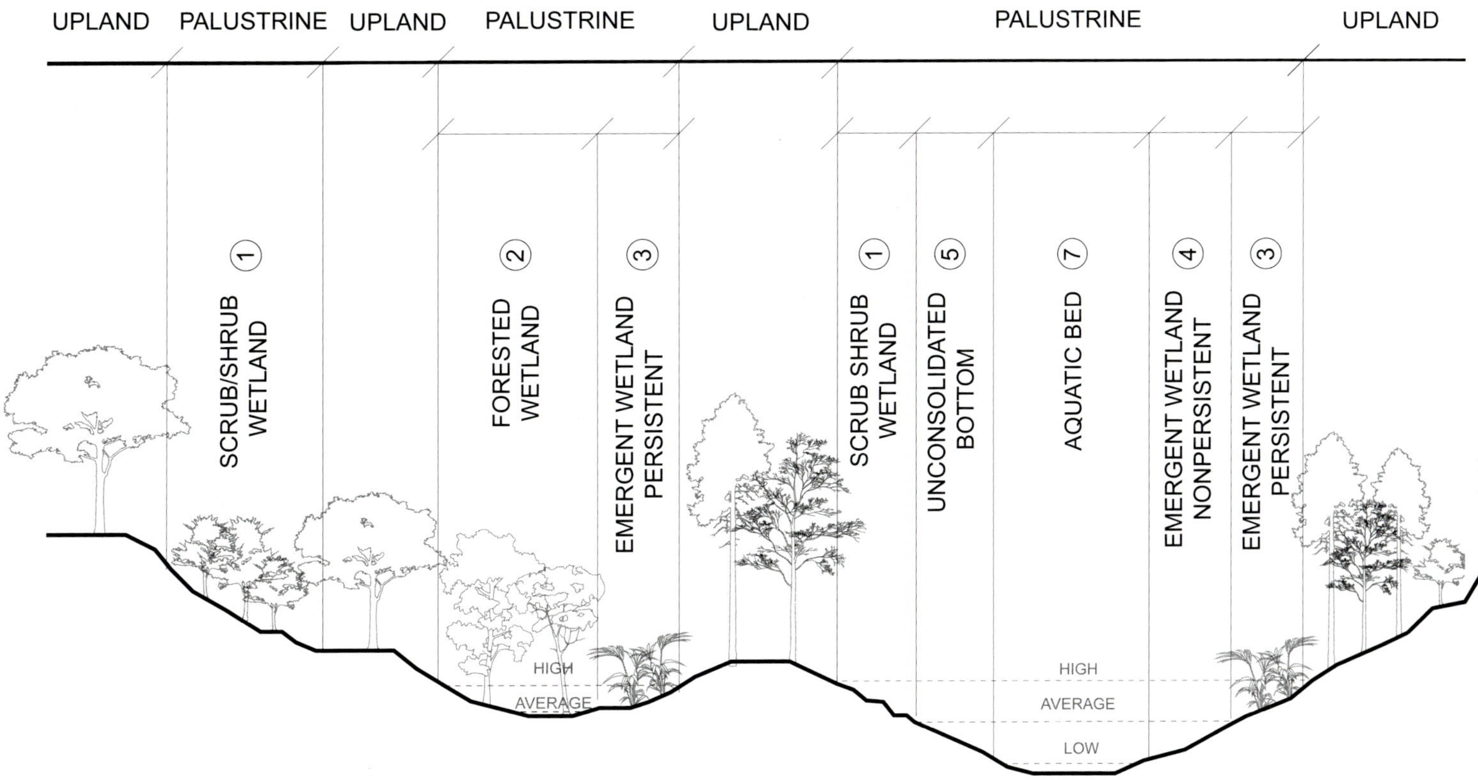

(1) SCRUB/SHRUB WETLAND

(2) FORESTED WETLAND

(3) EMERGENT WETLAND PERSISTENT

GAME ON | SUBURBAN

LEARNING TO LIVE WITH NATURE: WETLANDS

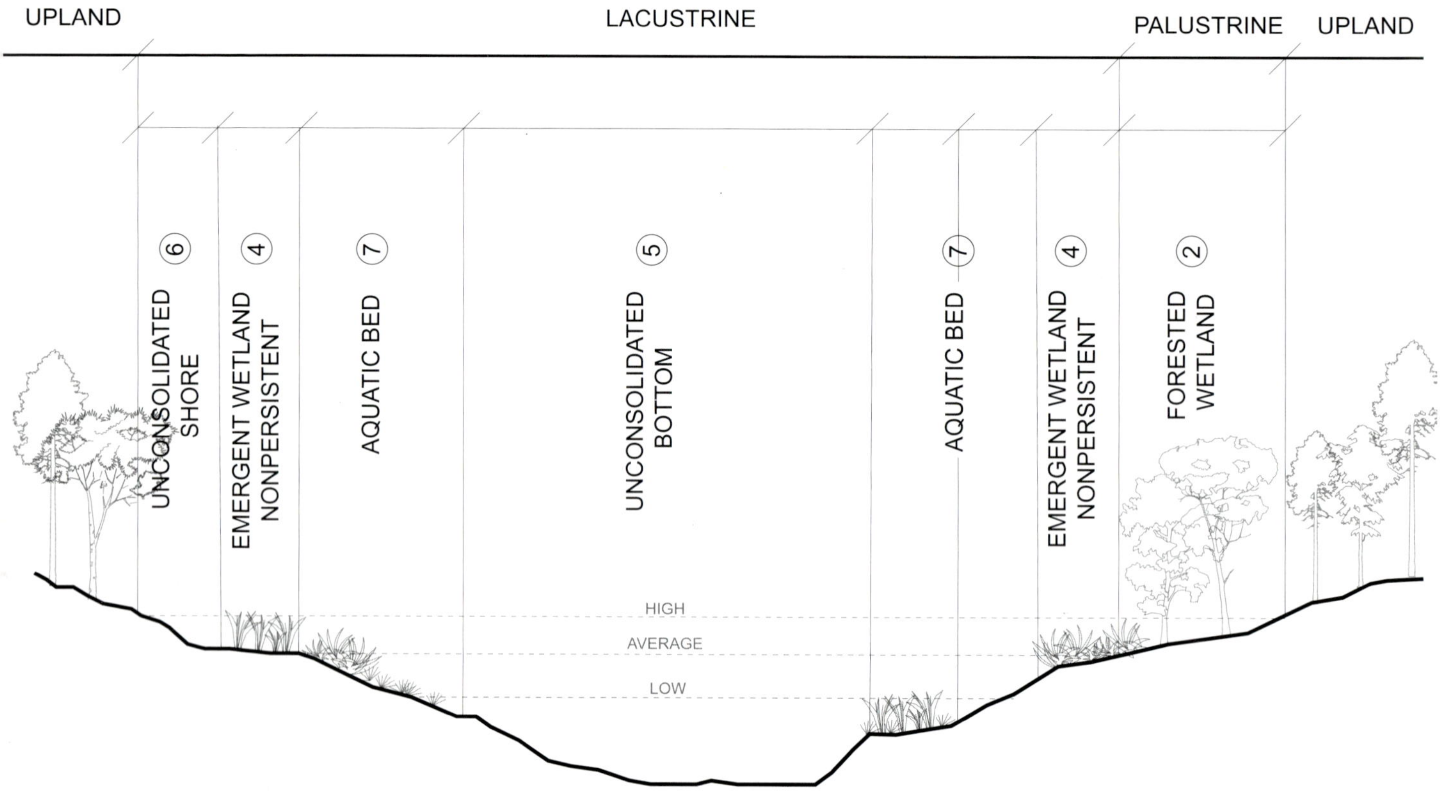

(4) EMERGENT WETLAND NONPERSISTENT

(5) UNCONSOLIDATED BOTTOM

(6) CONSOLIDATED SHORE

The design of the development, placed in the prairie wetlands, has become an integral part of SUBTOPIA. By using indigenous plants and creating lakes on each side of the project with a controlled waterway, it is possible to collect flooding overruns that come with hurricanes in a hot-humid region. The sculpted land will be zoned for living, growing, working, and living with nature.

2020

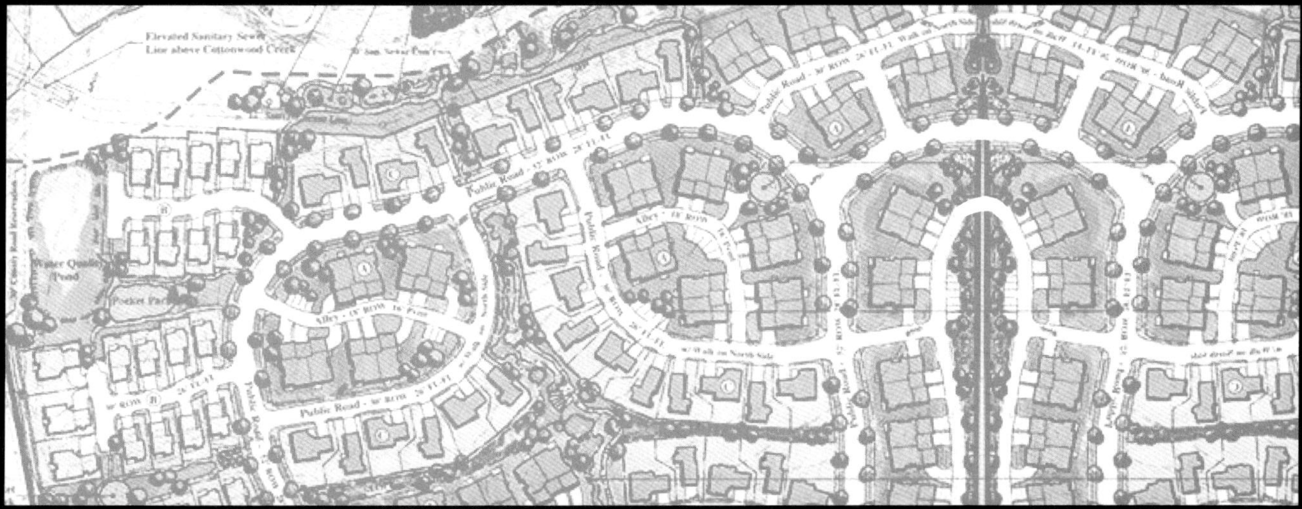

WATER

WATER USAGE

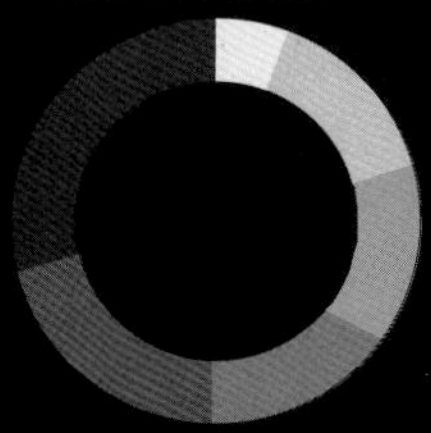

TOILET	30%
WASHER	21%
FAUCETS	16%
SHOWER	17%
LEAKS	16%
OUTDOOR	5%

WATER REUSED

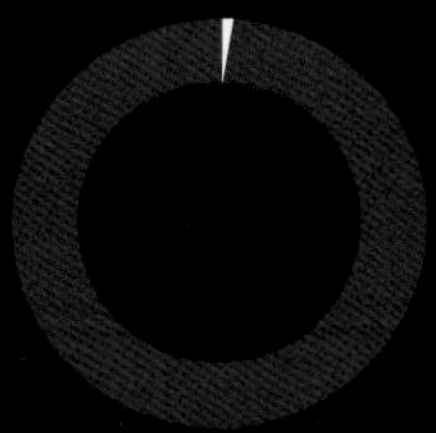

RECYCLED	1%
WASTED	99%

WATER CONSUMPTION PER HOUSE (FAMILY OF 4):

300 GALLONS PER DAY / 4 PEOPLE

FOOD

SPENDINGS

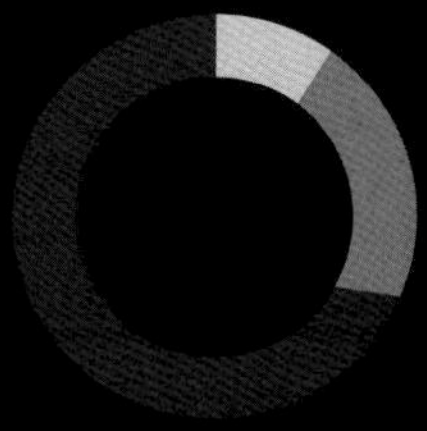

GROCERY	64%
FAST FOOD	26%
LOCAL FOOD	10%

HOME GROWN

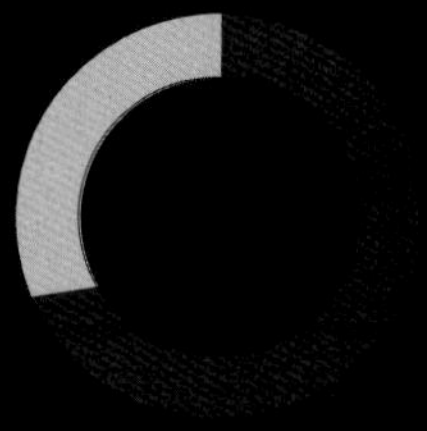

HYDROPONIC TOWERS	0%
COMMUNITY GARDEN	35%
NONE	65%

FOOD COST PER YEAR (FAMILY OF 4):

$7,700 PER YEAR

WASTE

WASTE

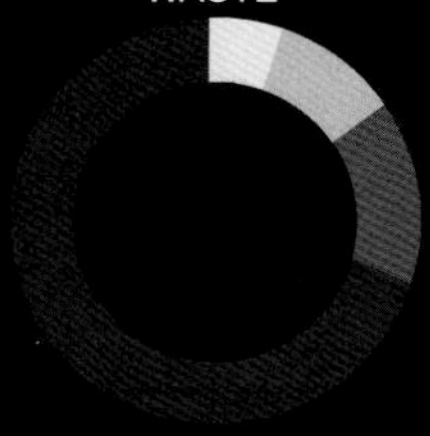

FOOD	70%
FECES	15%
YARD	10%
HAZARDOUS	5%

RECYCLING

COLLECTION MAINTENANCE:	65%
BIOGESTERS:	10%
COLLECTION RECEPTACLES:	25%

HOUSEHOLD WASTE PER DAY (FAMILY OF 4):

18 LBS / 4 PEOPLE

ENERGY

ELECTRICITY USAGE

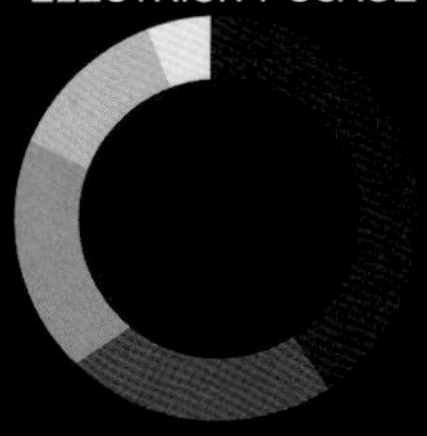

A/C AND HEATING	46%
WATER HEATING	14%
APPLIANCES	13%
LIGHTING	9%
TV AND MEDIA	4%

SOLAR

HOUSES WITH SOLAR	6%
NONE	94%

ENERGY CONSUMPTION PER HOUSE (FAMILY OF 4):

10,649 kWh PER DAY

2060

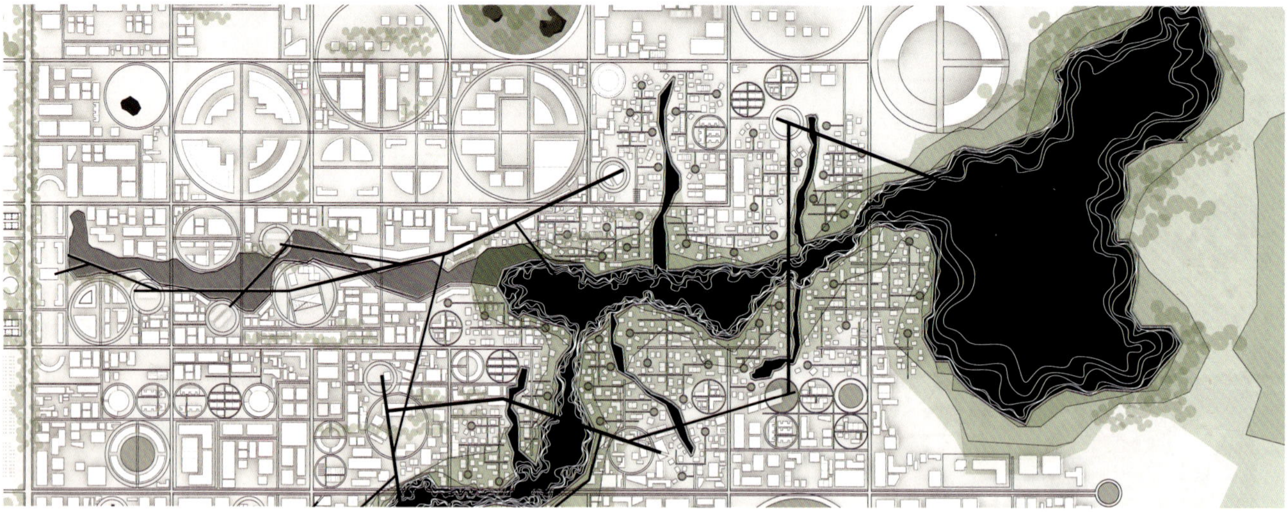

WATER

WATER USAGE

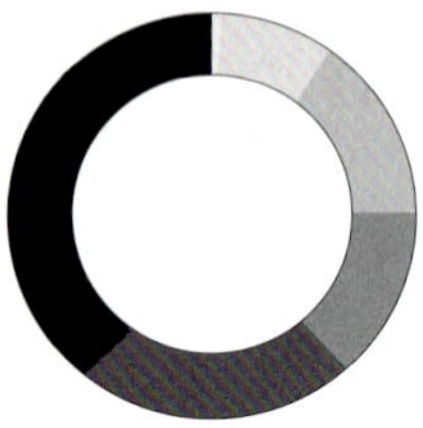

TOILET	39%
WASHER:	21%
FAUCET:	15%
SHOWER:	15%
LEAKS:	9%
OUTDOOR:	1%

WATER REUSED

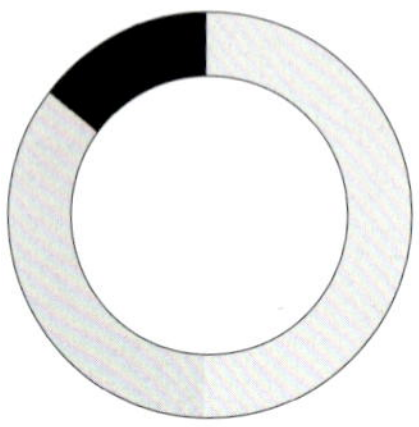

RECYCLED:	85%
WASTED:	15%

WATER CONSUMPTION PER HOUSE (FAMILY OF 4):

50 GALLONS PER DAY / 4 PEOPLE

FOOD

SPENDINGS

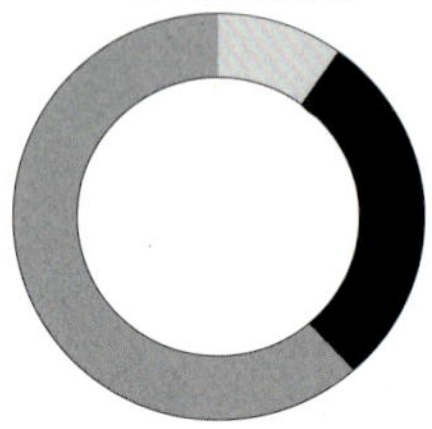

GROCERY:	30%
LOCAL FOOD:	60%
FAST FOOD:	10%

LOCAL FOOD GROWN

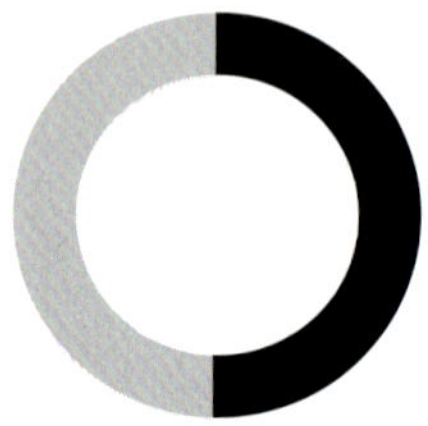

HYDROPONIC TOWERS:	50%
COMMUNITY GARDEN:	50%
NONE:	0%

FOOD COST PER YEAR (FAMILY OF 4):

$1,540 PER YEAR

WASTE

WASTE

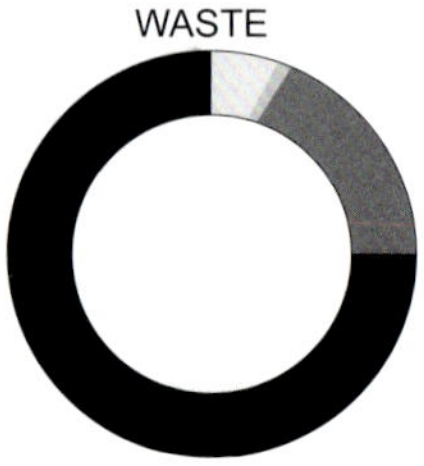

FOOD:	75%
FECES:	19%
YARD:	1%
HAZARDOUS:	5%

RECYCLING

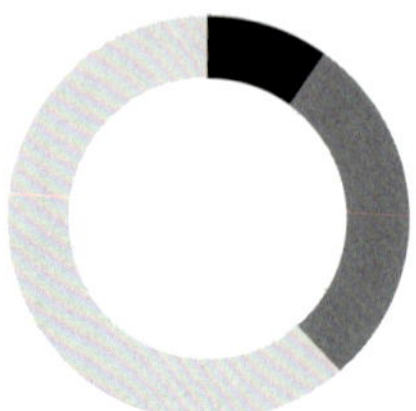

COLLECTION MAINTENANCE:	10%
BIOGESTERS:	30%
COLLECTION RECEPTACLES:	60%

HOUSEHOLD WASTE PER DAY(FAMILY OF 4):

4 LBS / 4 PEOPLE

ENERGY

ELECTRICITY USAGE

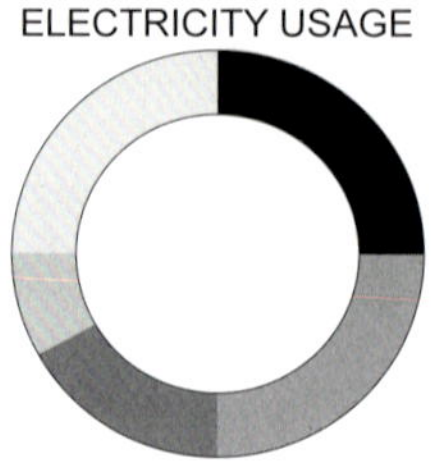

A/C AND HEATING	25%
WATER HEATING:	15%
APPLIANCES:	25%
LIGHTING:	10%
TV AND MEDIA:	25%

SOLAR

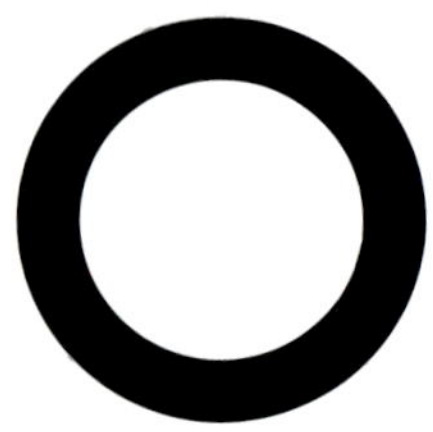

HOUSES WITH SOLAR:	100%
NONE :	0%

ENERGY CONSUMPTION PER HOUSE (FAMILY OF 4):

6389 kWh PER DAY

The net zero vision for SUBTOPIA involves initiating a water, power, food, and waste program that will reduce the water intake for people and crops. The infrastructure will have compost bins for waste, solar collectors, energy efficient HVAC systems to reduce power consumption, and hydroponics and gardens for food production.

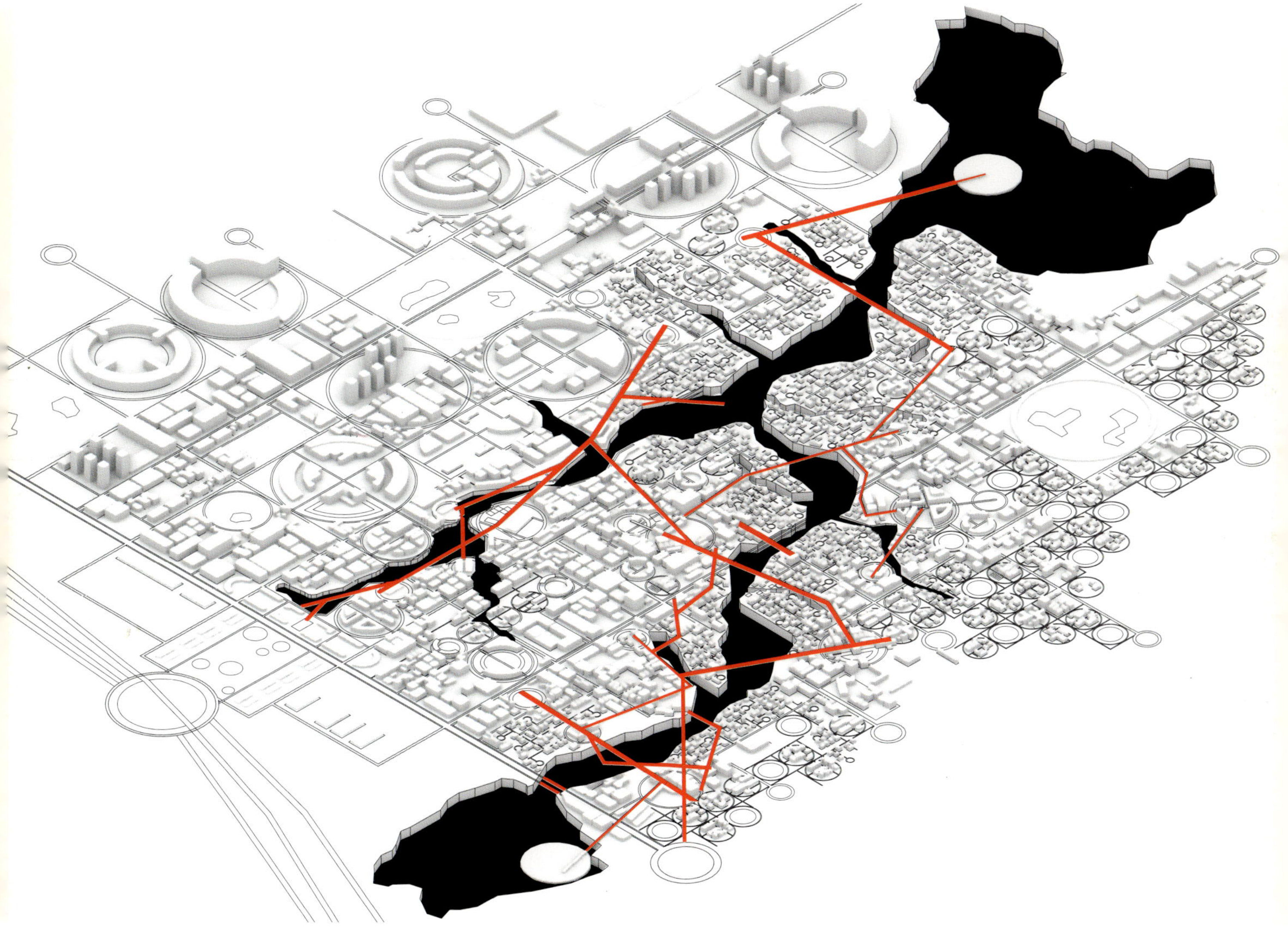

GAME ON | SUBURBAN

LIVING WITH NATURE

SUBTOPIA is designed to live with nature by using the prairie landscape to absorb the inevitable flooding and having a central waterway with boardwalks that give space for the water runoff to go during heavy rains. The natural landscape is allowed to come into the neighborhood, which heightens the respect for the land and ultimately gives a sense of place to SUBTOPIA.

Architecture has always been framed in history as having the ability to represent its own time. We are currently living in an explosive time facing existential threats of climate change, political upheaval, cyberattacks, and nuclear warfare.[184] EXploded tries to embody its own time by exploring the abstract idea of exploding and imploding forms that can make a neighborhood out of the fragments. Located in Berlin, Germany along the Berlin Wall, the historic reference to the collapse of the wall in 1989[185] is symbolic of the divided city in the twentieth century and the current division and uncertainty in the world that remains 60 years later. The neighborhood, called DOOMSBURY, is a homage to the troubled times we are living through as the stakes for the world increase. The central monumental space is a gallery that has etched glass floors as part of a visual exhibition that places observers within the context of the Berlin Wall. A central plaza of shards and lighted posts provide

LIVING INBETWEEN A MONTAGE OF EXPLOSIONS

the community with a backdrop for gathering and for a healing space. The 2,000 inhabitants create a neighborhood composed of mixed-use offices, apartments, residential living, and small commercial spaces with a ring of parking for cars with only golf carts allowed within the community. Forests, playgrounds, and edible gardens are communally shared by the residents with solar energy and geothermal systems used to achieve net zero. The green roofs and rooftop gardens provide additional insulation from the cold winters, as well as food for the local markets. The walking neighborhood is a prototype for how to integrate housing at all economic levels. The tension between the very real concept of living during explosive times can be contrasted to the integrated use of the sophisticated, sustainable technology in the neighborhood. The ambiguity of a central space becoming the center of a green neighborhood, yet also being the source of an implosion and explosion sets up the experience of the push and pull of living INbetween.

EXPLODED: HEALING THE BERLIN WALL

CUBE	IMPLOSION	EXPLOSION	IMPLOSION + EXPLOSION

NATURAL DISASTERS

830,000 DEATHS	655,000 DEATHS	230,000 DEATHS	316,000 DEATHS

SHAANXI EARTHQUAKE JANUARY 23, 1556	TANGSHAN EARTHQUAKE JULY 28, 1976	BOXING DAY TSUNAMI DECEMBER 26, 2004	HAITI EARTHQUAKE JANUARY 12, 2010

DISASTERS: NATURE VS. MANMADE[186]

DOOMSBURY: A FRAGMENTED CITY

MAN-MADE DISASTERS

23,568,559 DEATHS

WORLD WAR I

1914 – 1918

85,000,000 DEATHS

WORLD WAR II

1939 – 1945

800,000 DEATHS

RWANDA GENOCIDE

1994

2,383,000 DEATHS

VIETNAM WAR

1955 – 1975

When comparing the history of natural and human-made disasters it becomes clear that the number of deaths by human-made causes far outweighs those of natural disasters. As the world faces climate change, the number of deaths will increase by natural disasters, however, since history often has a way of repeating itself, it's likely that we will have more human-made disasters as well.

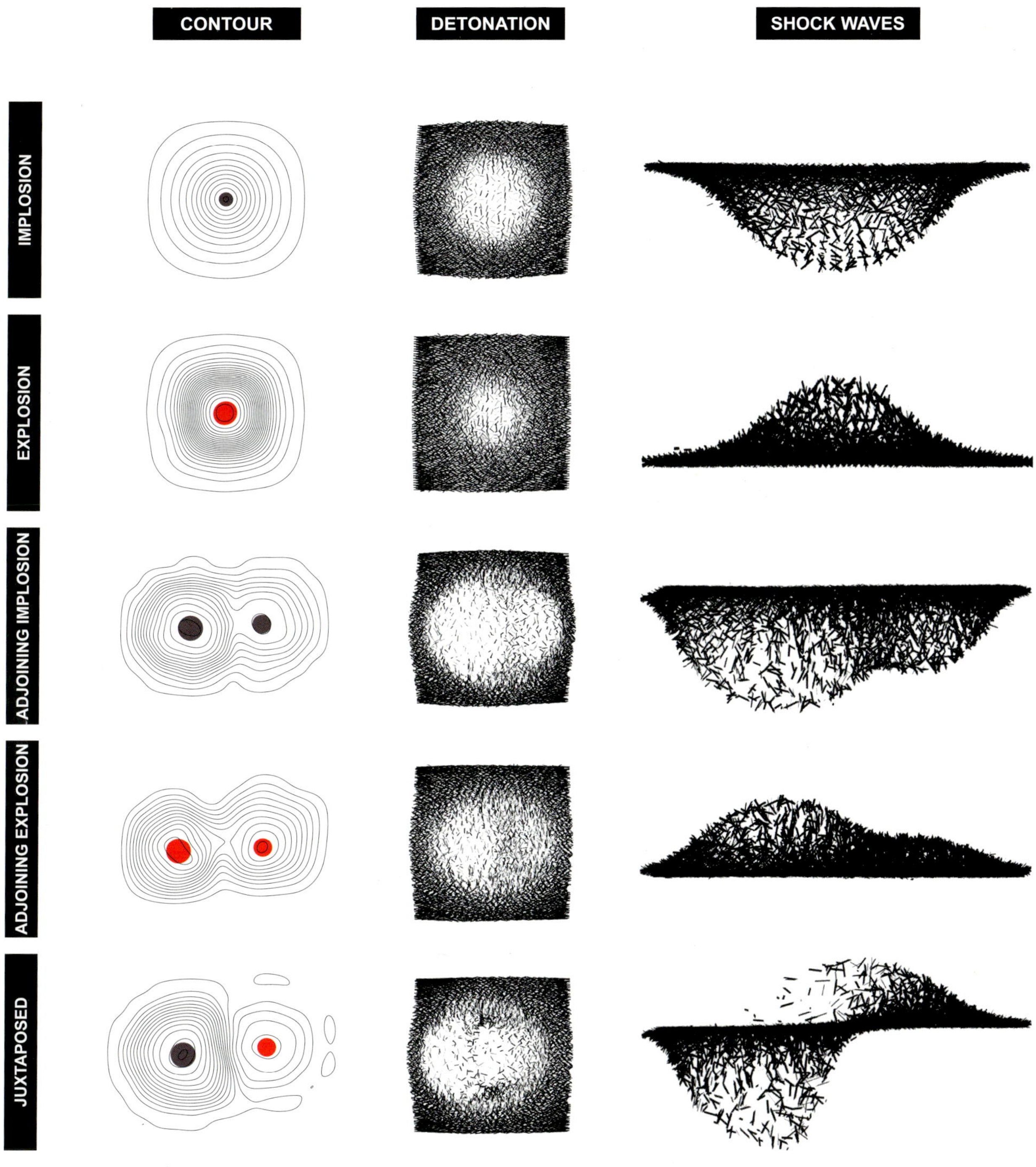

GAME ON | EXPLODED

FORM: IMPLOSION + EXPLOSION

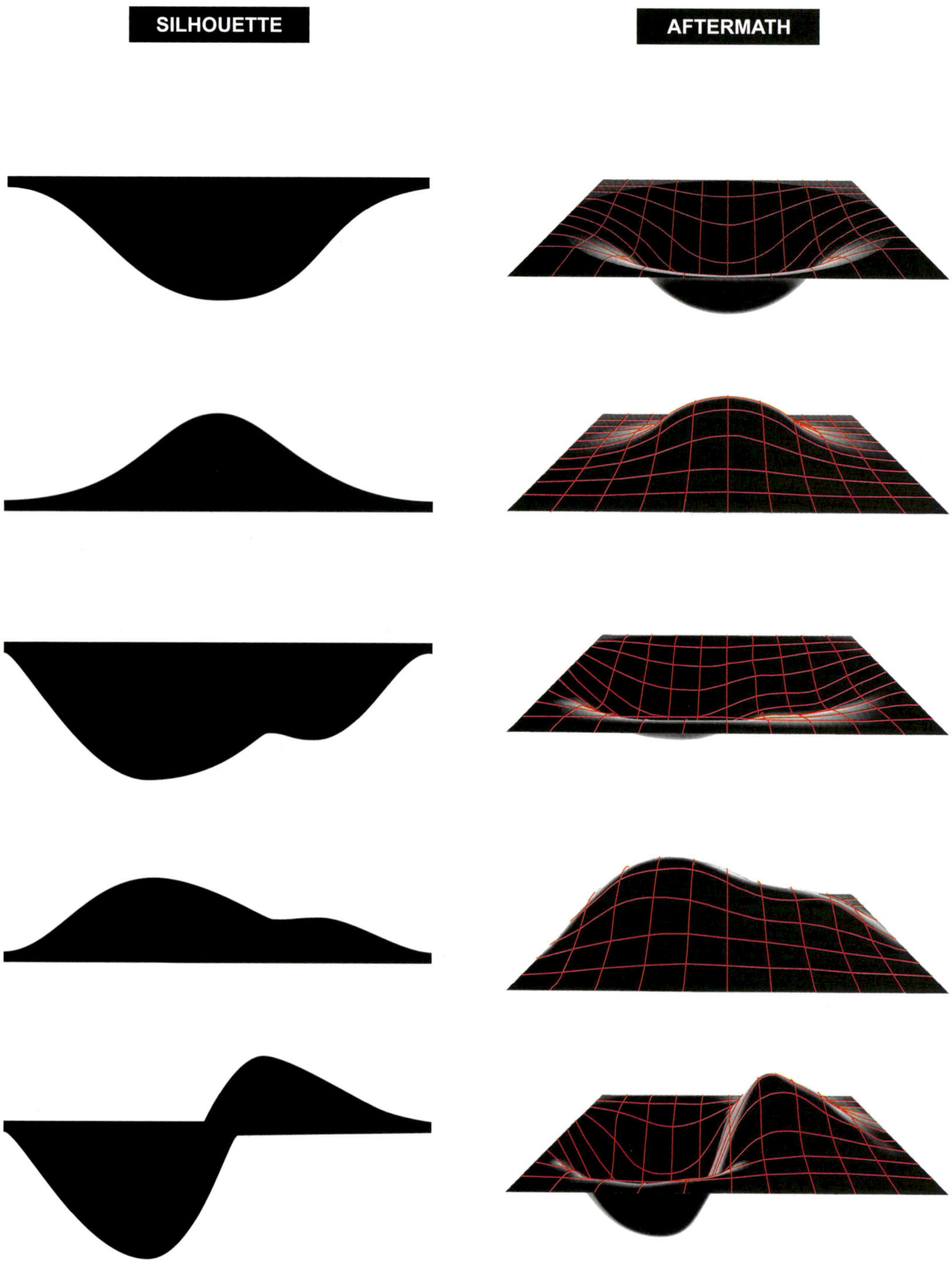

By investigating the properties of explosion and implosion it is possible to create computer modeling of the shock waves produced by the interaction of the two architectural forms. The silhouette and aftermath of the explosions are the result of the studies becoming the groundwork for exploring the new formal concept of the architectural spaces.

GAME ON | EXPLODED

BERLIN WALL: 1961-1989

DOOMSBURY

DOOMSBURY, a neighborhood of 2,000 inhabitants, takes its cue from the adjacent Berlin Wall. Erected in 1961 and taken down in 1989,[187] the divided city was a metaphor for a world divided. Since the historic attack on September 11, 2001 more than 70 politically motivated walls have been erected around the world.[188] This division is reflected in the concept for the neighborhood.

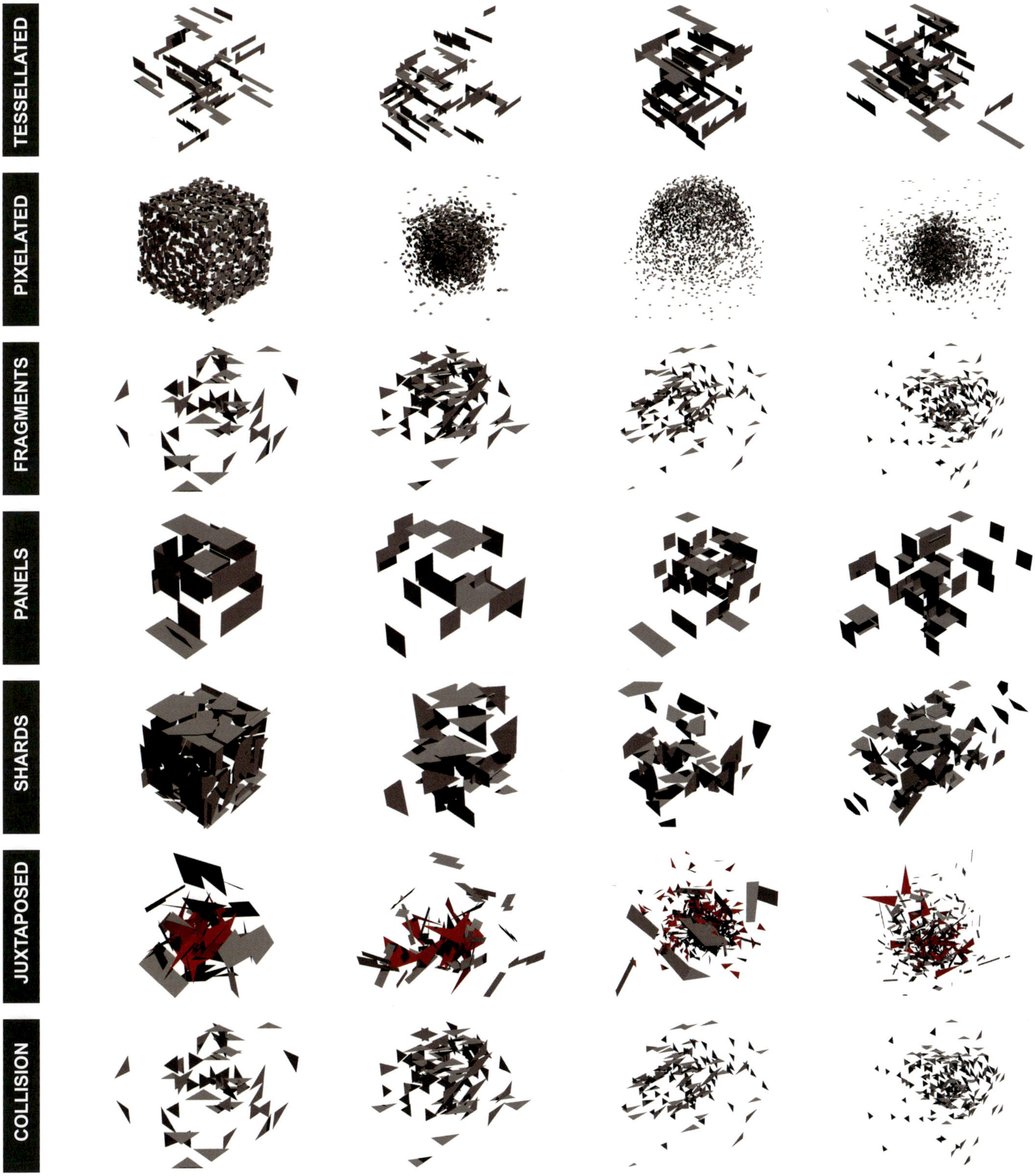

EXPLOSIONS: INBETWEEN SPACE AND TIME

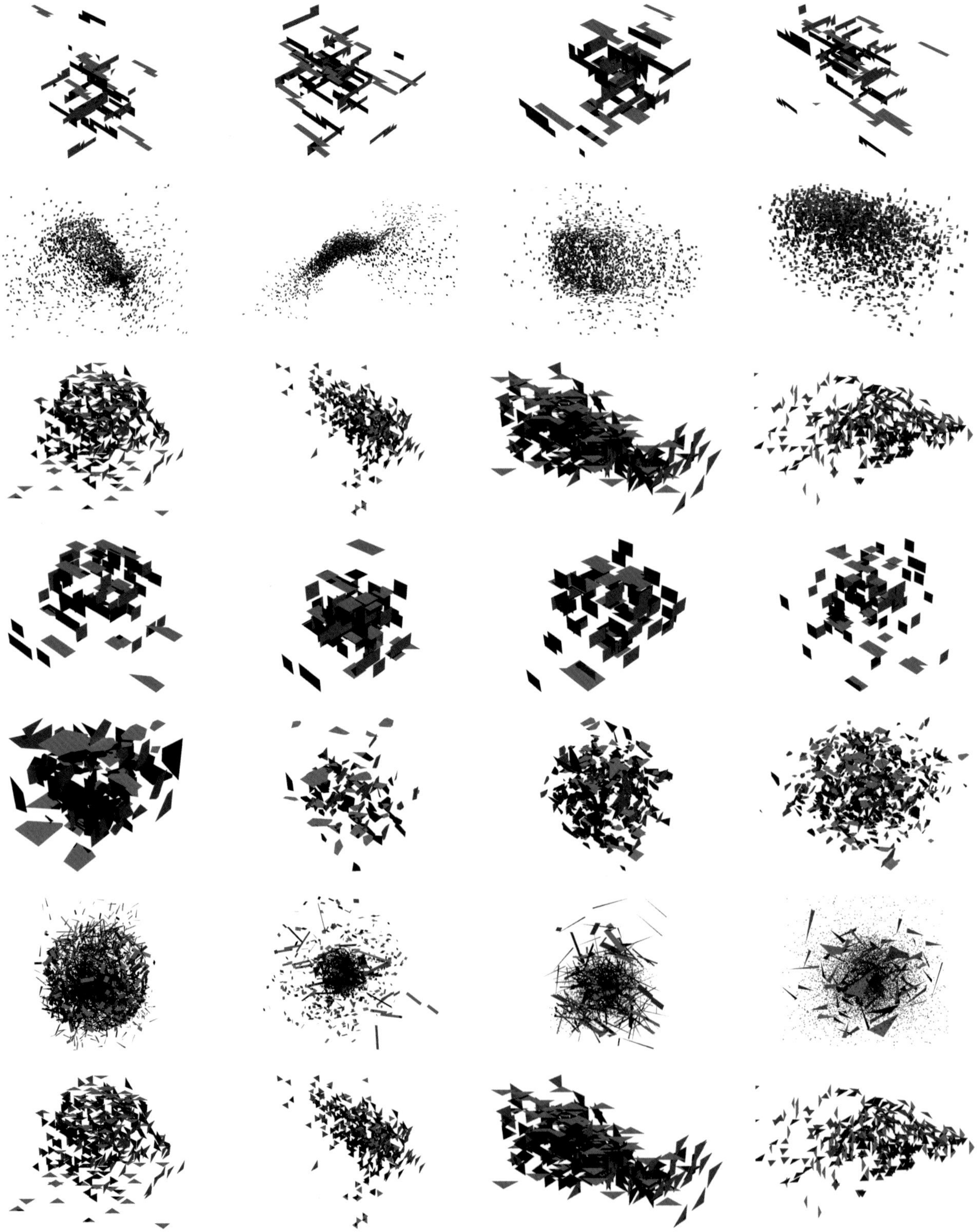

By exploring the architectural forms of implosion and explosion through the lens of pixels, shards, fragments, panels, and tesselations, it is possible to combine and juxtapose the two formal moves into simultaneous forms of explosion and implosion occuring in one monumental structure. This suspended-in-animation shape becomes the template for DOOMSBURY.

EXPLOSION

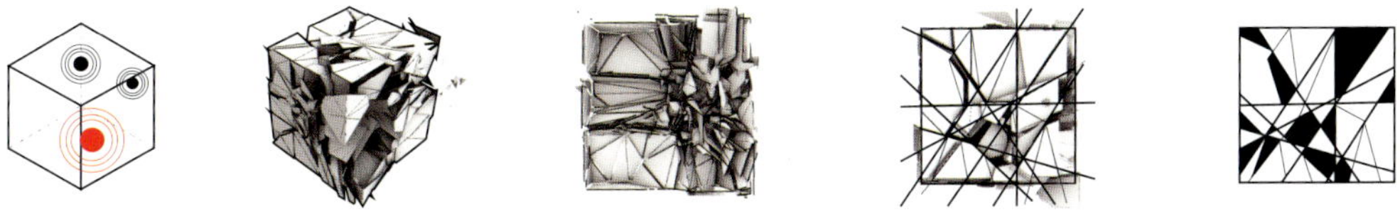

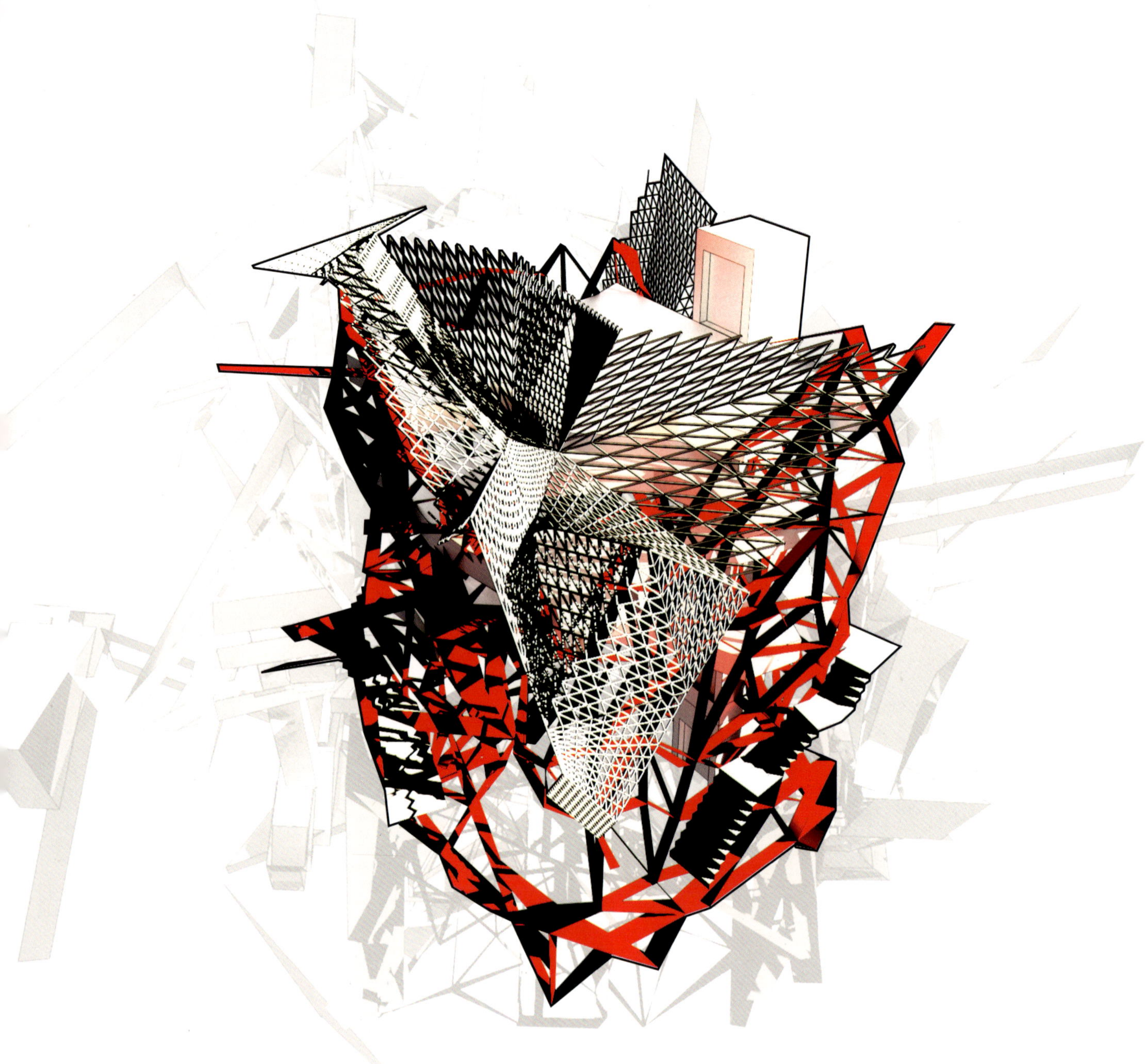

IMPLOSION

The fractured layers of red, steel plates imploding into the space are contrasted by metal stairs and a glass elevator wedged between the explosive façade of anodized metal and the interior red, imploding steel. During the day the building appears to explode, but at night light emanates from the gallery that highlights the concept of forms exploding in space.

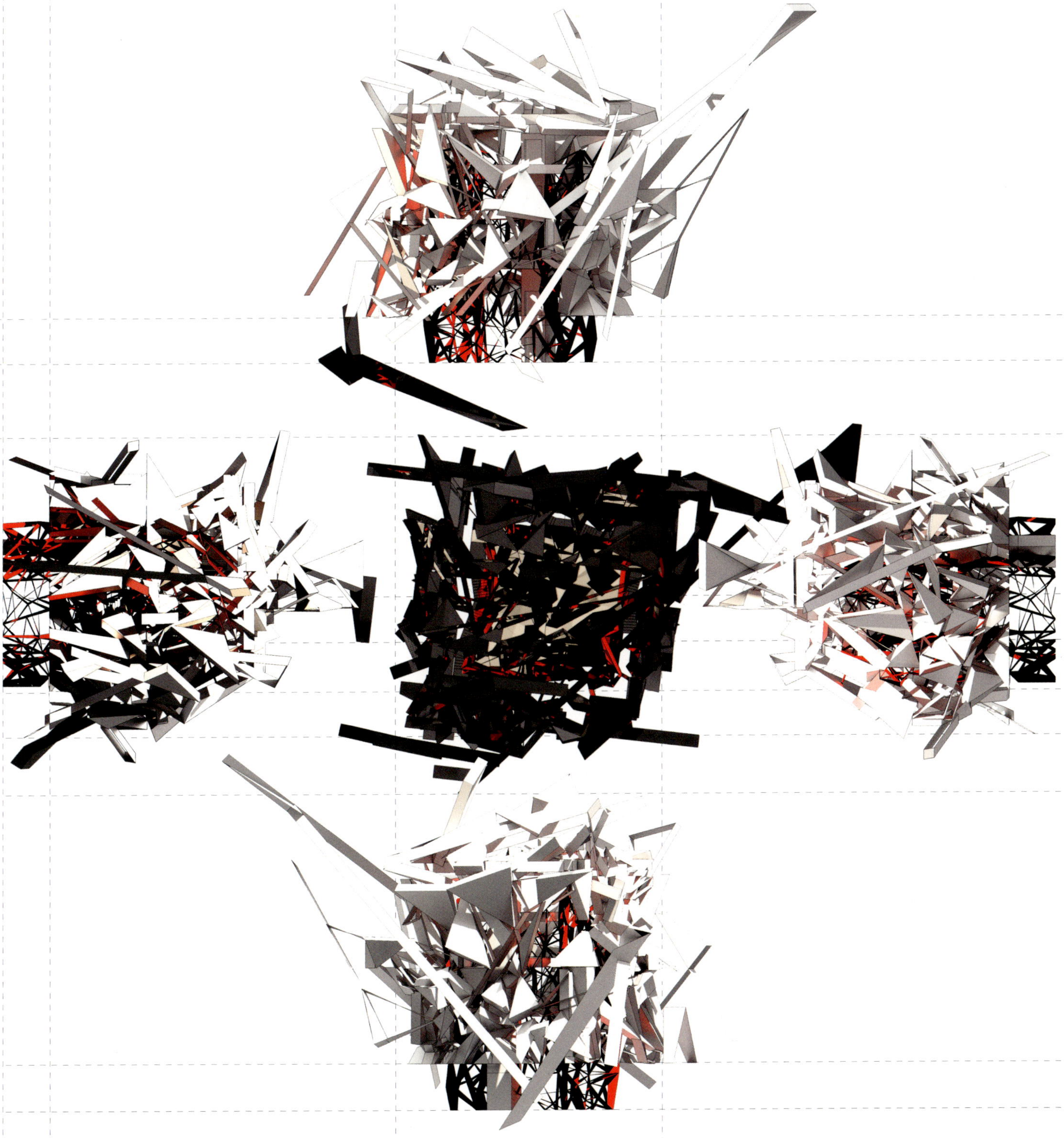

GAME ON | EXPLODED

MUSEUM: EXPLODED DRAWINGS

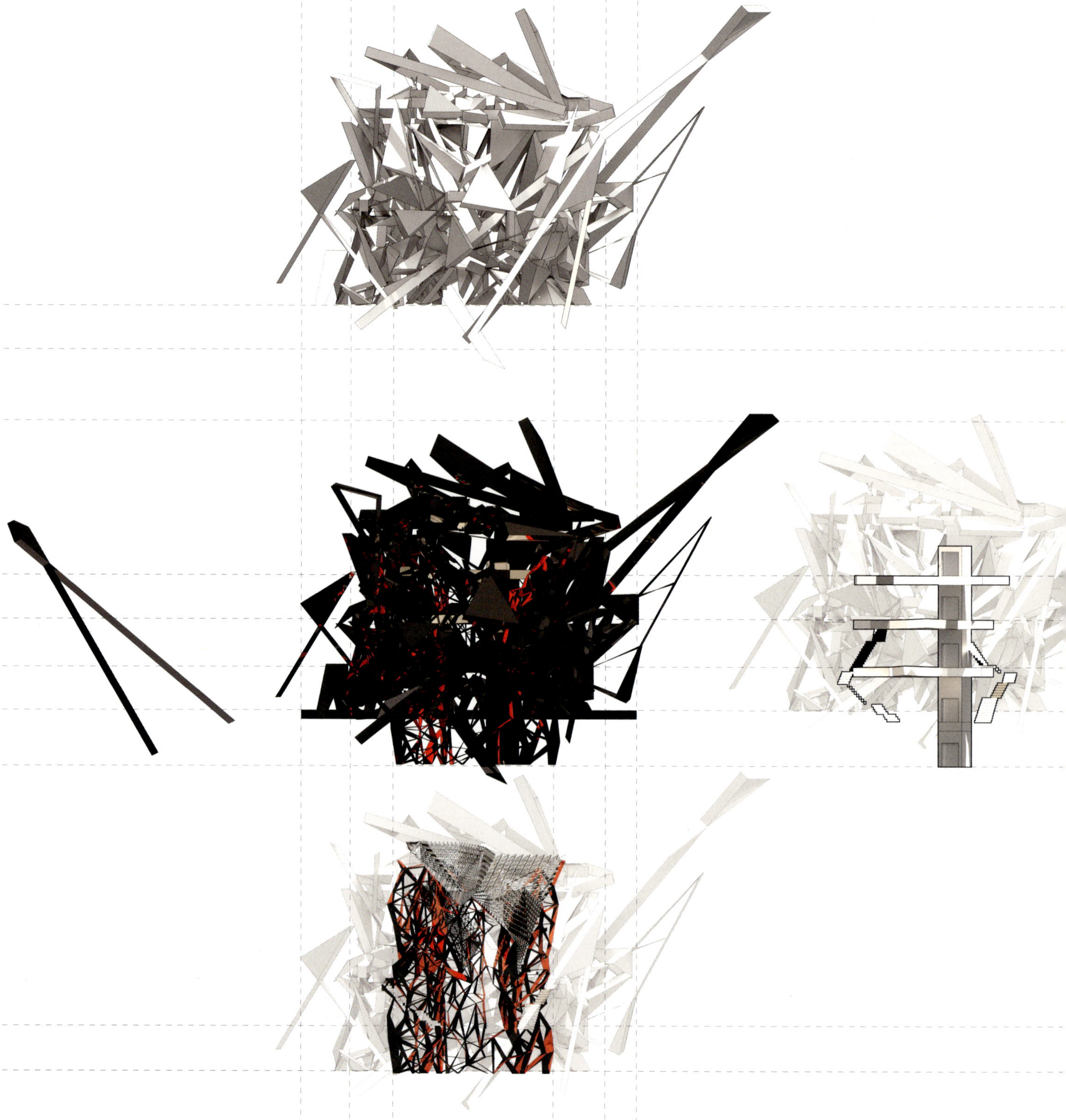

As the structure is analyzed, the steel metal plates extend the façades that replicates the simultaneous explosion and implosion of the museum. INbetween the plates, lighting is shaped into a reveal that minimizes the light source, but manages to highlight the movement at night. The glass elevator and stairs sandwiched between the two structures mediate the experience of the pathways.

GAME ON | EXPLODED

MUSEUM: INSIDE OUT

The museum is designed from the Inside / Out. The architecture is suspended in time by existing at the same moment as the form explodes and implodes. The red steel fragments can be seen from the interior pushing and pulling simultaneously while looking up. The circular plaza erupts in both the ground plane and the buildings projecting out in space.

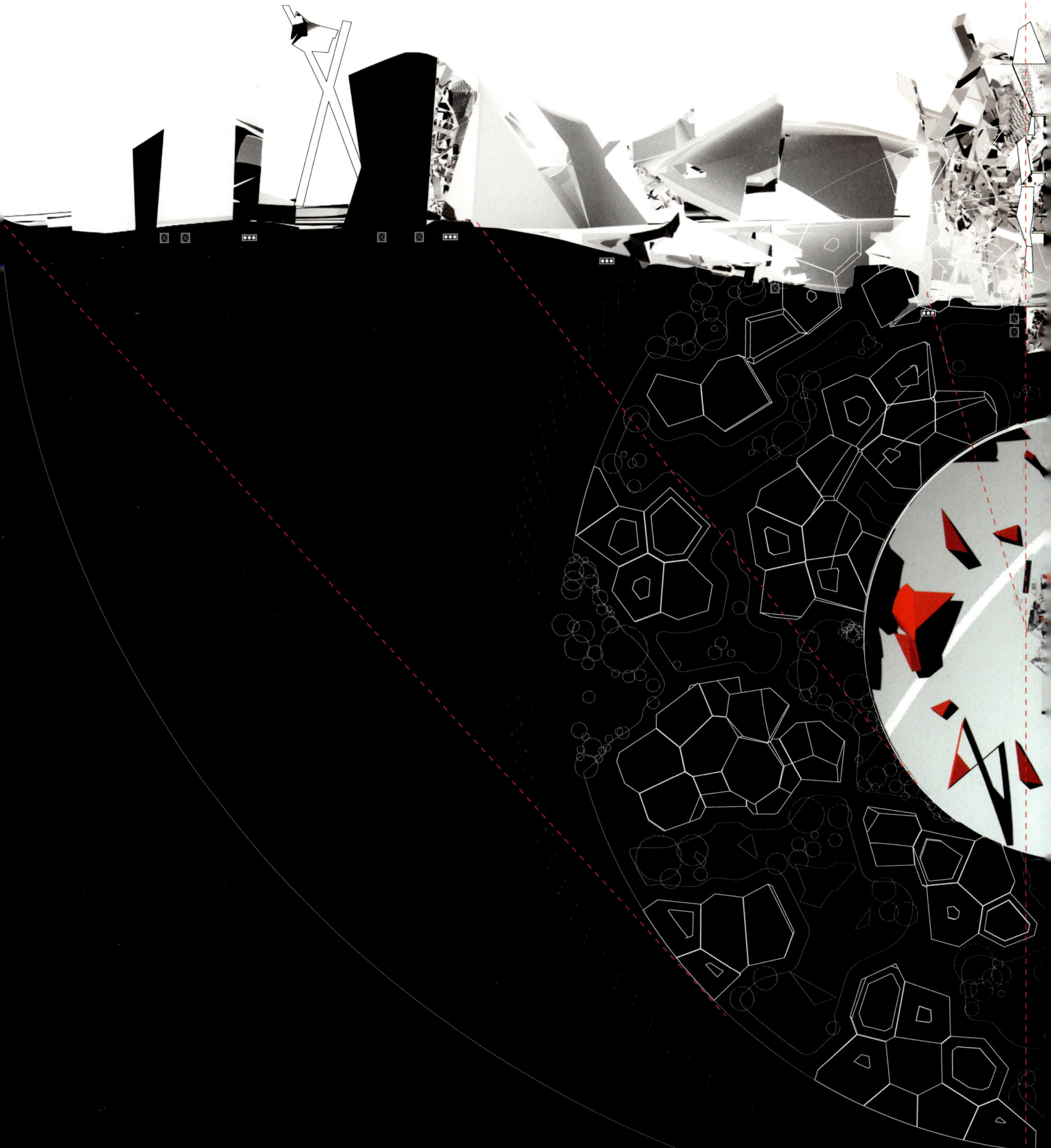

GAME ON | EXPLODED

DOOMSBURY: CENTRAL PLAZA

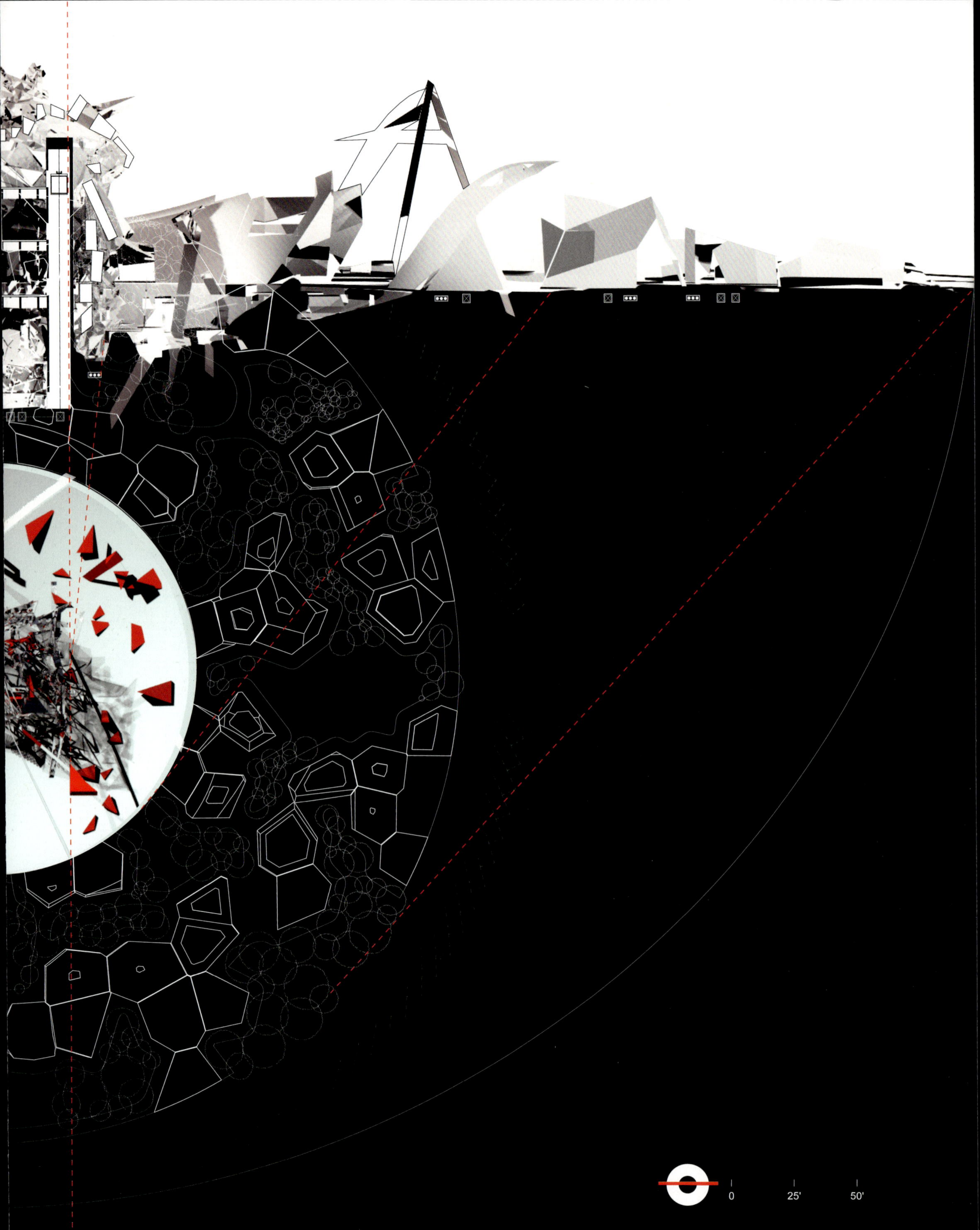
0
25'
50'

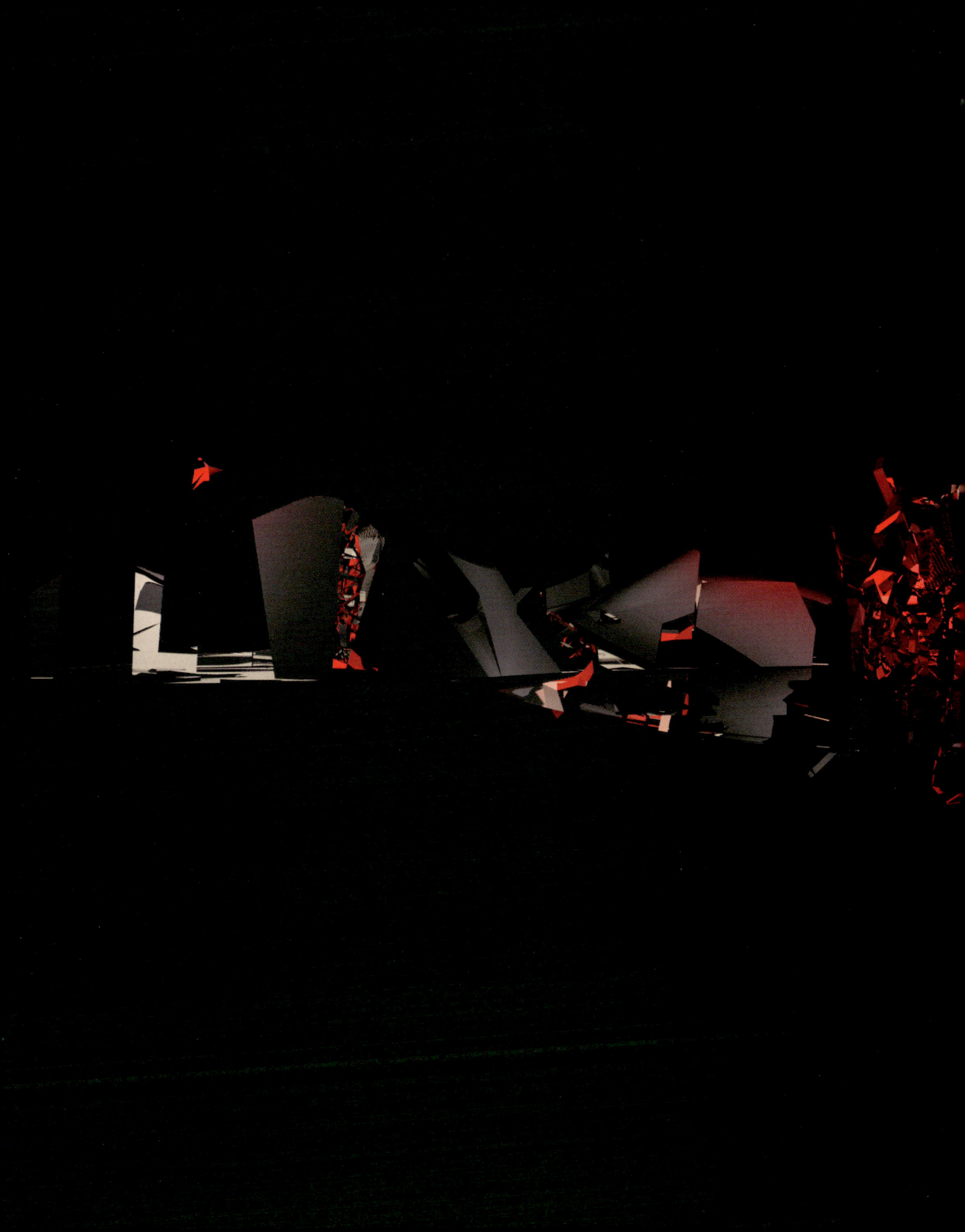

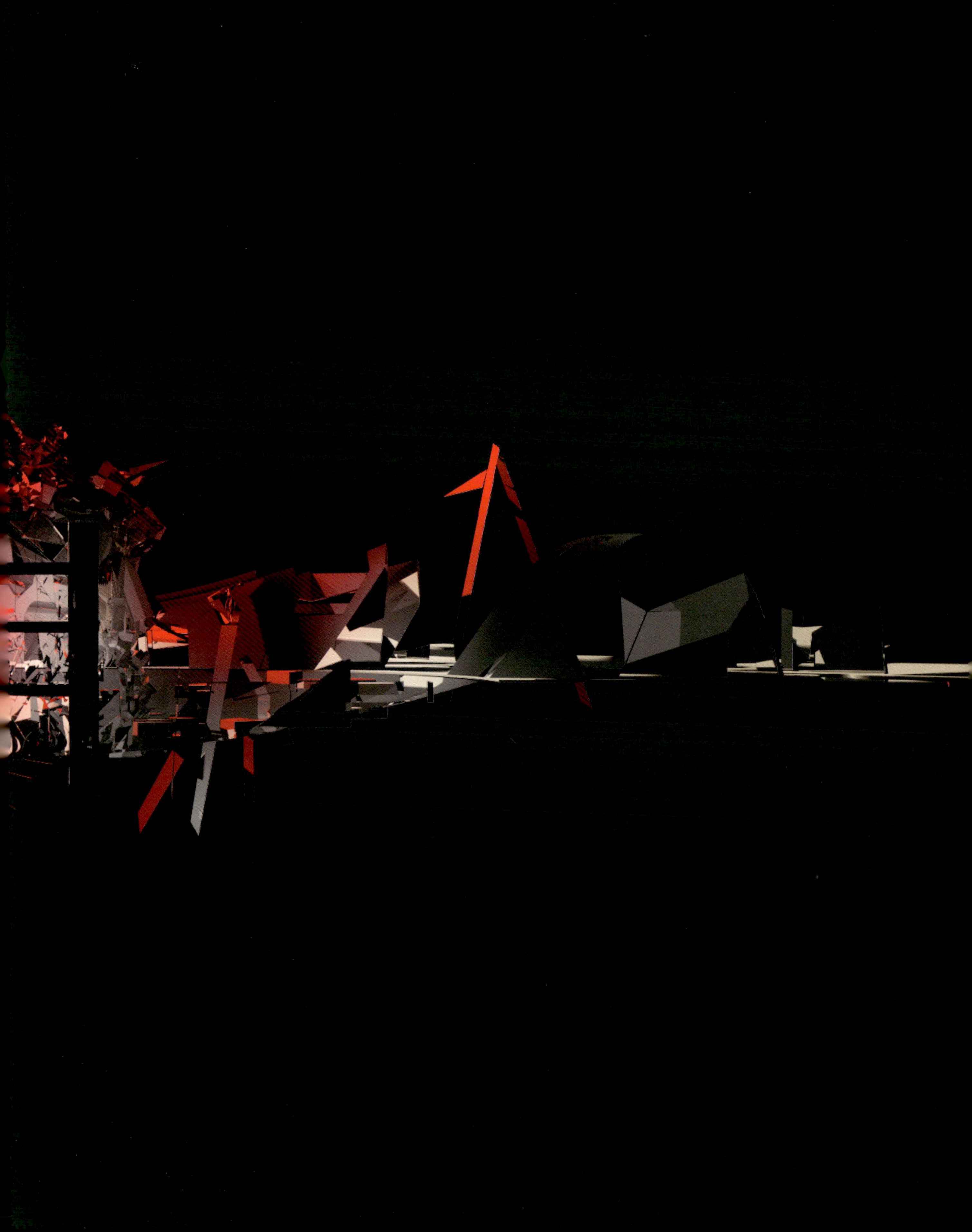

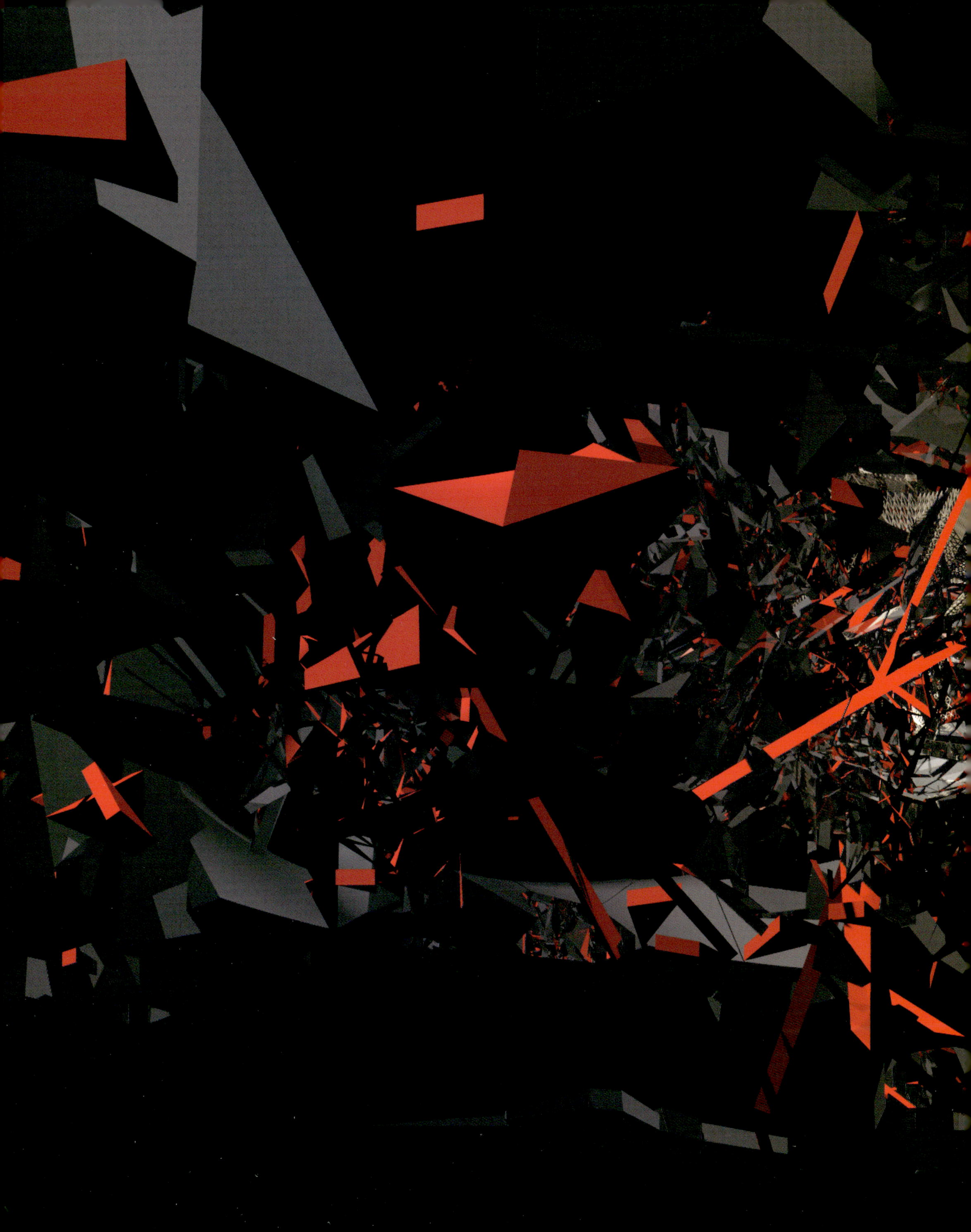

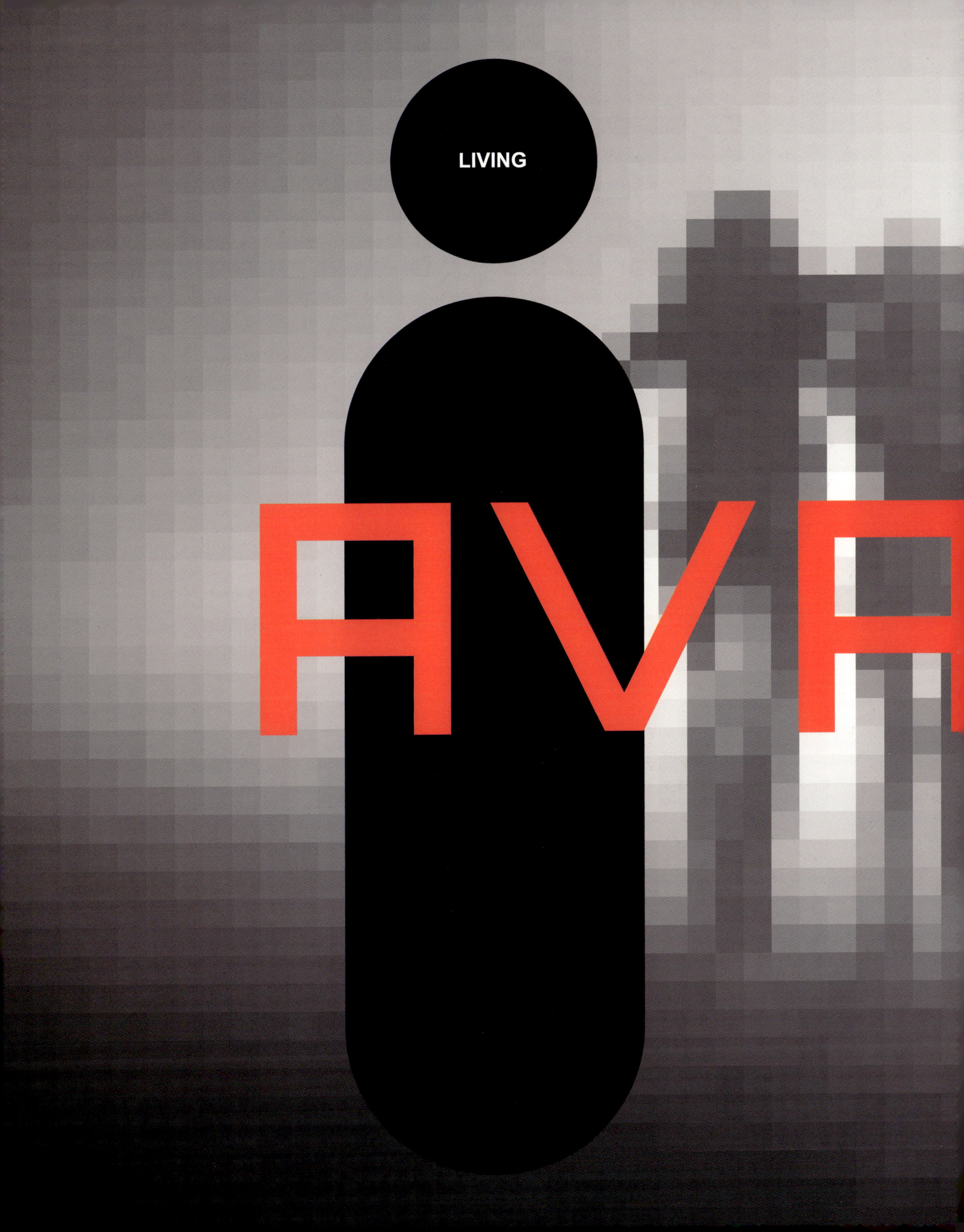
LIVING
AVA

DYING

TAR

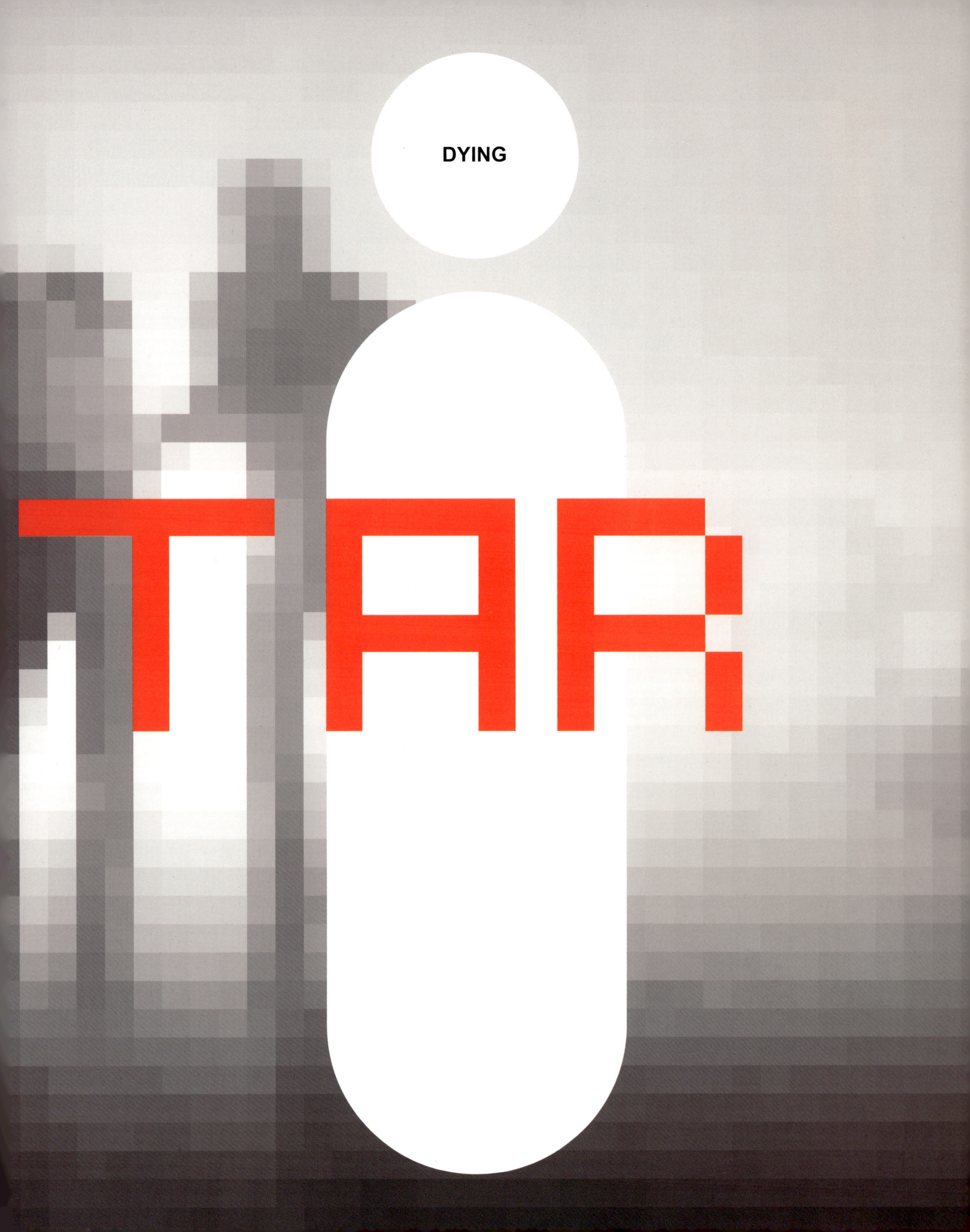

Social media and the digital age have spotlighted the increasing sense of isolation[189] and unleashed some base instincts in humans. The globalization of the internet held the promise of bringing the world closer together, however, it turned out there is a dark side, depression and a lack of self-confidence in teenagers are on the rise with the invention of the smart phone and social media.[190] A Happy Place, located north of Atlanta, Georgia, creates a safe haven for teenagers in a camp setting to "turn off" the digital world and try to reconnect with the real world. The addiction of the virtual world[191] is understandable because of the advent of avatars that dress better, are thinner, and can change their appearance to be what they can only dream of in the real world. The pandemic has further taken youths down the rabbit hole of creating this other life due to the increased time online.[192] By escaping technology and returning to a natural setting with their peers for a limited time, the allure of the digital world is kept in check and redefined as a positive force for connecting with others around

SURVIVING DEPRESSION IN THE DIGITAL AGE

the world and with history. Gaming is put in perspective as a form of entertainment, not a lifestyle. With the invention of the Metaverse the ability to go into the web and experience the third dimension will only enhance the attractiveness of losing oneself[193] and highlights the need for a Happy Place to go in order to reconnect with the real world. The camp is organized in circular rings from the individual to the collective with an eating area shaped around a circular garden, with cabins create a housing campus with a lawn for sports, gatherings, and a central fire pit to share the night time communal experience. Understanding reality and one's position in the world is a discovery time for teenagers and the question is how to evaluate this global social experiment with the increasing use of technology and the negative side effects of social media with teenagers connecting through devices instead of human contact. The Happy Place represents an alternative, a time to take a breadth, a moment to reconnect with each other directly without the interference of a device.

A HAPPY PLACE: ESCAPING TECHNOLOGY

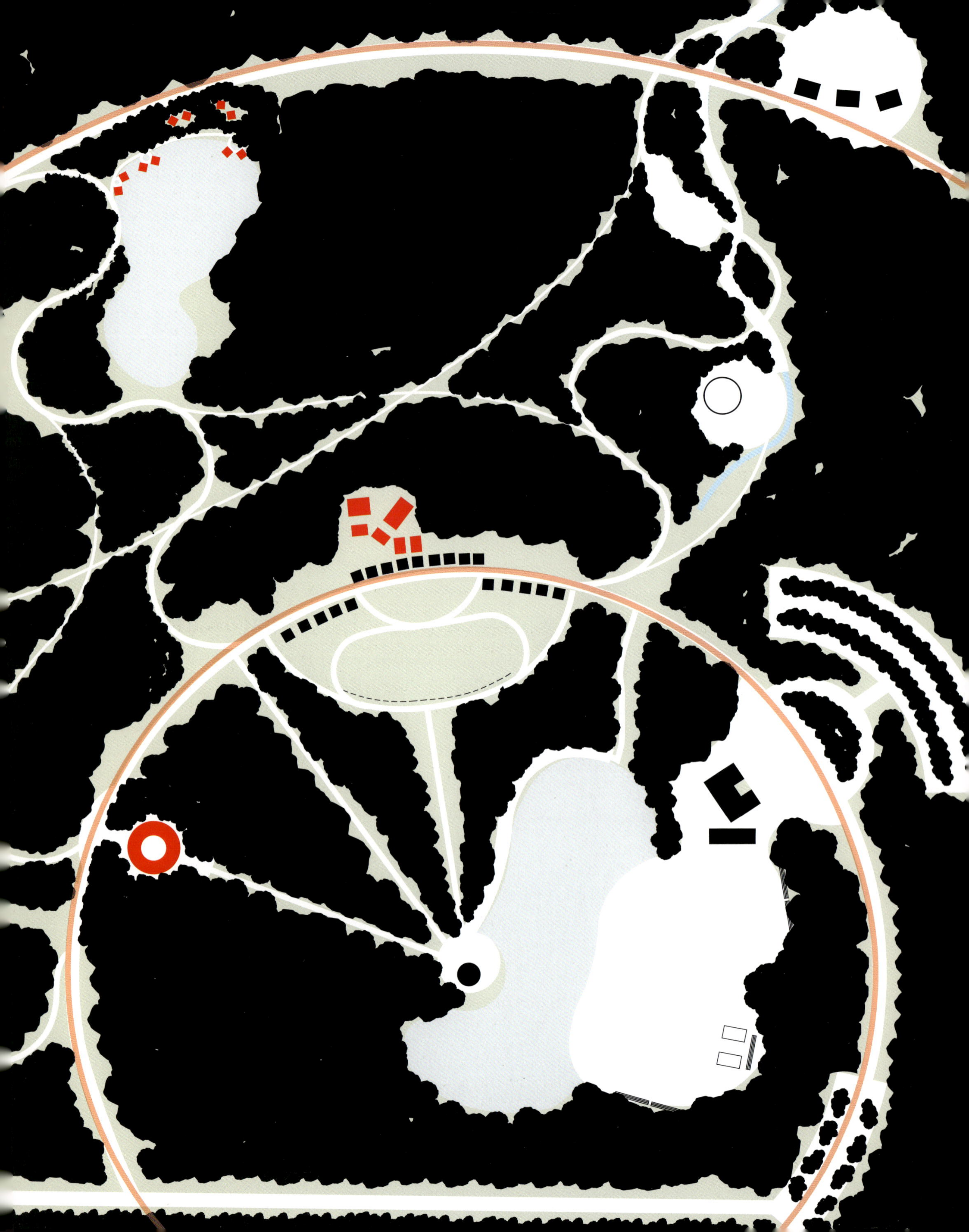

6.3 MILLION

CHILDREN WITH DEPRESSION + ANXIETY[194]

2.1 MILLION

30% ARE TREATED[195]

AVATAR | DEPRESSION

LIVING WITH NATURE: MASTER PLAN

COMMUNAL RING

1 WELCOME CENTER

2 HOME

3 THE CIRCLE

4 FIRE RING

5 EVENT FIELD

HEALING RING

6 GUEST PARKING

7 AMPHITHEATER

8 THE SQUARE

9 ADMINISTRATION

INDIVIDUAL RING

10 ART HUTS

11 MEDITATION GARDEN

12 BOAT DOCK

13 STAFF DORMITORY

14 ANIMAL FARM

0' 250' 500'

AVATAR DEPRESSION

CABINS: A SAFE PLACE TO GO

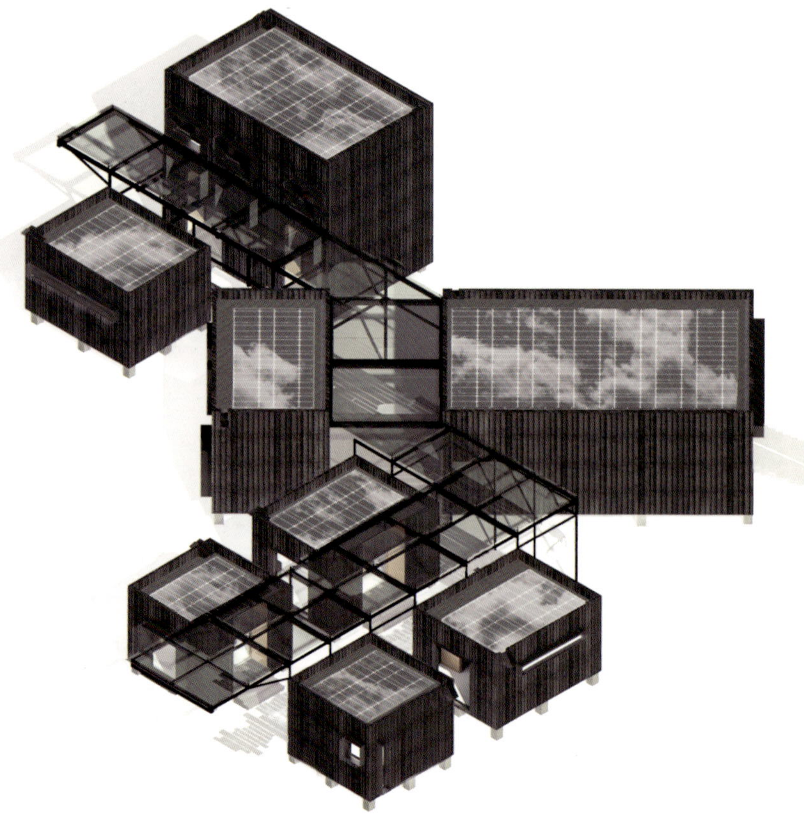

The cabins are loosely organized around a circular path with a lawn for communal activities. The cabins reflect a safe place to go and creates a public pathway with porches that promote interaction among the teens. The buildings made of charred wood don't weather and act as a backdrop to the forest and internal gardens.

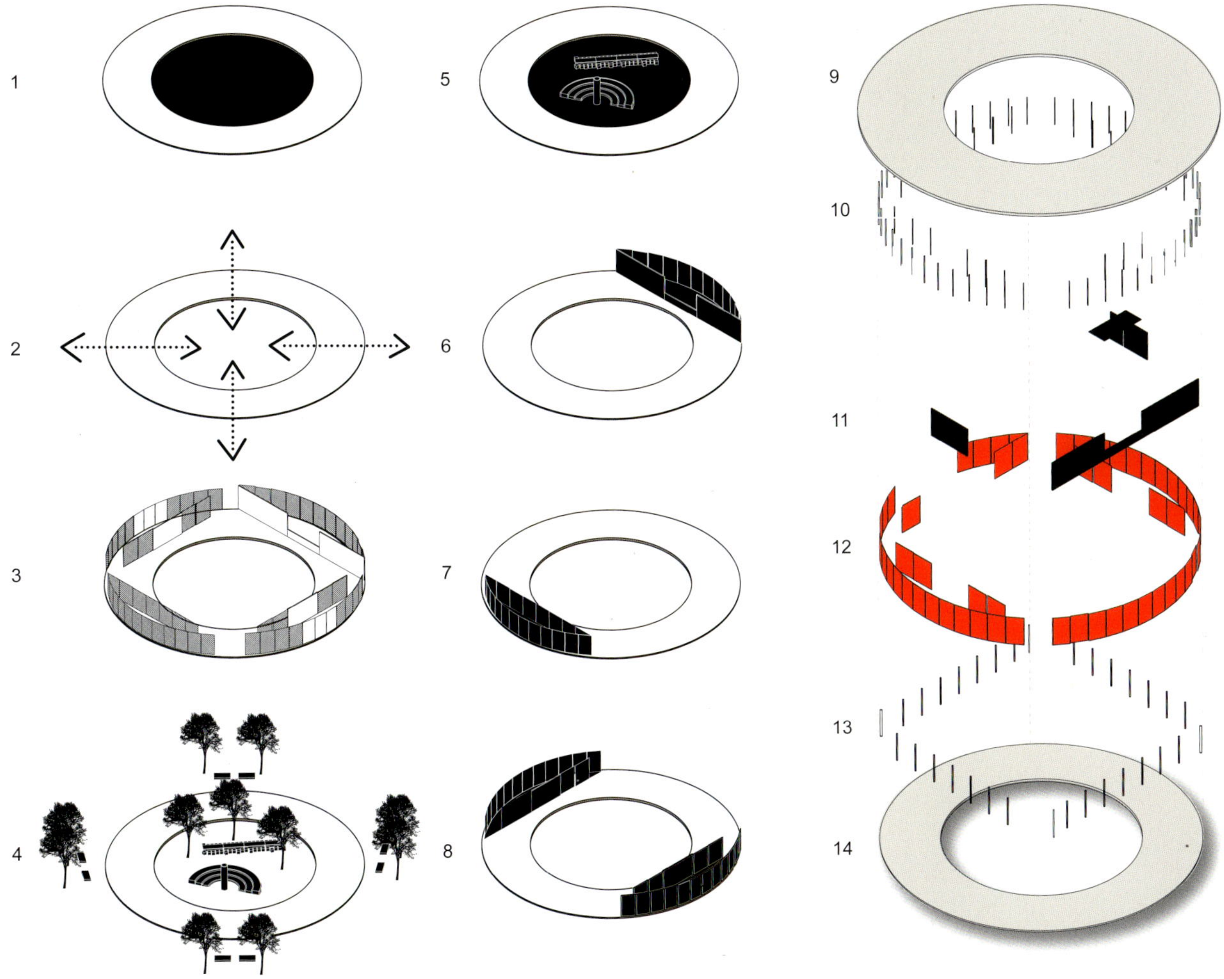

1 INTIMATE SPACE
2 AXIAL ACCESS
3 MESH SCREENS
4 GARDENS
5 FIRE PIT
6 KITCHEN
7 ENCLOSED DINING
8 RESTROOMS
9 ROOF
10 MULLIONS
11 PARTITIONS
12 SCREEN PANELS
13 STEEL COLUMNS
14 FOUNDATION

AVATAR | DEPRESSION

THE CIRCLE: COLLECTIVE DINING

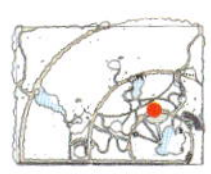

THE CLEARING

GATHERING AROUND A PRIMAL CIRCLE

The Circle is a collective eating area that has a contained garden in the center with a barbecue area located around an amphitheater. The indoor / outdoor seating can be used in all the different weather conditions. The rotating, perforated siding panels bring the forest into the building, while the kitchen and storage have a controlled space for cooking.

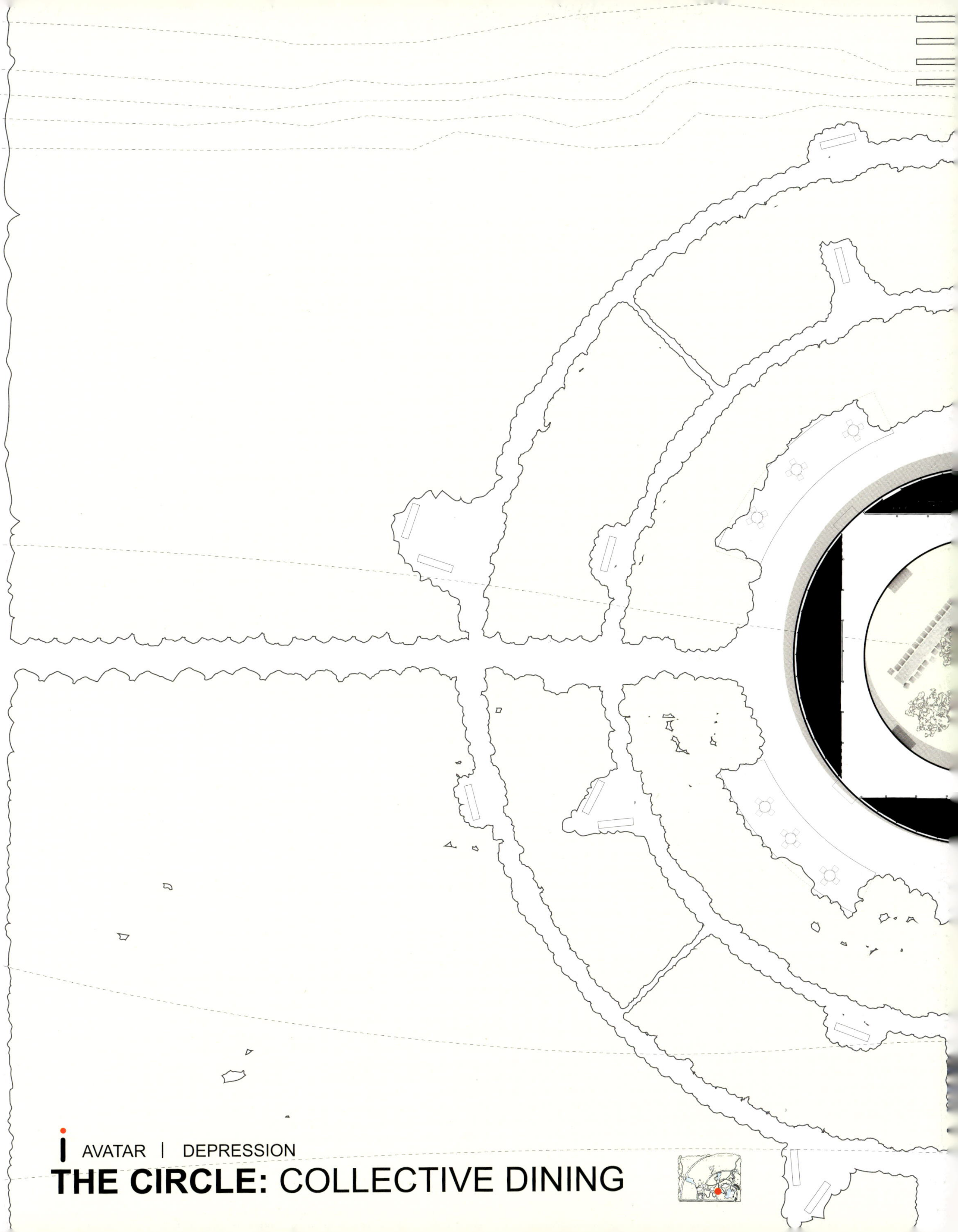

AVATAR | DEPRESSION

THE CIRCLE: COLLECTIVE DINING

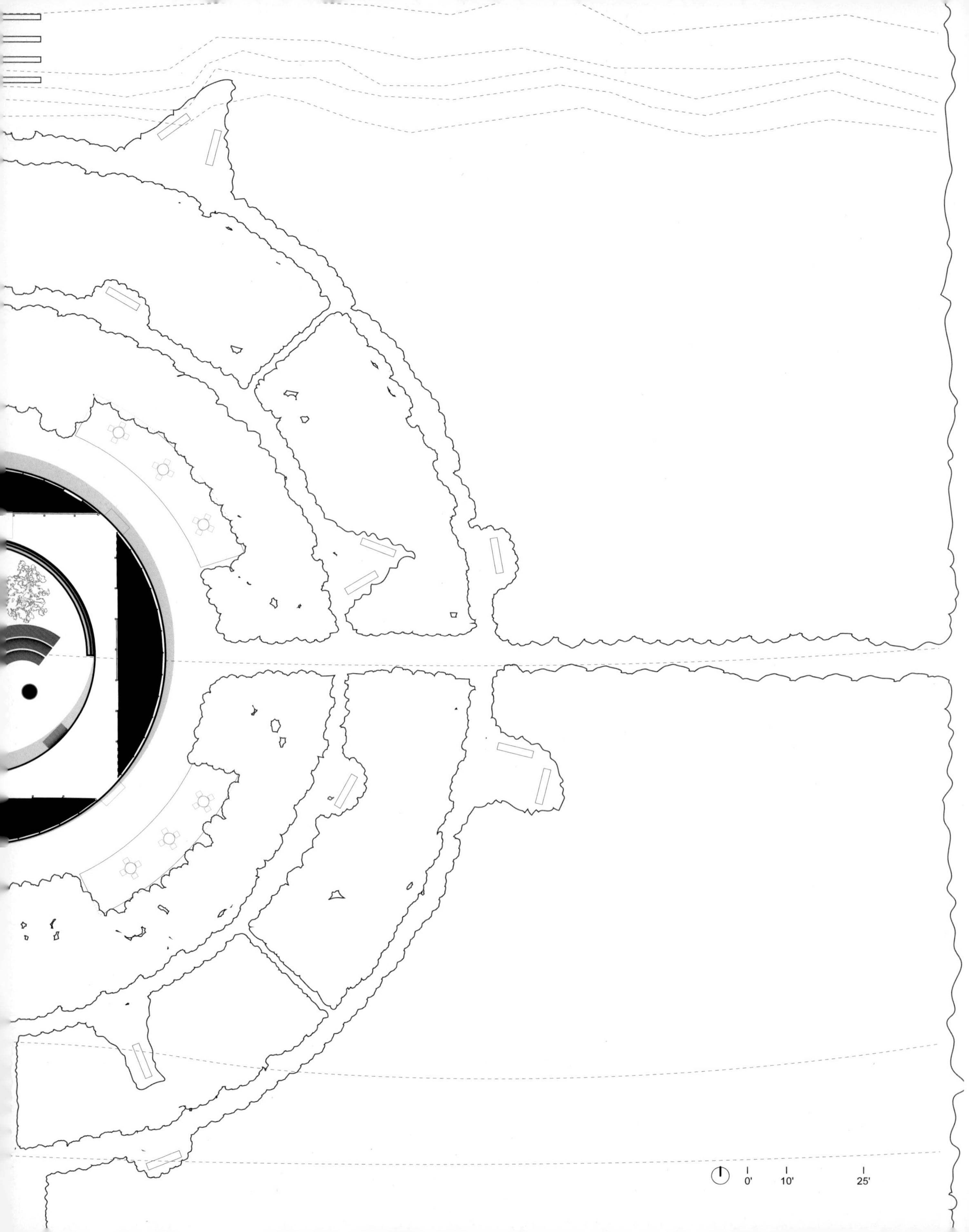
0'
10'
25'

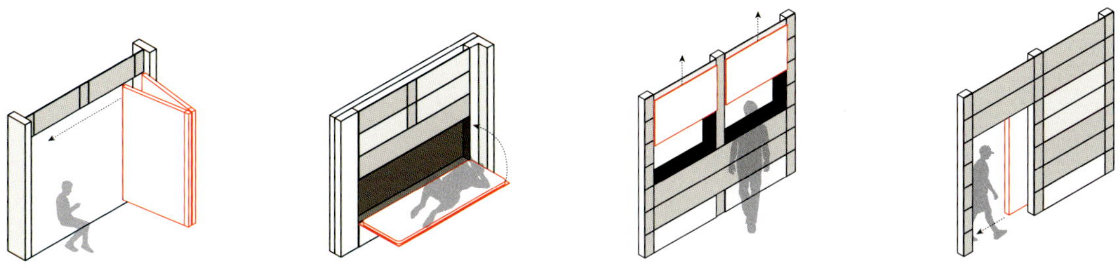

AVATAR | DEPRESSION

THE CIRCLE: DINING IN NATURE

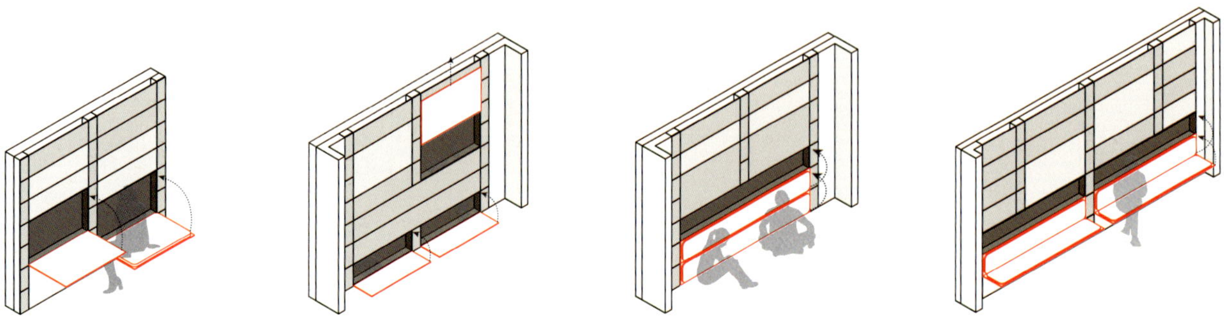

The panel systems extends throughout the grounds of the Happy Place. The perforated sliding panels of the Circle and the adaptive wall panels of the therapy cabins are all designed to allow the individual to be in control of their environment by being able to change the architectural spaces at will by moving the panels.

AVATAR | DEPRESSION

THE CIRCLE: CLOSED | OPEN

The transparency of the circle can be seen in section with the central location of the fire pit and the floating circular form lifted above the ground appears to hover over the landscape. With deep overhangs, the protection from the rain and the sun creates a sense of place where those going to the circle can feel like they are eating in the woods.

INDIVIDUAL

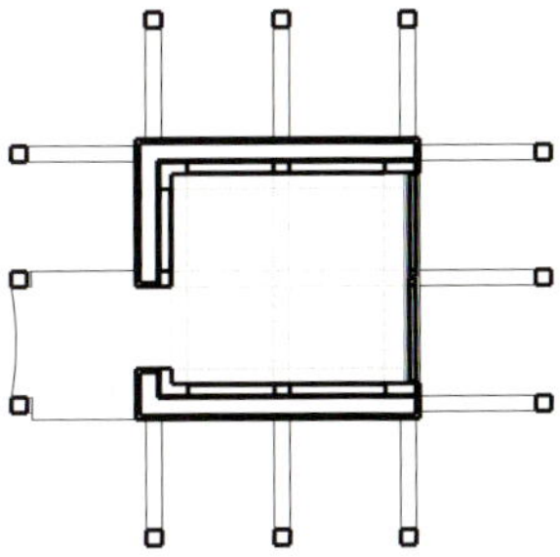

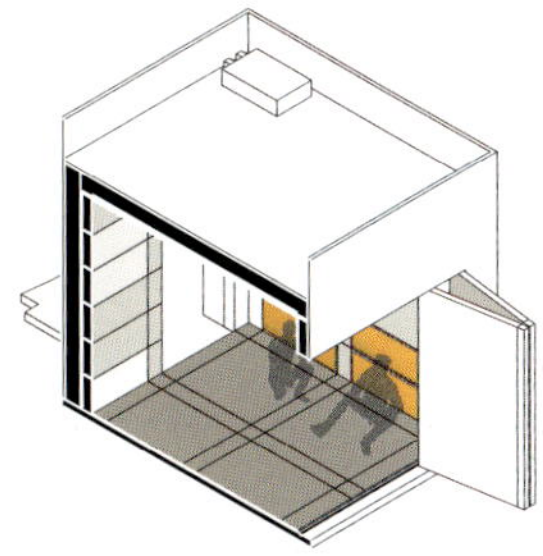

SMALL GROUP

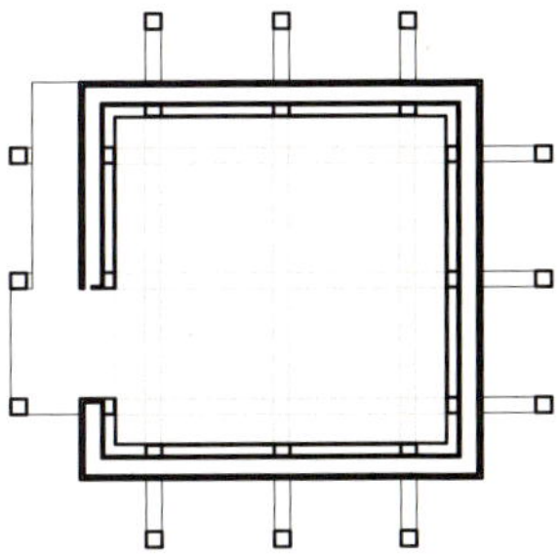

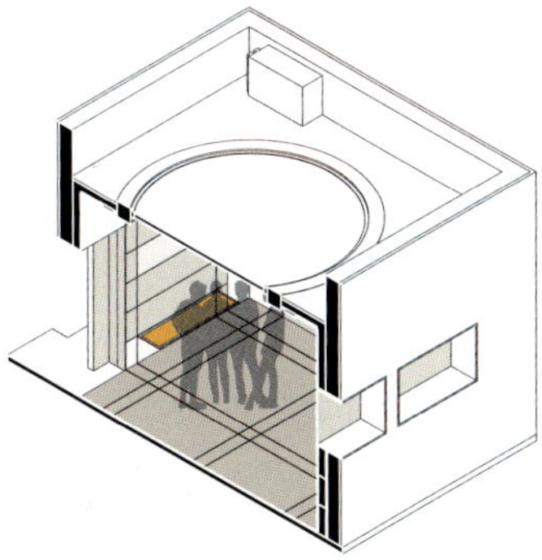

LARGE GROUP

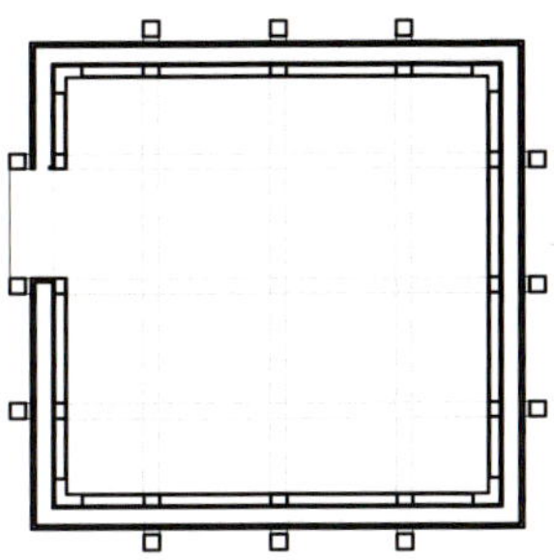

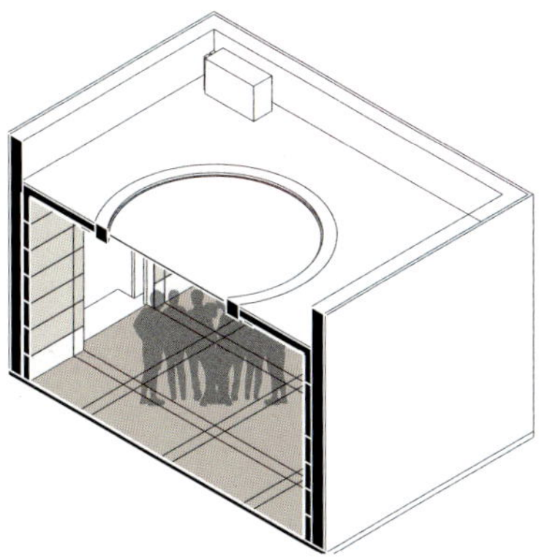

i AVATAR | DEPRESSION

COUNSELING: INDIVIDUAL | GROUP HEALING

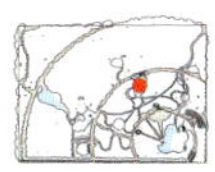

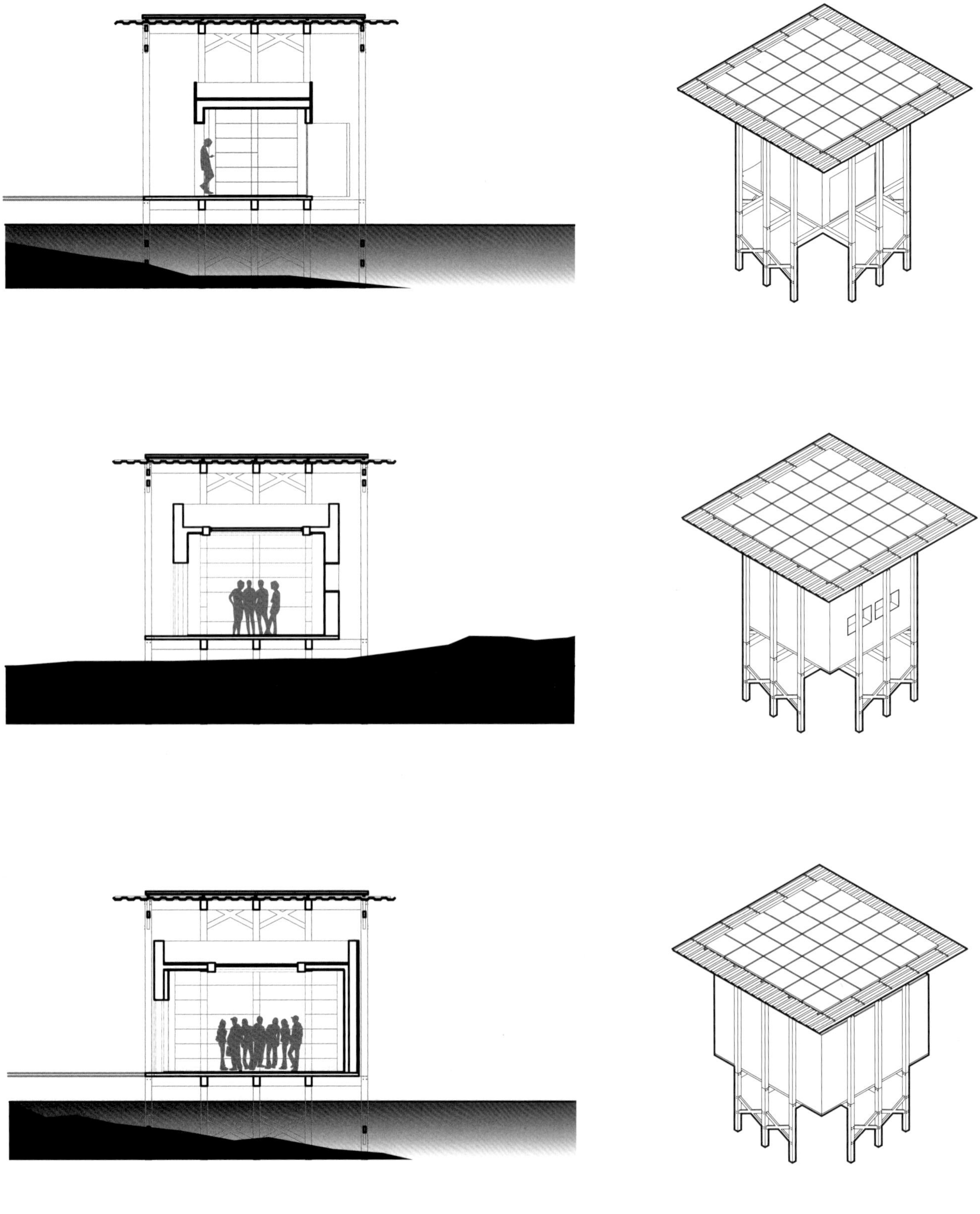

The three types of counseling cabins are designed as small, medium, and large for each type of therapy session and are located in different relationships to the lake. The circular skylight above has reflected light from the metal hovering roof structure that creates negative pressure for breezes to go through the cabins.

The contemporary fear of the oncoming effects of irreversible climate change by 2030[196] has created the basis for a Garden of Extinction. With the approaching global disasters, the need to protect a variety of seeds to support human life is becoming ever more important. As so many of the animal species on earth have become endangered, so too is the plant life that humanity depends upon to survive. This project creates a working museum of endangered and poisonous plants that are open to the public with exhibitions located below the circular forms of the gardens. The tilted garden is oriented to the path of the sun for the most efficient use of the available sunlight. Below the surface is a bunker housing a controlled environment of a seed farm that is stored safely and only to be used after a global disaster and when the atmosphere is once again conducive to growing the plants. This prototype will be replicated around the world to produce multiple buildings that

SURVIVING EXTINCTION IN THE AGE OF GLOBAL DISASTERS

are designed for the local sites and plant conditions. The very existence of this museum is meant to be a warning to the public that time is running out and humanity is facing an existential dilemma. The organization of the museum is designed using the typologies of the labyrinth, maze, and stepwell. Taken from different historical references and used for a variety of purposes the strong architectural forms are meant to highlight the circuitous paths humans have taken to get at this crossroads in time. The stepwells in history date to the 7th century CE in India[197] and are now used to collect rainwater in cisterns to be used in the submerged bunker of stored seeds and humans. The central descending stair acts as a fire escape for the building below grade and as a visual filter for the museum. The bunker is equipped for housing people and seeds and uses technology as a way to transport those living there into different real, augmented, and virtual realities.

A GARDEN OF **EXTINCTION**

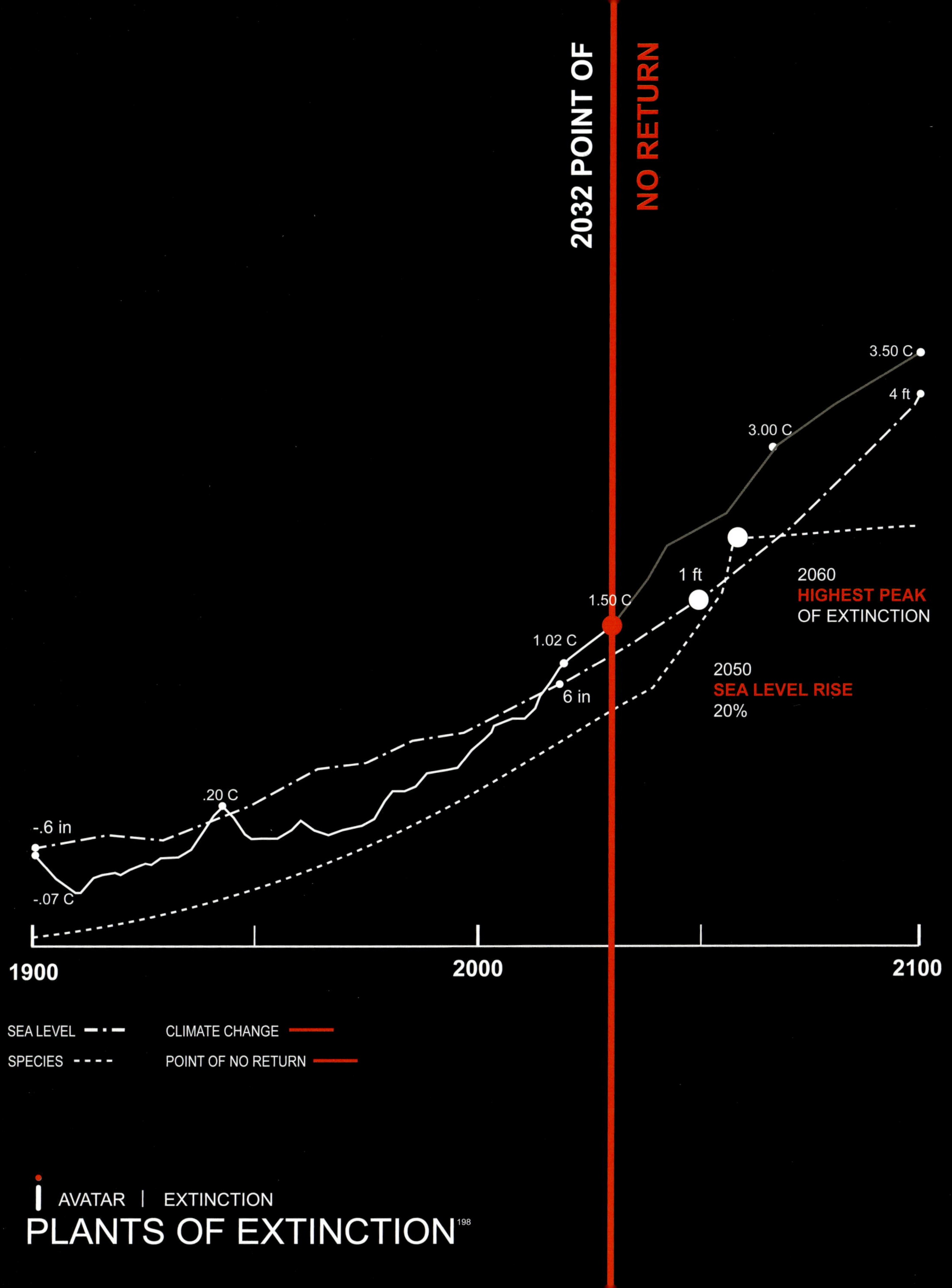

AVATAR | EXTINCTION

PLANTS OF EXTINCTION[198]

GARDEN OF EXTINCTION: A HIDDEN SEED BANK

0' 25'

1. GARDEN ENTRY
2. TEMPERATE FOREST
3. LOTUS POND
4. STORAGE
5. FLOWER GARDEN
6. FERN GARDEN
7. SHRUB GARDEN
8. RESTROOMS
9. LOADING DOCK
10. CACTI GARDEN
11. ARENA
12. SUBTROPICAL FOREST
13. POISON GARDEN
14. MUSEUM ENTRY
15. STEPWELL

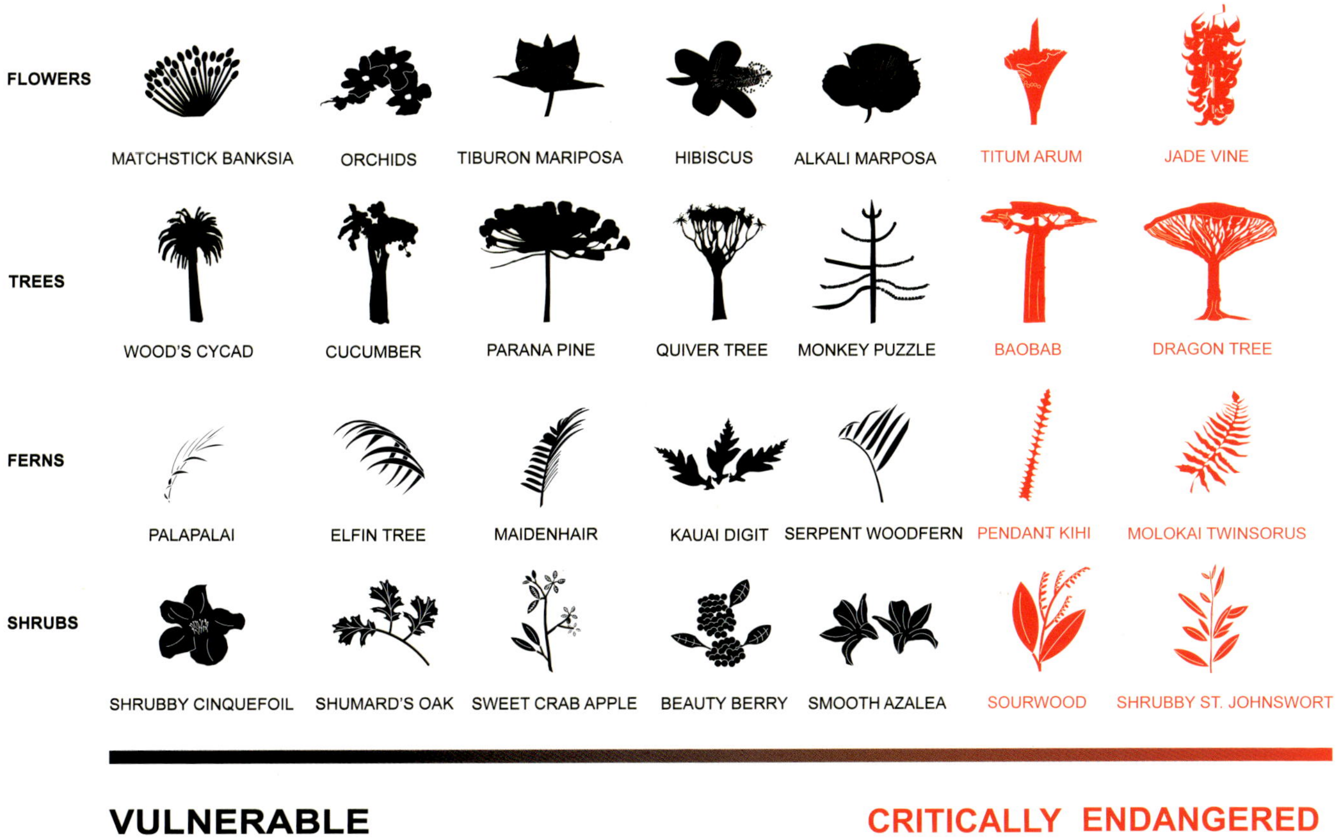

ENDANGERED PLANTS

AVATAR | EXTINCTION

ENDANGERED | POISONOUS PLANT SPECIES[199]

POISONOUS SPECIES

The point of no return has now been projected as irreversible by scientists[200] *and will mark the beginning of catastrophic weather conditions of sea rises that will threaten the very existence of human's capacity to live in the earth's atmosphere. The museum gardens reflect this threat by displaying the plants that are endangered, as well as poisonous plants.*

LABYRINTH A SINGULAR PATH LEADING TO A CENTER

MAZE A NETWORK OF PATHS WITH DEAD ENDS

27,508 SF
TEMPERATE FOREST

20,130 SF
SUBTROPICAL FOREST

12,426 SF
CACTI GARDEN

13,821 SF
FLOWER/ SHRUB/ FERN GARDEN

5,858 SF
POISON GARDEN

GARDENS

STEPWELLS A SUBTERRANEAN STRUCTURE WITH A SET OF STEPS LEADING TO A WATER SOURCE

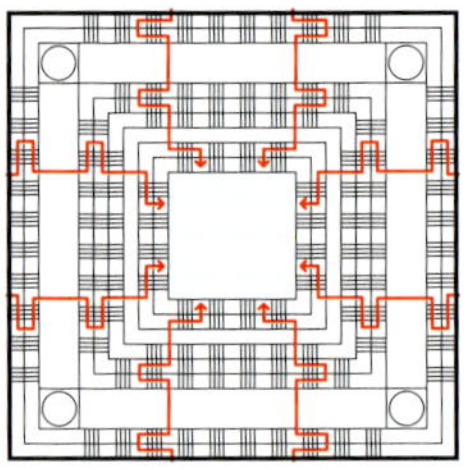

CHAND BAORI (RAJASTHAN)
800 CE

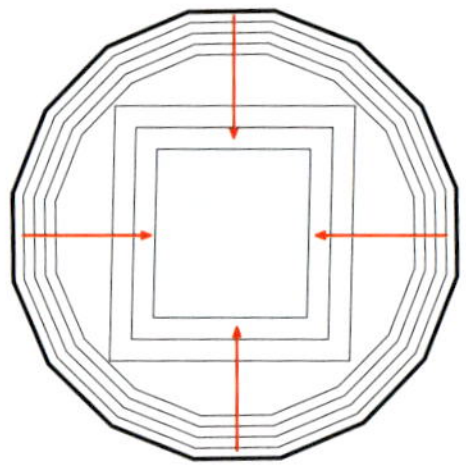

ADALAJ STEPWELL, CHAMPANER
1400 CE

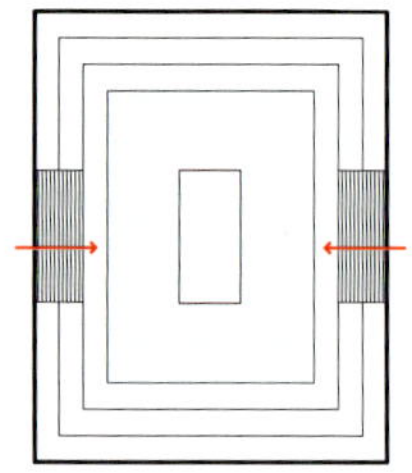

AMPI STEPWELL (KARNATAKA)
1400 CE

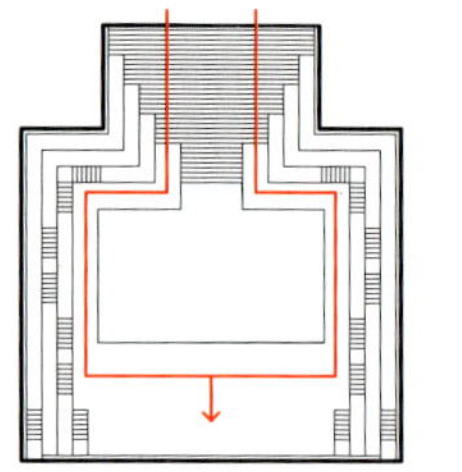

PERALASSERY TEMPLE POND
1000 CE

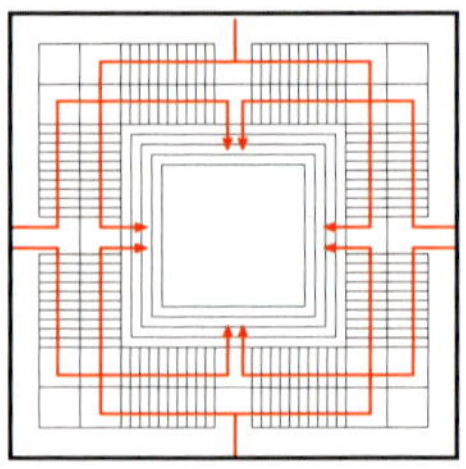

RAMKUND (BHUJ)
1026 CE

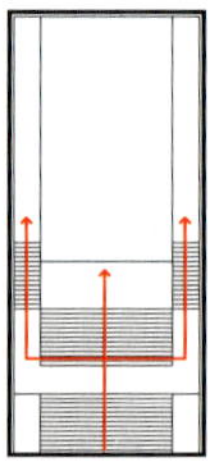

AGRASEN KI BAOLI (DELHI)
1300 CE

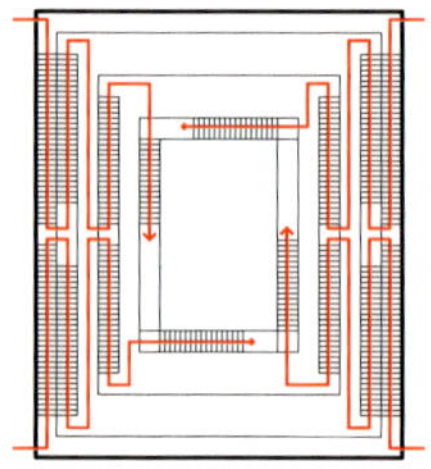

KHEDAMATA STEPWELL
800 CE

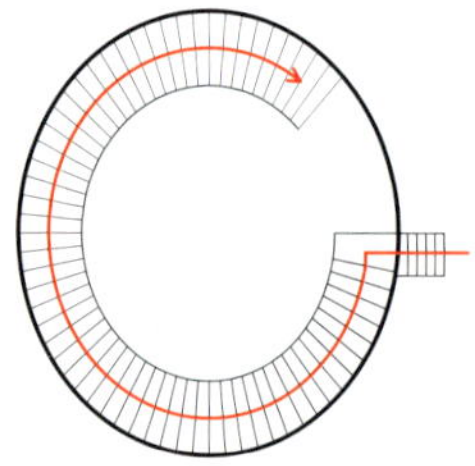

HELICAL STEPWELL, CHAMPANER
1500 CE

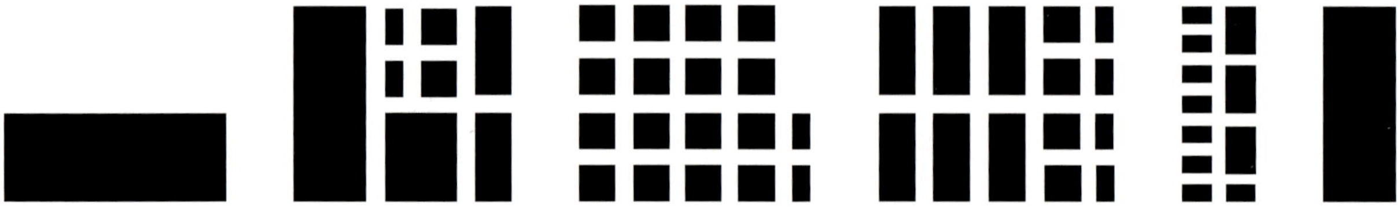

2,074 SF
FIRST FLOOR - F1
EXHIBITION

17,993 SF
F0
LOBBY
AUDITORIUM
CAFE
GIFTSHOP
RESTROOMS
MECHANICAL/ ELECTRICAL

15,068 SF
F-2
ADMINISTRATION
PRIVATE OFFICES
OFFICES
CONFERENCE ROOMS
RESTROOMS

15,985 SF
F-3
LABORATORY
RESEARCH LAB
TISSUE CULTURE LAB
MEETING ROOMS
COMPUTER LAB
COAT ROOM
RESTROOMS

17,450 SF
F-4
LIVING VAULT
BEDROOMS
LIVING ROOM
KITCHEN
FOOD STORAGE
RESTROOM
GYM
MECHANICAL/ ELECTRICAL

17,450 SF
F-5 + F-6
SEED VAULTS

MUSEUM

The collection of vulnerable, endangered plants is part of a unique exhibition displaying the last known seed of a particular plant. The scholarly approach to identifying and cataloging the plants is part of the research initiated by the museum staff and the collaboration of other institutions.

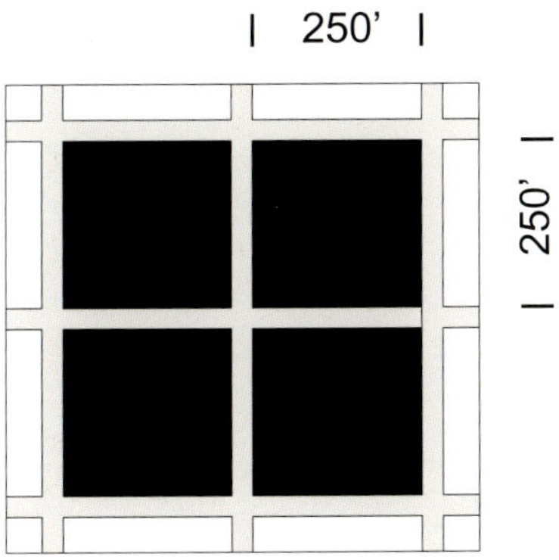

4 CITY BLOCKS

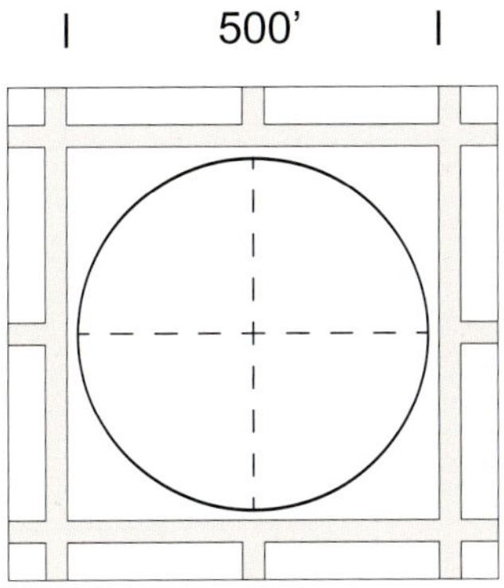

TOPOGRAPHY

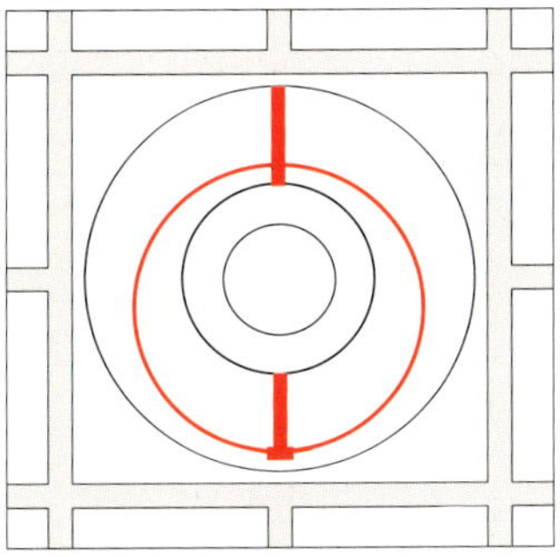

ENTRANCES

TUNNEL WALL ENTRANCE

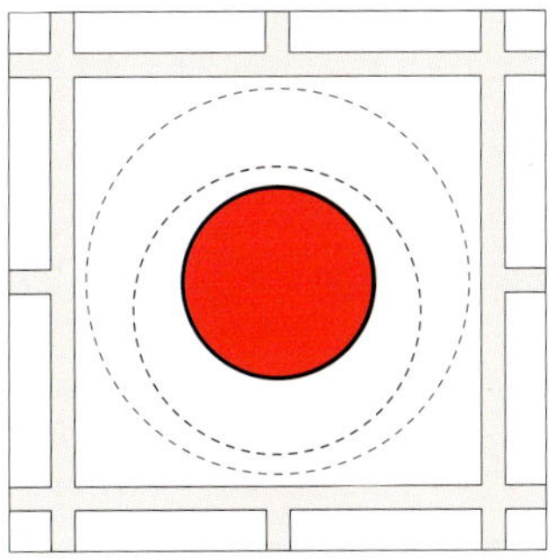

MUSEUM

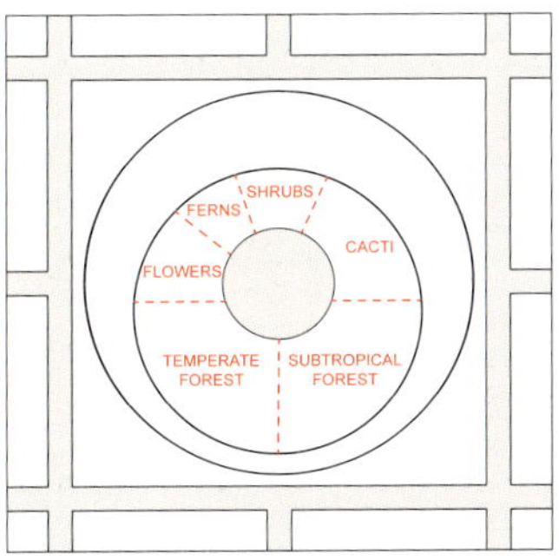

ENDANGERED PLANTS

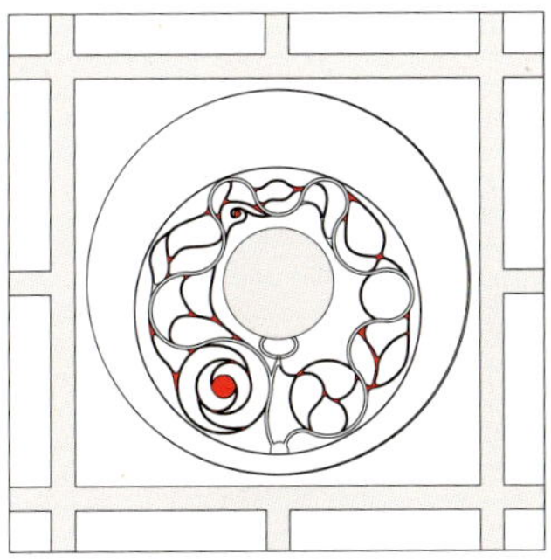

PATHS

A GARDEN OF EXTINCTION: A PROCESS

| 150'|

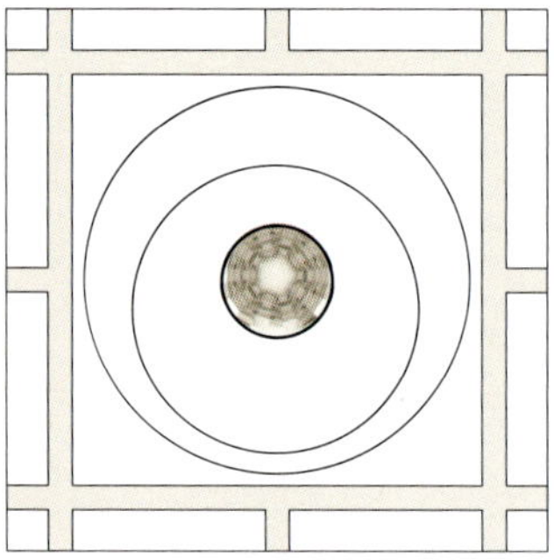

STEPWELL

LABYRINTH

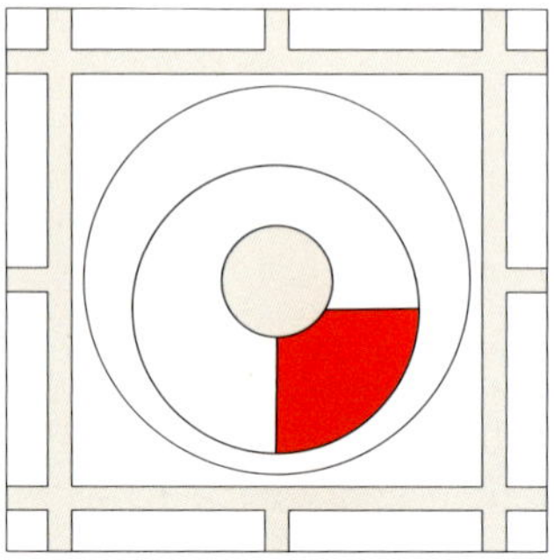

POISON GARDEN

ENDANGERED GARDEN

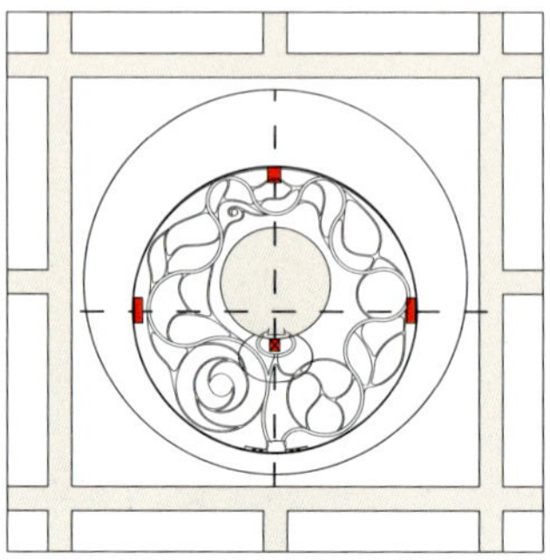

AXIS

GARDEN OF EXTINCTION

The design of the museum came from multiple sources to reflect the desire to be inclusive in sharing the need to save the planet. The labyrinth organizes the garden in a tilted circular form, while the stepwell brings the visitor below the surface perceiving the museum as geological forms descending to a pool of water.

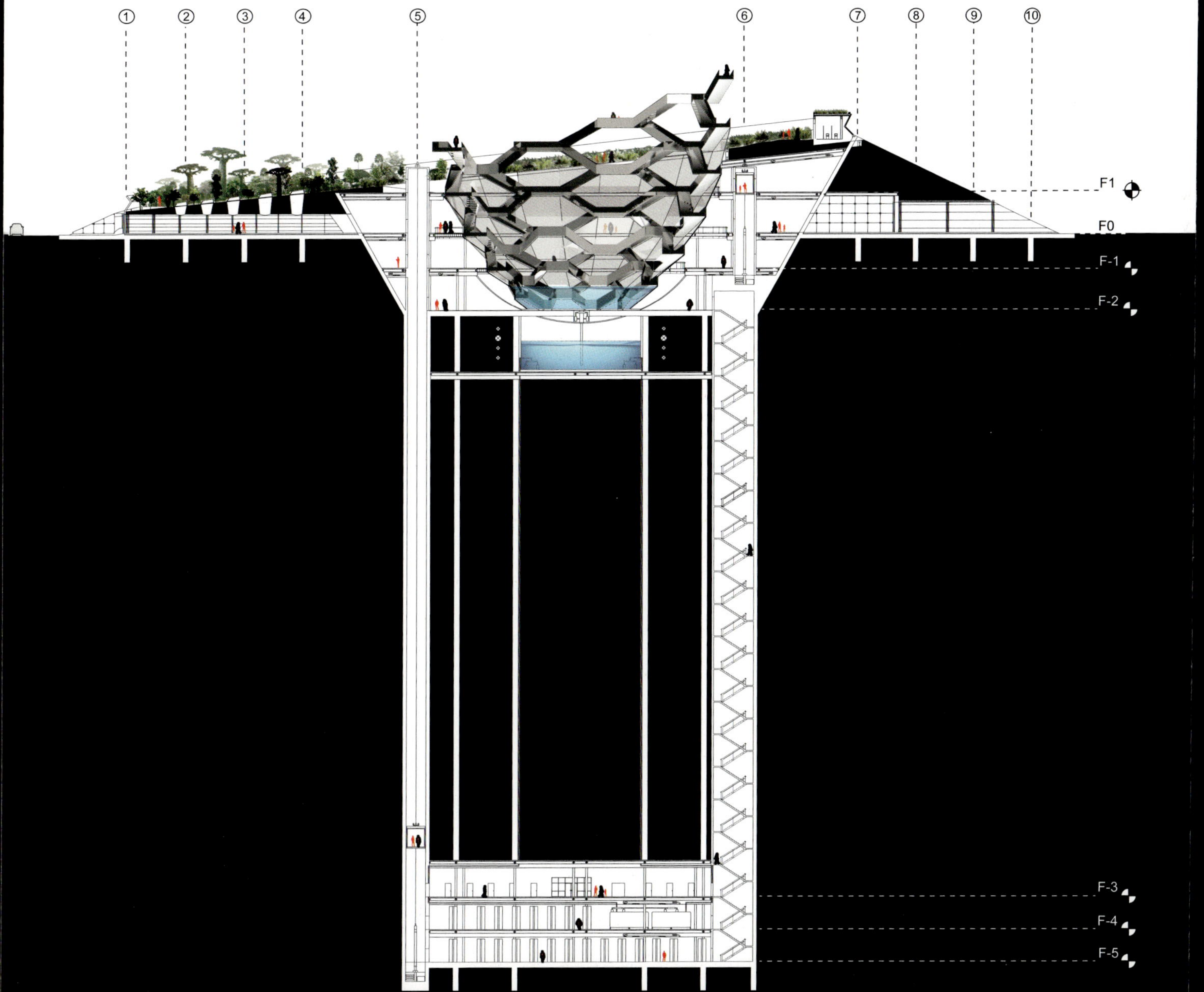

1
2
3
4
5
6
7
8
9
10
F1
F0
F-1
F-2
F-3
F-4
F-5

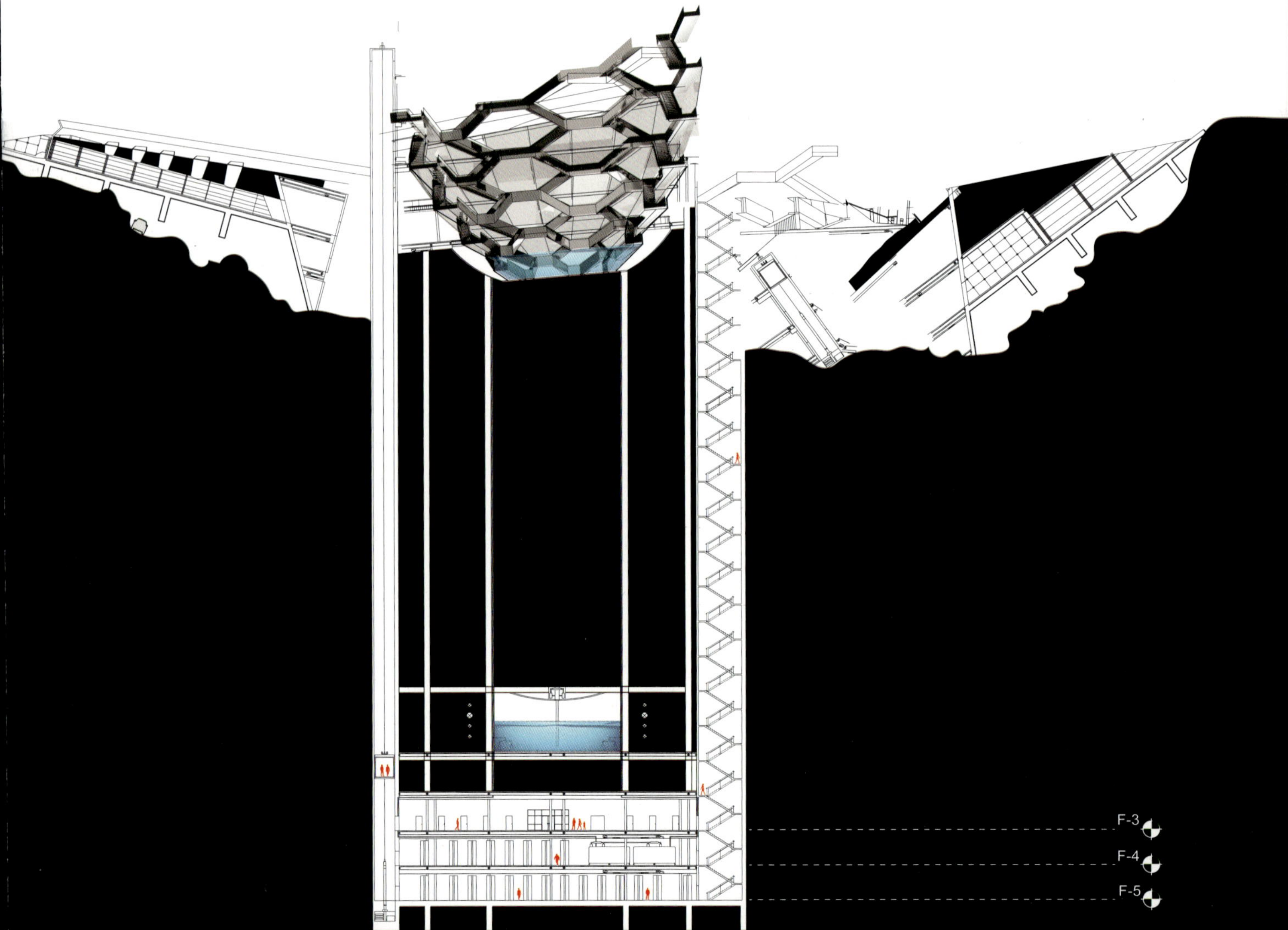

2040

0 25' 50'

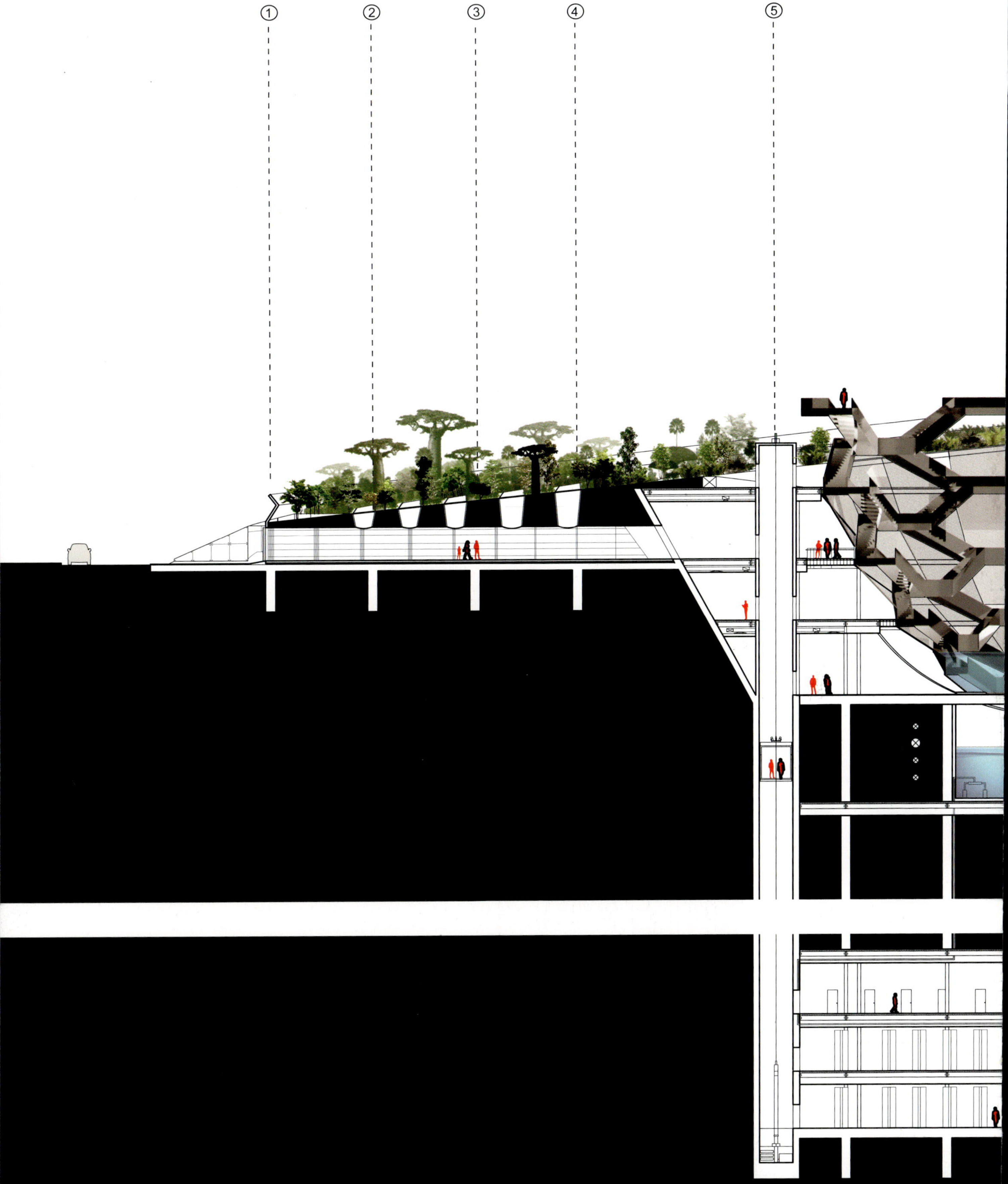
1
2
3
4
5

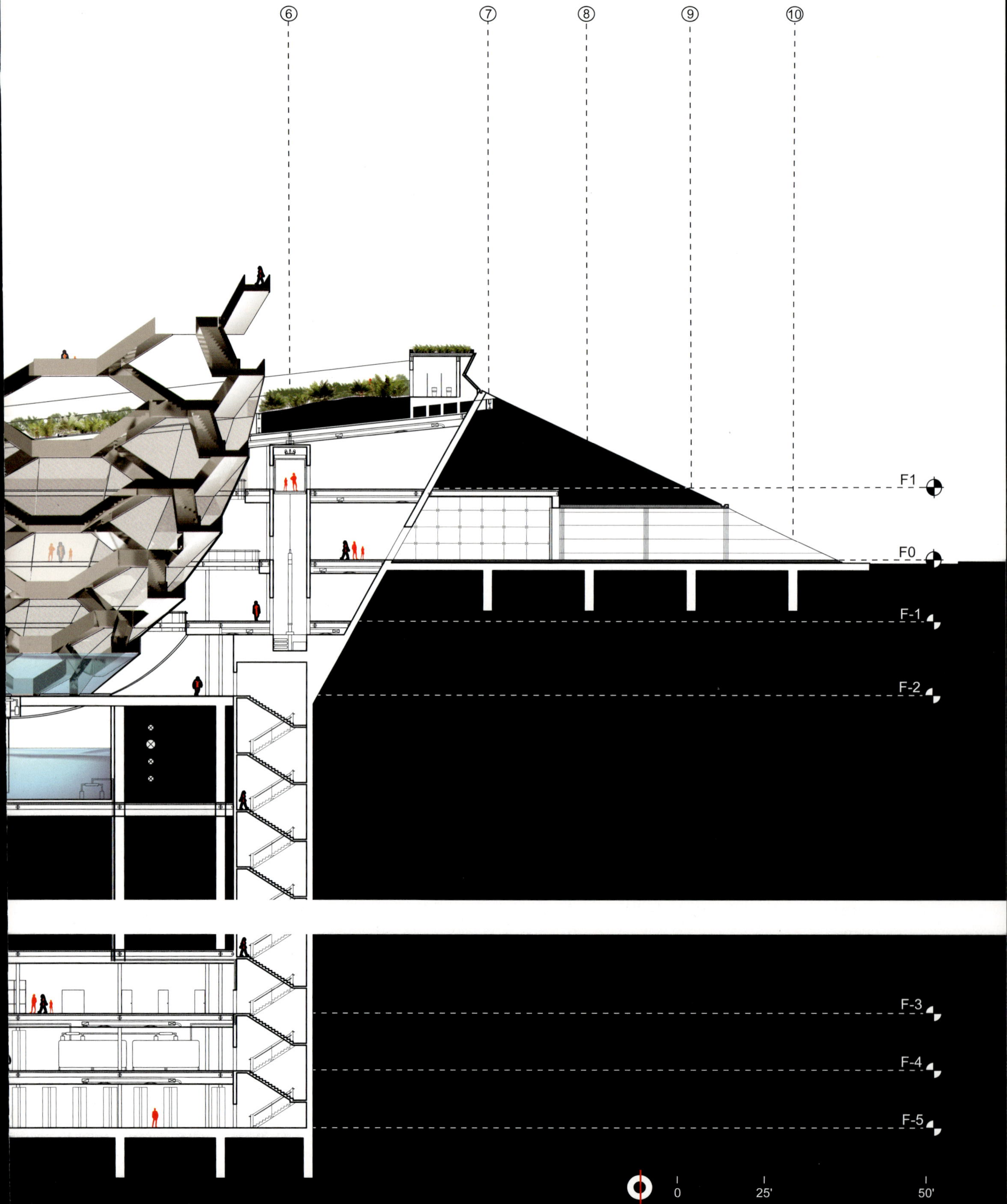

6
7
8
9
10
F1
F0
F-1
F-2
F-3
F-4
F-5
0
25'
50'

2025 MUSEUM OPENING DAY

2040 GLOBAL DISASTER

AVATAR | EXTINCTION

GLOBAL DISASTER 2.0

2100 EARTH 2.0

2080 VIRTUAL ESCAPISM

AVATAR | EXTINCTION

SEED GROWTH: HUMANS

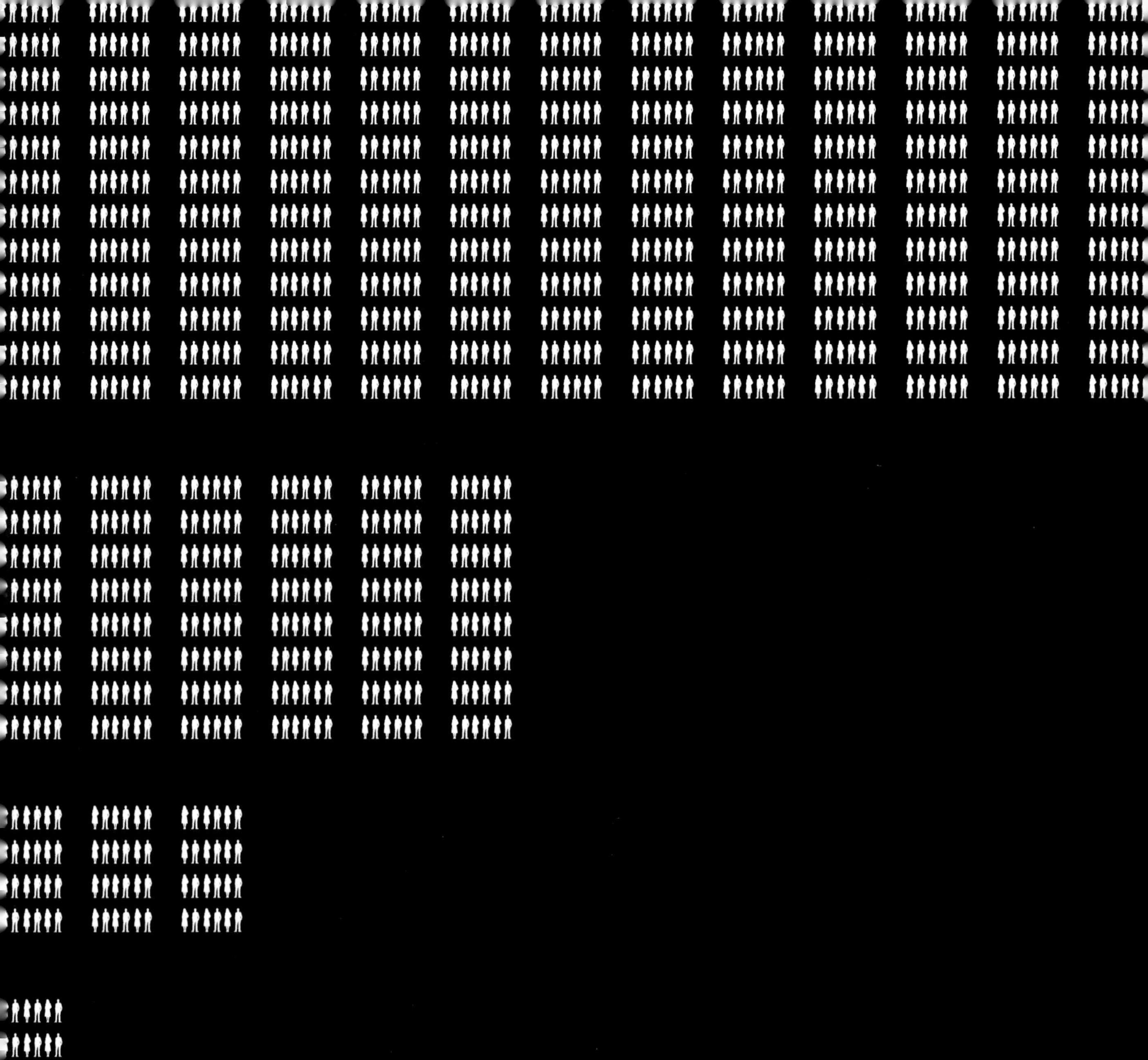

The concept of the Garden of Extinction not only represents the saving of the seeds of plants, but also the seeds for humans. The chart of repopulating humans on earth after a catastrophic event illustrates the exponential growth of people over time.[201] The

As the world approaches code red with global change predicted reaching temperatures of 1.5 degrees centigrade by 2032,[204] the need for creating an image of what humanity is losing has become increasingly critical. Before + After establishes a datum of Ground Zero in NYC as the centerpiece of an international museum with satellite exhibitions in eight communities around the world. Each museum represents a different catastrophic event caused by global warming in the locations chosen by the museum curators. These environments are becoming increasingly extreme as the temperatures change the fauna, the weather conditions, the animal population, and have consequences unforeseen by the locals. From wildfires, tornadoes, hurricanes, rising sea levels, heatwaves, and drought, each location houses an exhibition, offices, restroom, and an interactive connection to all eight museums. A separate building acts a performance shrine both to the inevitable disaster and the after-effects of the catastrophic event. Conceived as Before + After

EXTREME ENVIRONMENTS

the museums are an architecture of a specific landscape, but also highlight the ecology after the disaster happens. The example illustrated is located in California where the threat of wildfires is dramatically increasing throughout the state. The museum is a memorial to the lost forests, property, and lives enshrined in an area carved out of the forest. The orientation building and the performance center are each perceived differently before and after the disaster. The visitor center located at the base of a hill has a suspended roof and exterior walls that are composed of a mirror finish stainless steel that reflects and expands the perception of the forest. Once a wildfire passes through, the burnt trees become totems with only the trunks remaining as a testament to the lost forest. The shrine located on the hill with a water roof and stainless steel overhang has protected trees under the umbrella with shutters coming down during a wildfire. This homage to the forest has the trees remaining healthy and growing after the wildfire has scarred the forest.

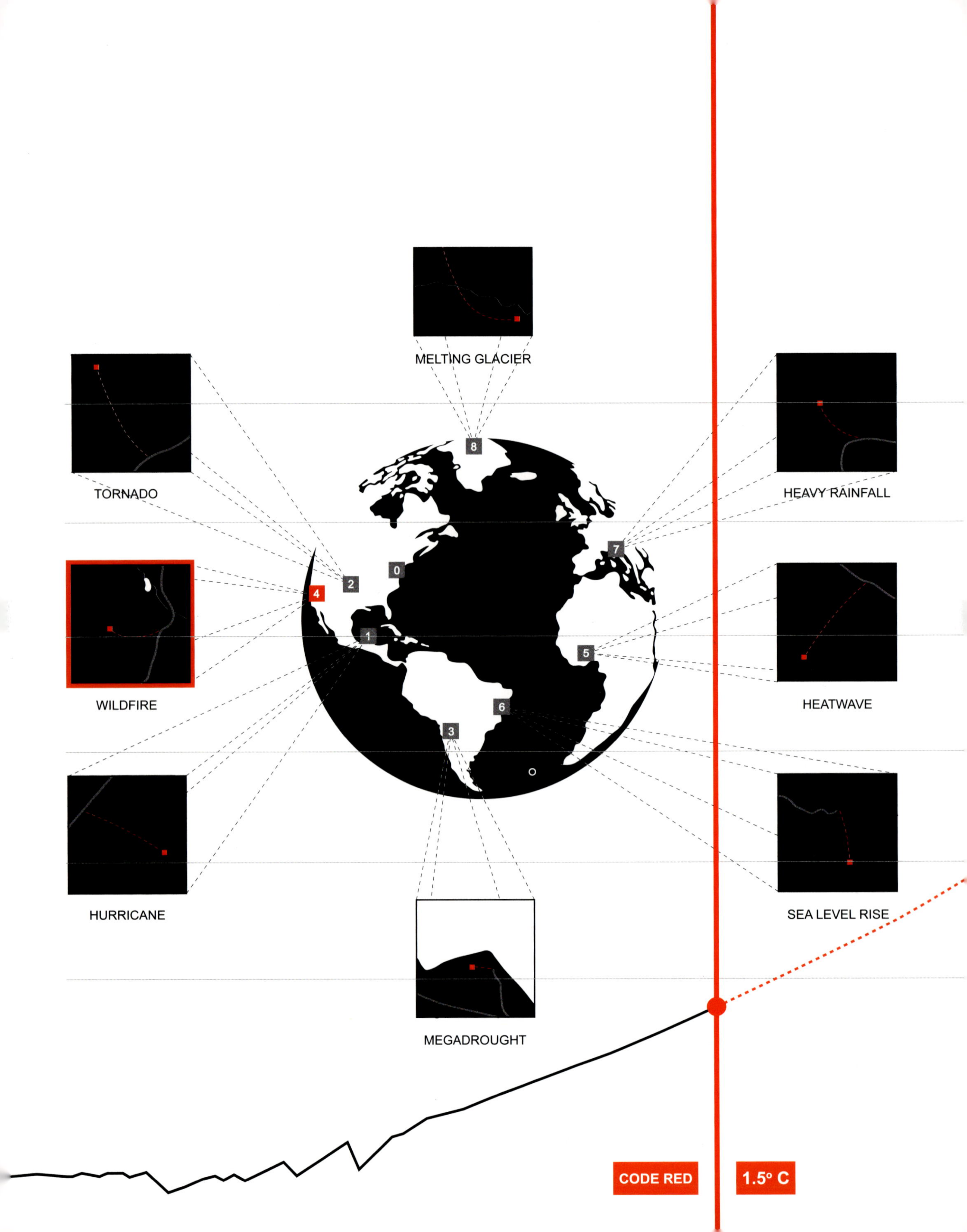

MELTING GLACIER
TORNADO
HEAVY RAINFALL
WILDFIRE
HEATWAVE
HURRICANE
SEA LEVEL RISE
MEGADROUGHT
0
1
2
3
4
5
6
7
8
CODE RED
1.5° C

AVATAR | EXTINCTION

LIVING BEFORE AND AFTER

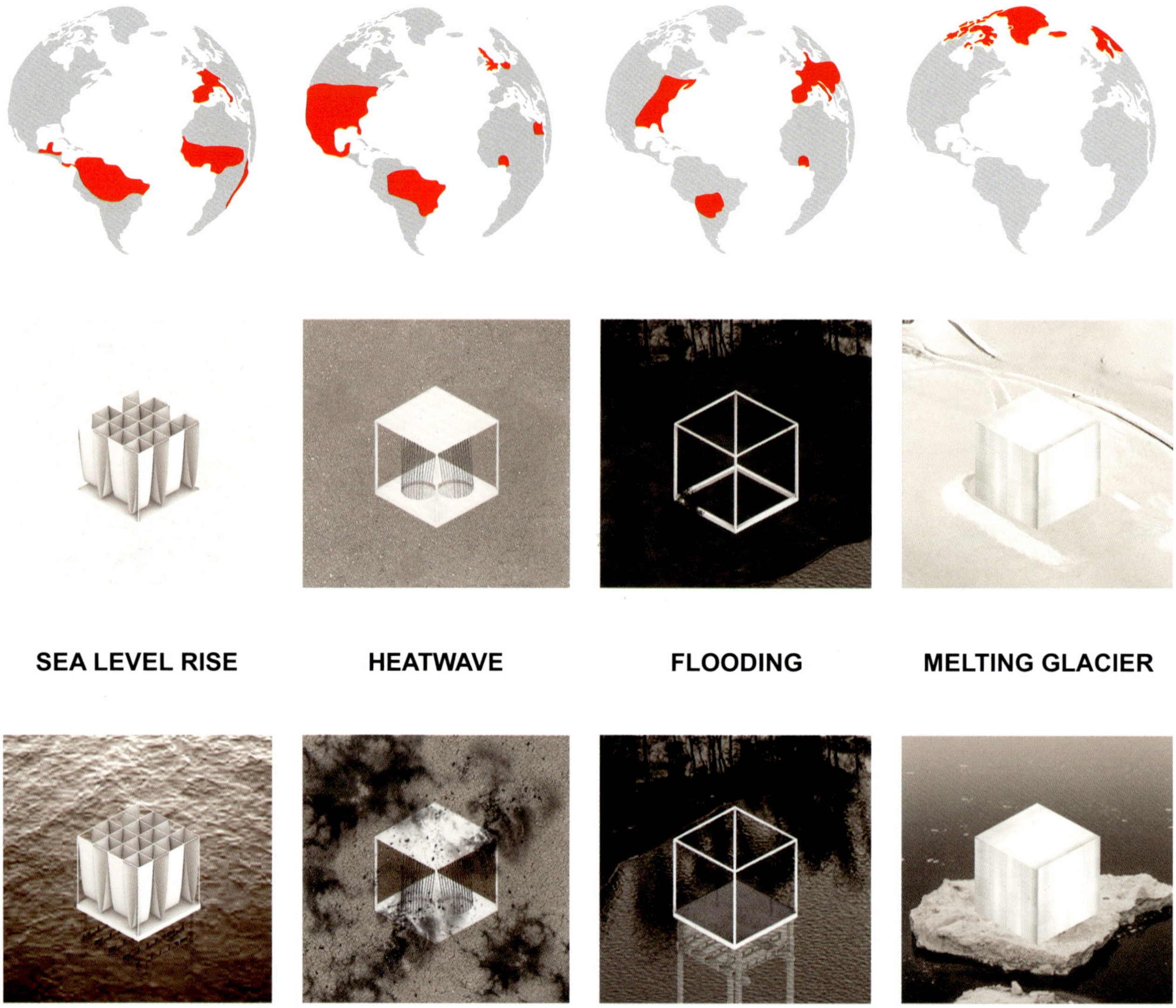

The strategic locations of the museums represent curators choosing areas that are most affected by the catastrophic disasters occurring due to global warming. The before and after effects are cataloged and viewed at the main museum in NYC near Ground Zero with interactive screens showing the changes in each location over time.

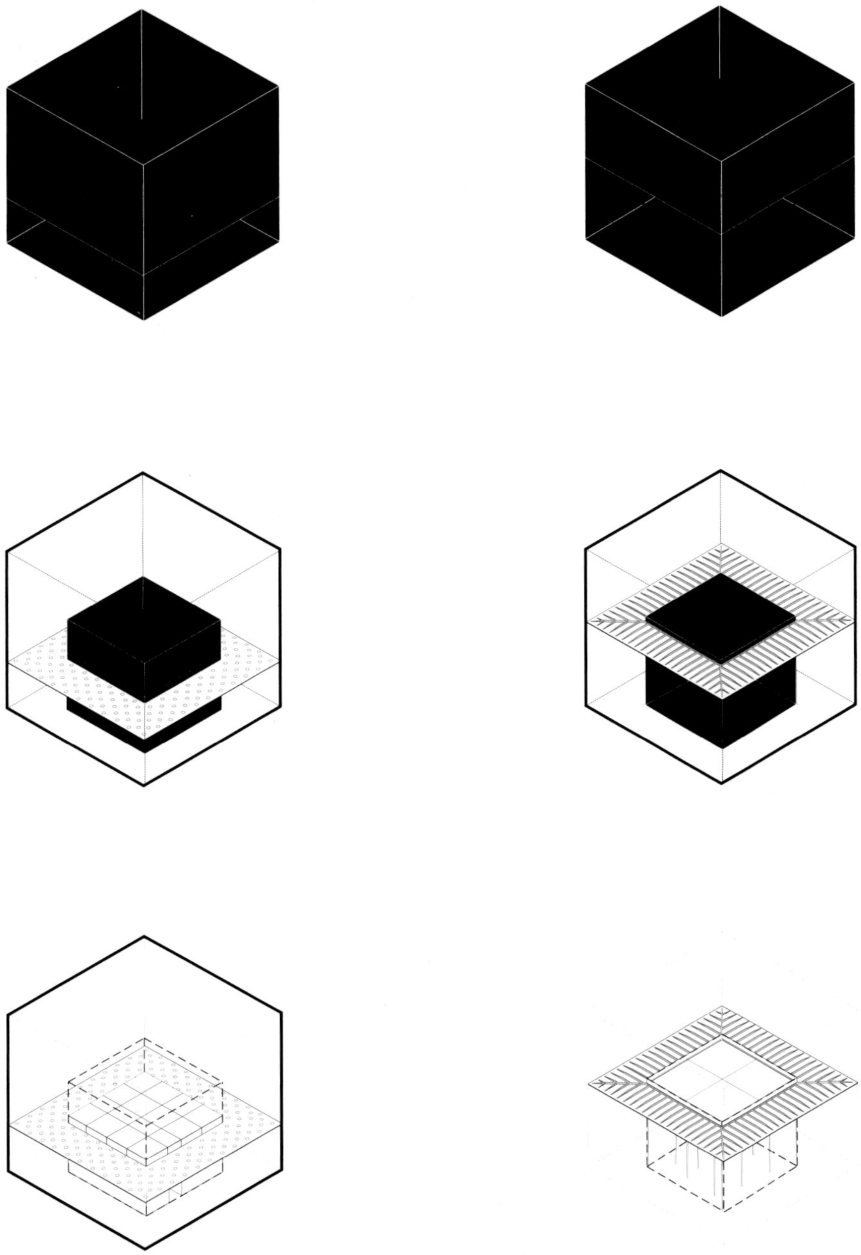

i AVATAR |

WILDFIRE MUSEUM: A VISITOR CENTER + SHRINE

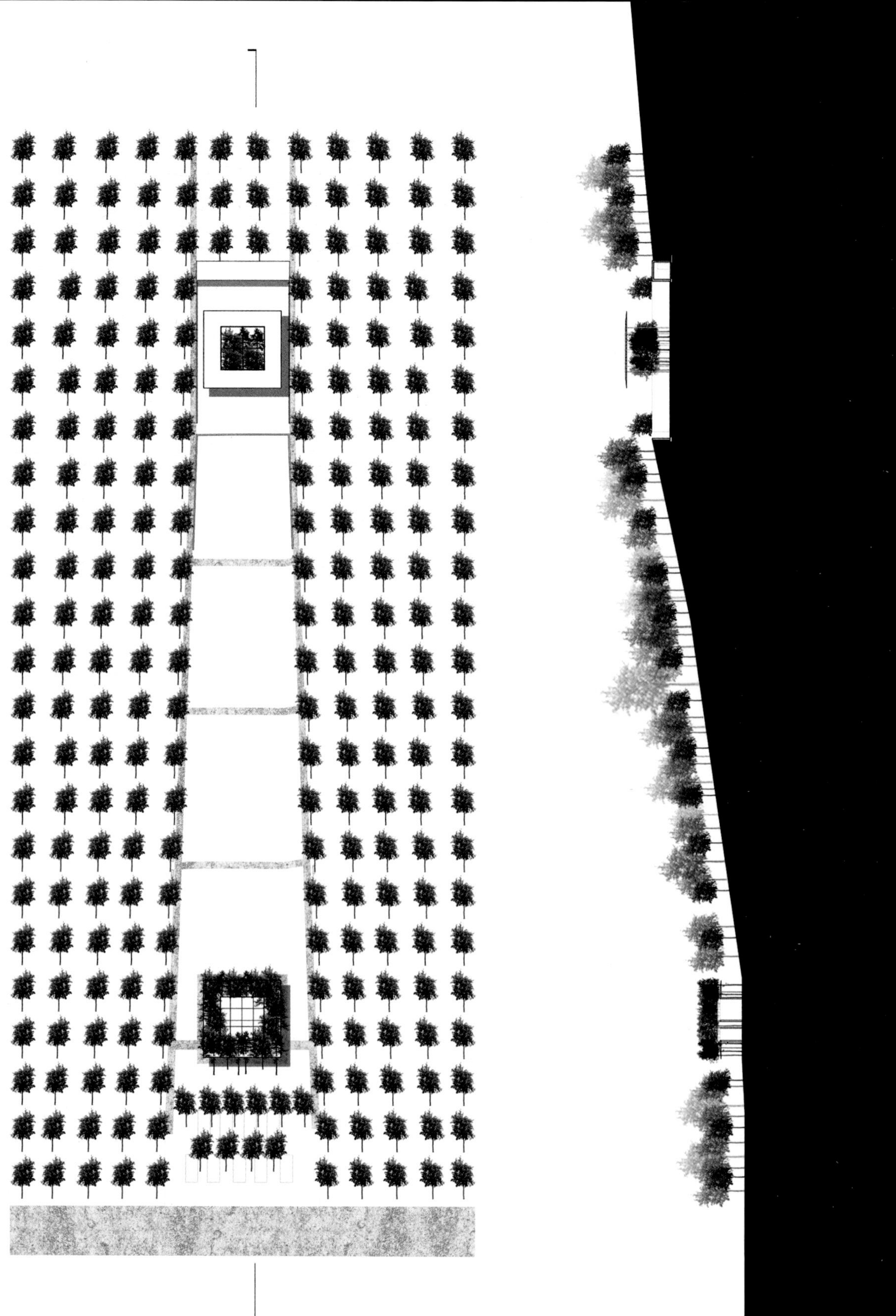

The geometries of the proportions of a cube create the organization of the two buildings: one based upon a roof of subtraction speaking to the earth and the other based upon levitation and looking upwards at the sky. The sectional quality of the hill heightens the visitor's perception of climbing to a shrine to those who were lost.

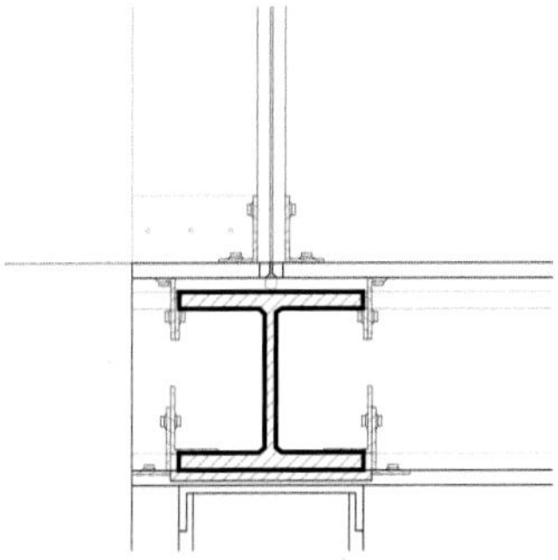

GLASS AND BEAM CONNECTION

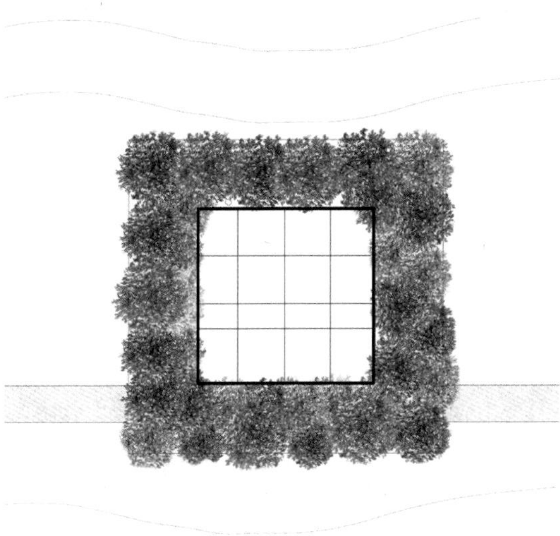

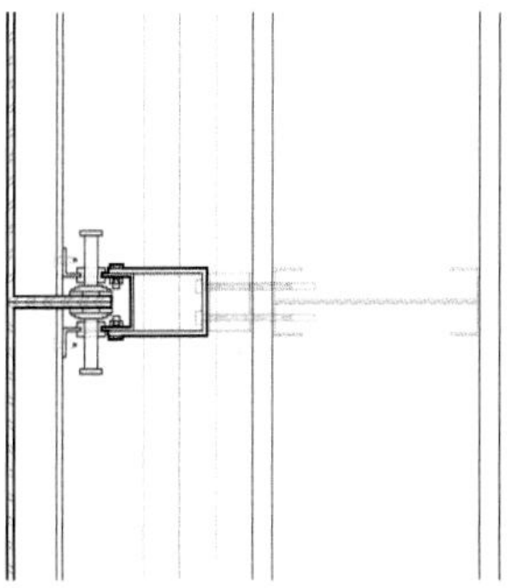

REFLECTIVE METAL FAÇADES

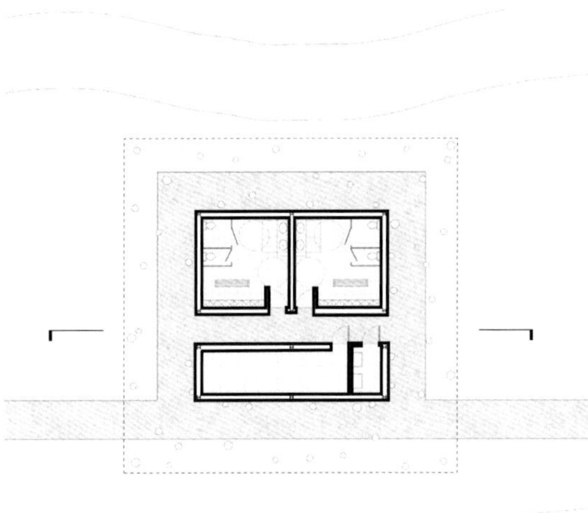

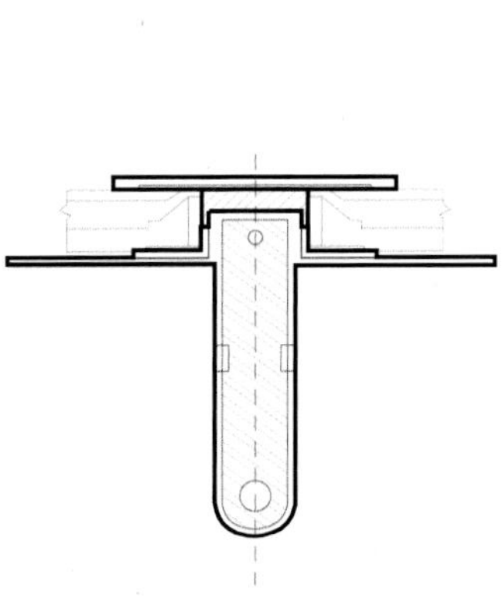

GLASS CONNECTION

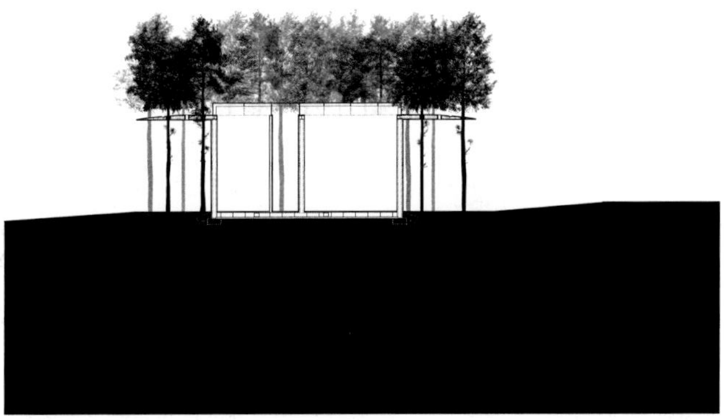

DECAY
(VISITOR CENTER)

0 80'

LIVING BEFORE AND AFTER: DECAY AND SHRINE

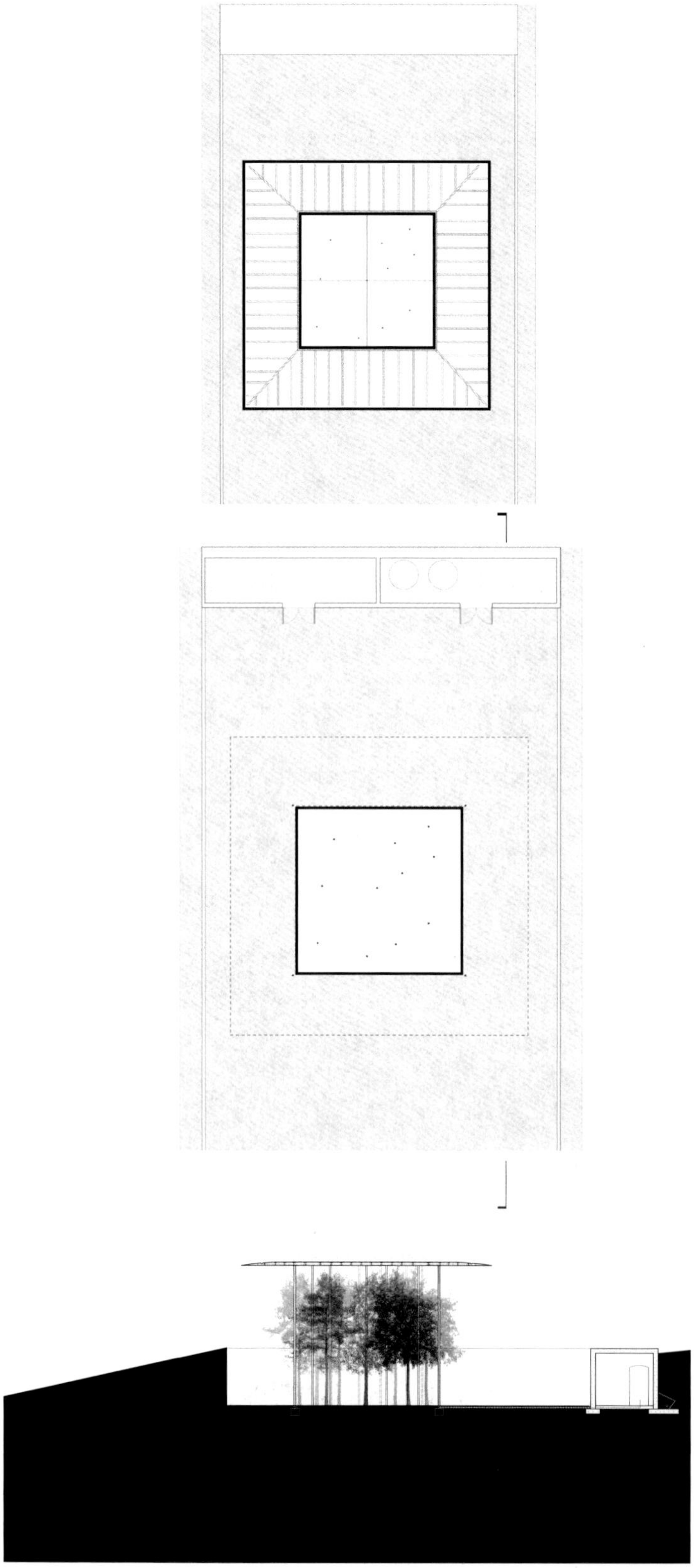

SHRINE
(EXHIBIT)

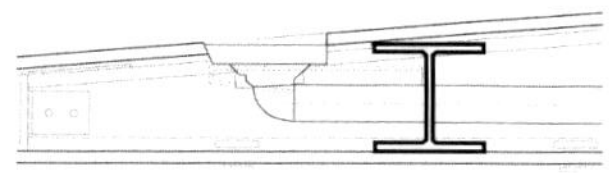

DRAINAGE

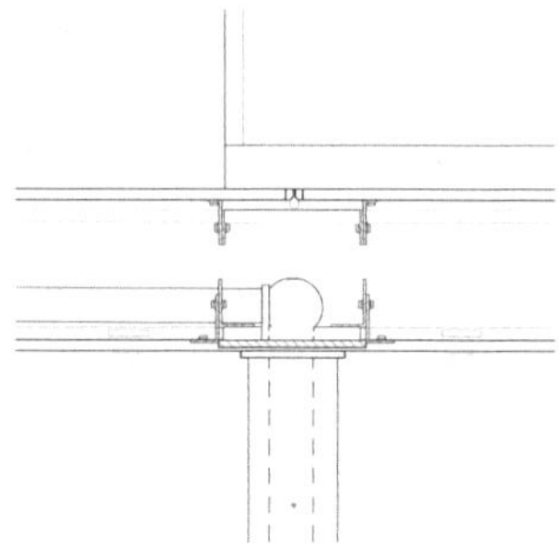

WATER PIPE

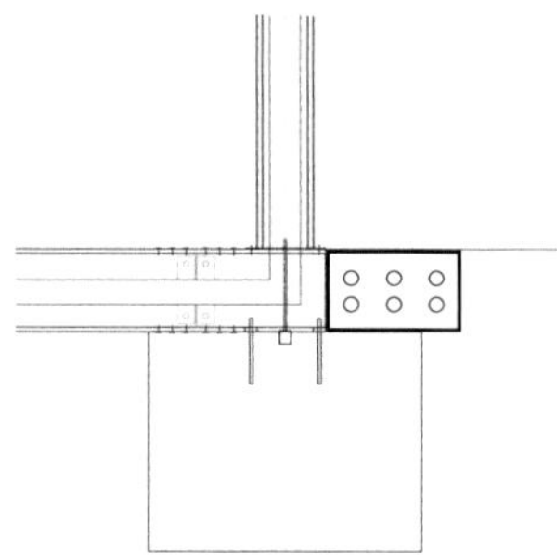

FOUNDATION

The two pavilions create a dialogue of decay and an act of renewal. The detailing of the stainless steel and water roof is used as protection from the fires and increased temperatures of the blazing forest. The reflections and extensions of the real trees are projected on the stainless roof and walls.

AVATAR | EXTINCTION

A VISITOR CENTER: BEFORE + AFTER

WILDFIRE

i AVATAR | EXTINCTION

A SACRIFICIAL EXHIBIT: BEFORE + AFTER

WILDFIRE

In a throw-away society driven by capitalism, parts of the business and housing communities have evacuated the neighborhoods that have outlived their usefulness. Mining and railroad towns, the shifting economies and demographics of a city, factories, and the new migration to online retail have all turned these sectors into ghost towns in the past hundred years.[205] Rooted in their past, ghost town memories are embedded in their history. This pathway to the renewal of an area very often involves artists moving into cheap, typically abandoned buildings. The danger facing these communities is the gentrification that causes the prices to rise so fast the resident artists can't afford the rent there anymore and then must find another area of the city to inhabit. This process can be short circuited by establishing housing co-ops, art galleries, and reusing a vacant railroad track that can be turned into a mass transit line linking downtown with this new evolving community. The program creates an economic plan with a stable housing stock

BRIDGING THE PAST + THE PRESENT

for taking the next step in establishing a vibrant community. ARTSTOP is about building a unique community. The garden seen through the train geometrically whisking by like a cornfield in motion, to an open atrium that has the sculpted random remnants of the demolished local steel structures playing against the rigidity of the open grid of the building conjure images of the past life of the area. The ARTSTOP acts as a gathering place with a community kitchen and open seating where the community comes together in a crossroads of circulating paths. The central galleries are organized about a regular geometric grid of fabric that defines smaller spaces that have a sense of transparency and reveals layered, ghost-like images of the three-dimensional sculptures on display. The surrounding rigid grid rotates with two-dimensional works that are hanging on panels. The garden, atrium, and gallery have bookend housing, workshops, and offices that compliment bridging the past into a new reality where the ghosts live comfortably in the present.

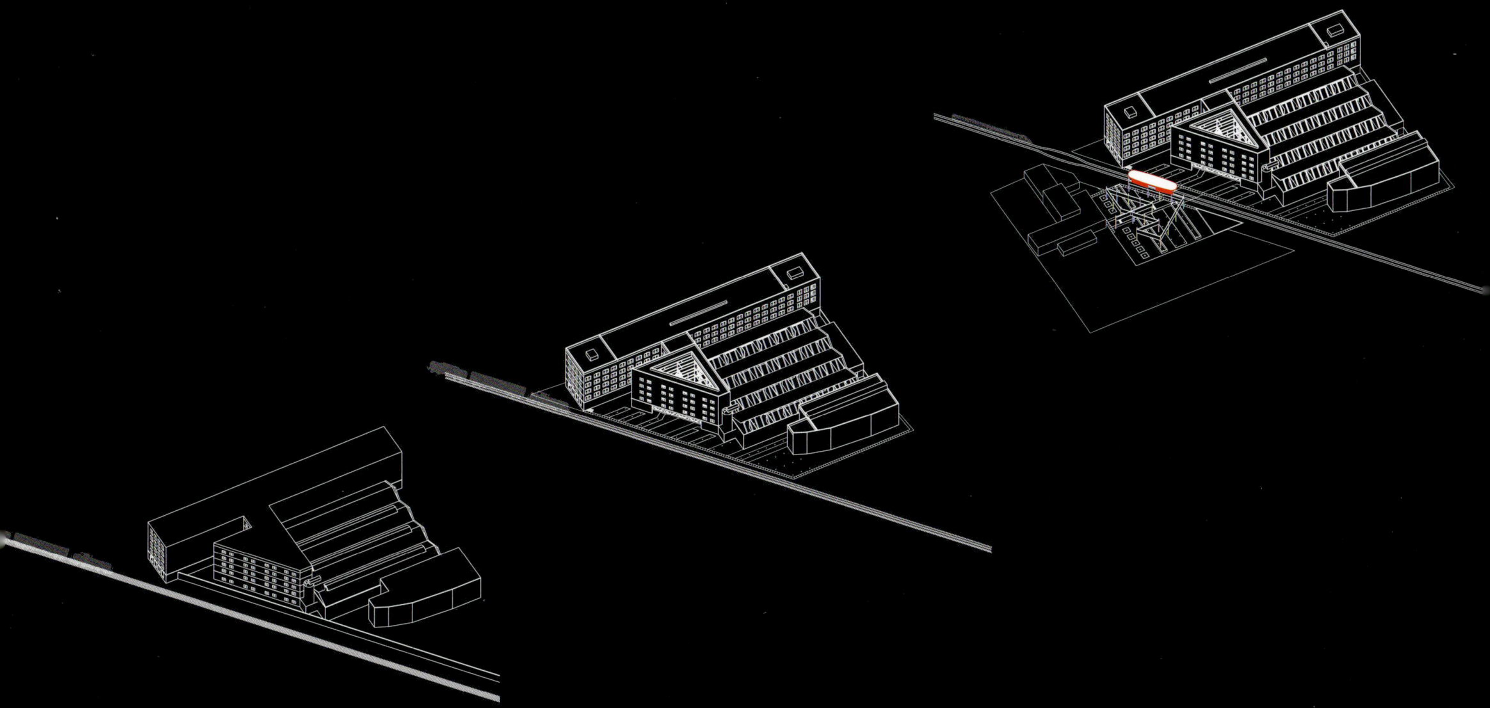

GHOST TOWN
1920

POSSESSION
2018

GROWTH
2022

1850 | 1860 | 1870 | 1880 | 1890 | 1900 | 1910 | 1920 | 1930 | 1940 | 1950 | 1960

MINING TOWNS **RAILROAD TOWNS**

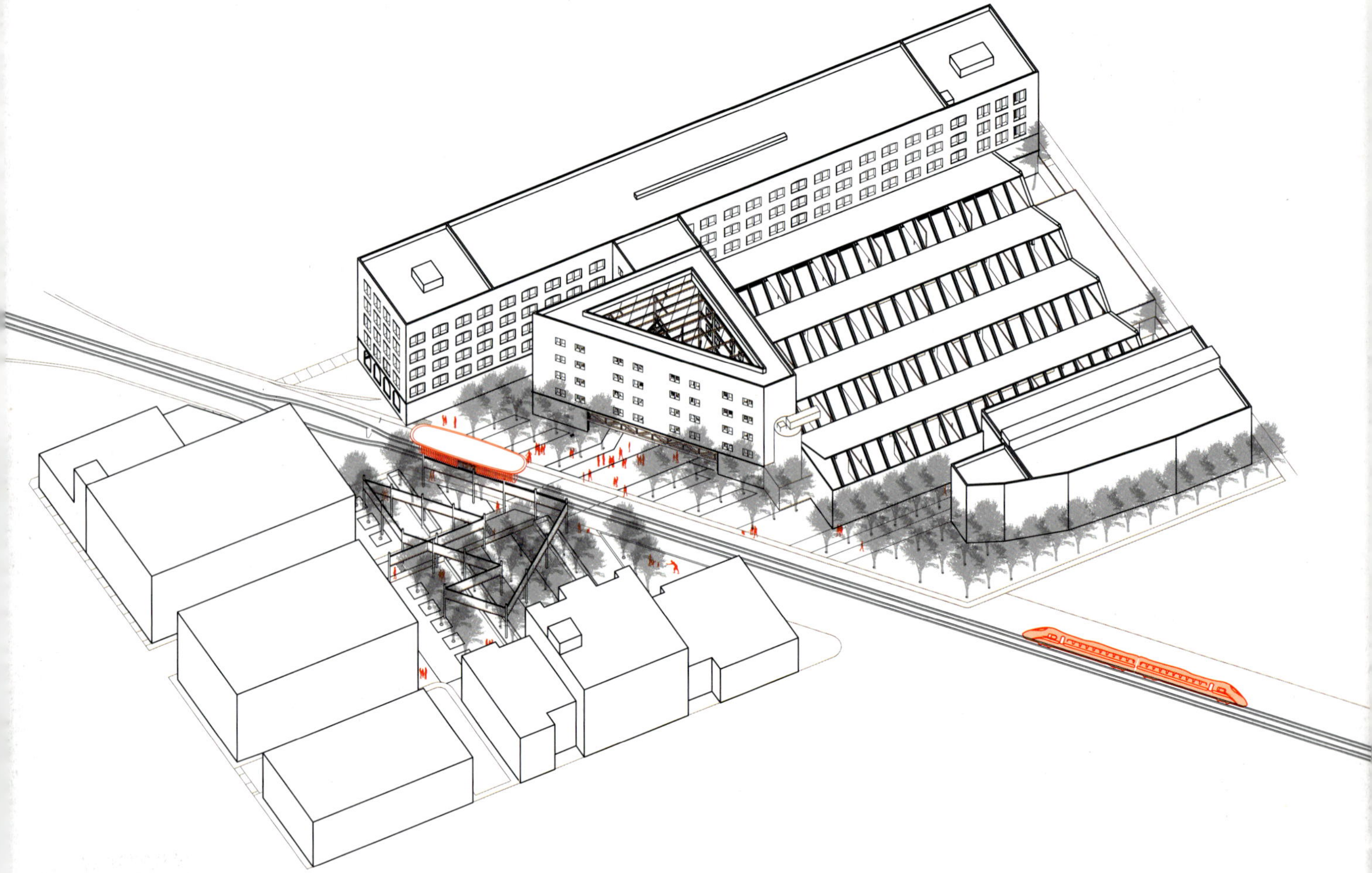

ART STOP
2025

1970 | 1980 | 1990 | 2000 | 2010 | 2020 | 2030 | 2040 | 2050 | 2060 | 2070 | 2080

FACTORIES **RETAIL**

Reviving abandoned railroad tracks and rundown factories, ARTSTOP rejuvenates a forgotten neighborhood by reconceptualizing the economy, the housing, and support facilities, by bridging the past with fragments of the ghosts coming back to life. The progression through time has allowed the neighborhood to have affordable housing, co-ops, and community centers.

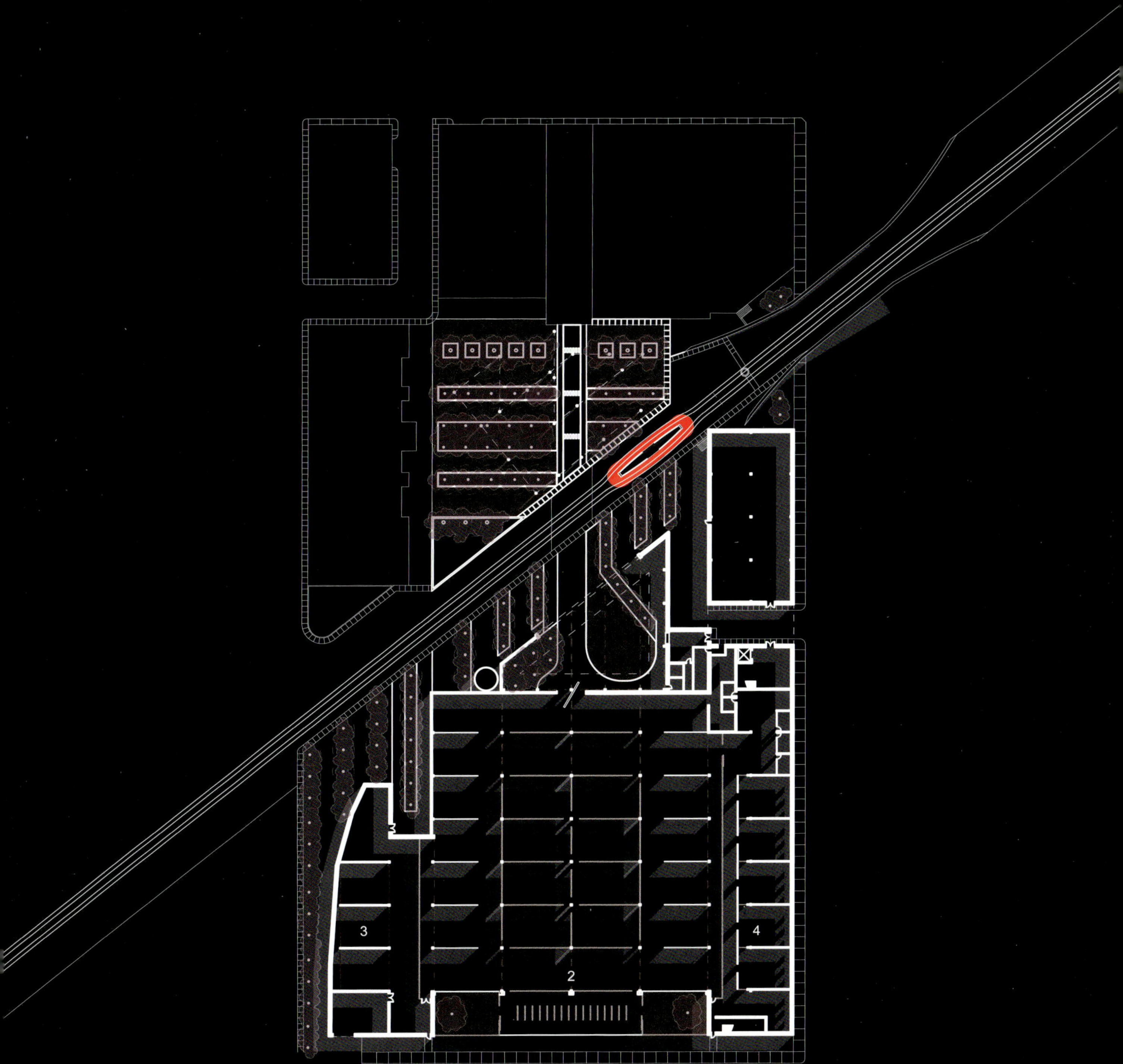

AVATAR | DEATH

ART STOP STATION ② GALLERIES ③ WORKSHOP ④ RESIDENT STUDIO ⑤ EVENTS ⑥ KITCHEN

GHOST TOWN: BRIDGING THE PAST

0' 10' 30' 50'

GALLERY

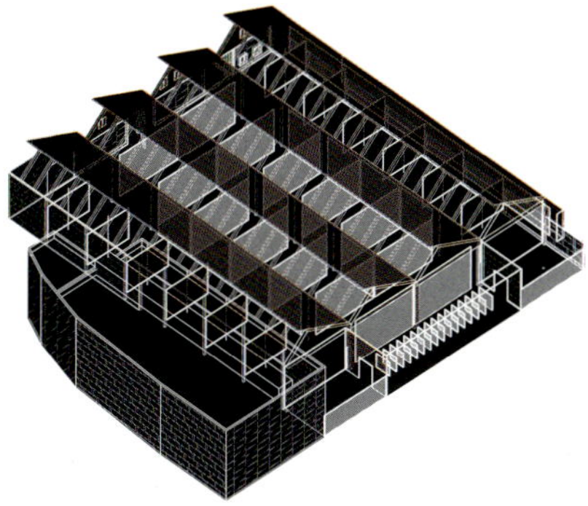

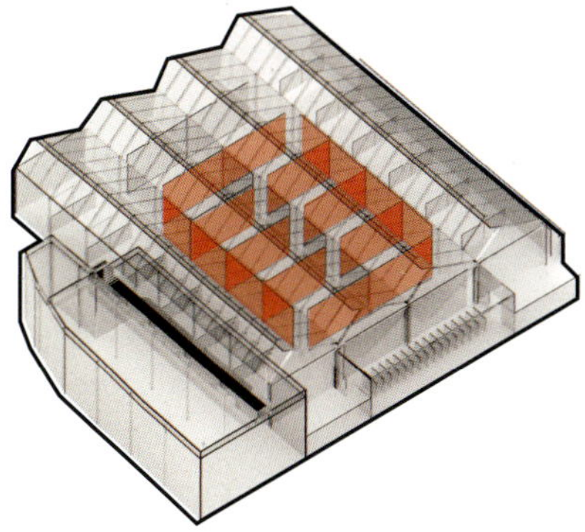

ATRIUM

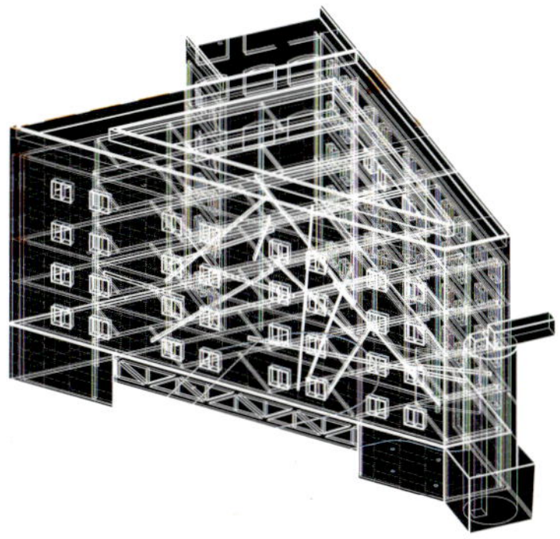

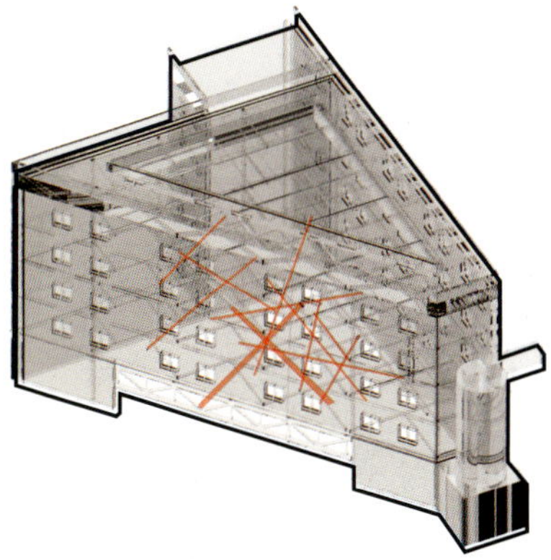

GARDEN

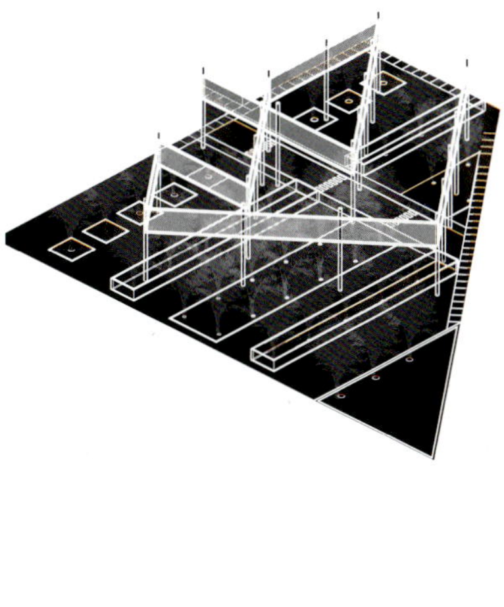

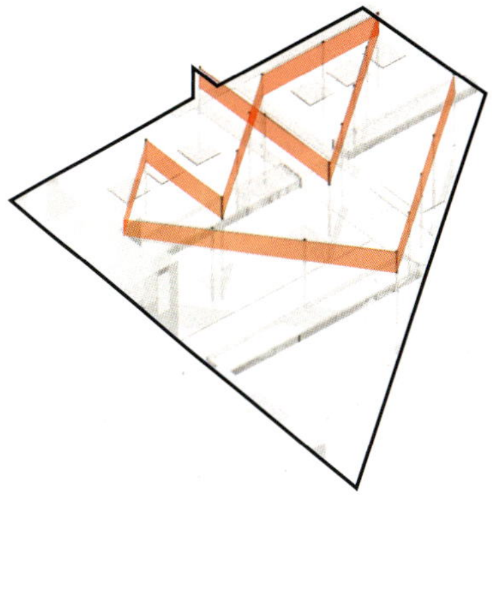

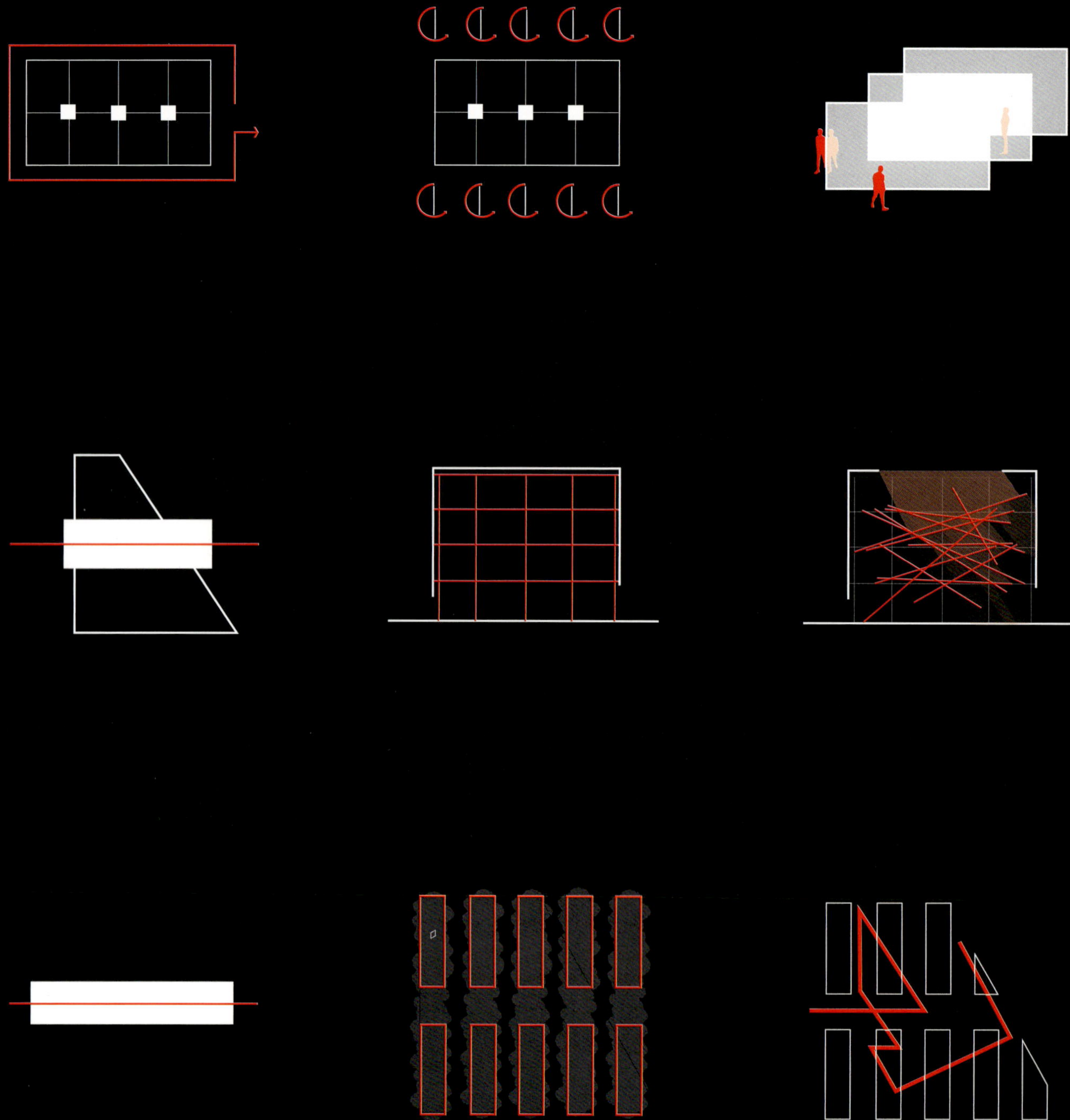

The ARTSTOP is composed of a garden, an atrium, and a gallery. The garden is pierced by a reflective glass that contrasts with the geometrical sculpture that plays with transparency in the trees. The atrium and gallery spaces similarly use memory and the dislocation of imagery to bring up forgotten layers of its past history as a ghost town.

DAY: TWILIGHT 6:32 AM EQUINOX

NIGHT: FULL MOON 7:32 PM SOLSTICE

The atrium of the ARTSTOP is refurbished by exposing the existing structure of the grid within a triangular space. The roof is retrofitted with a skylight that softens the exterior light during the day and night. The removed steel structure is used as a wayfaring sculpture throughout the large open space. The atrium is conceived as a public space with seating, a kitchen, and a garden.

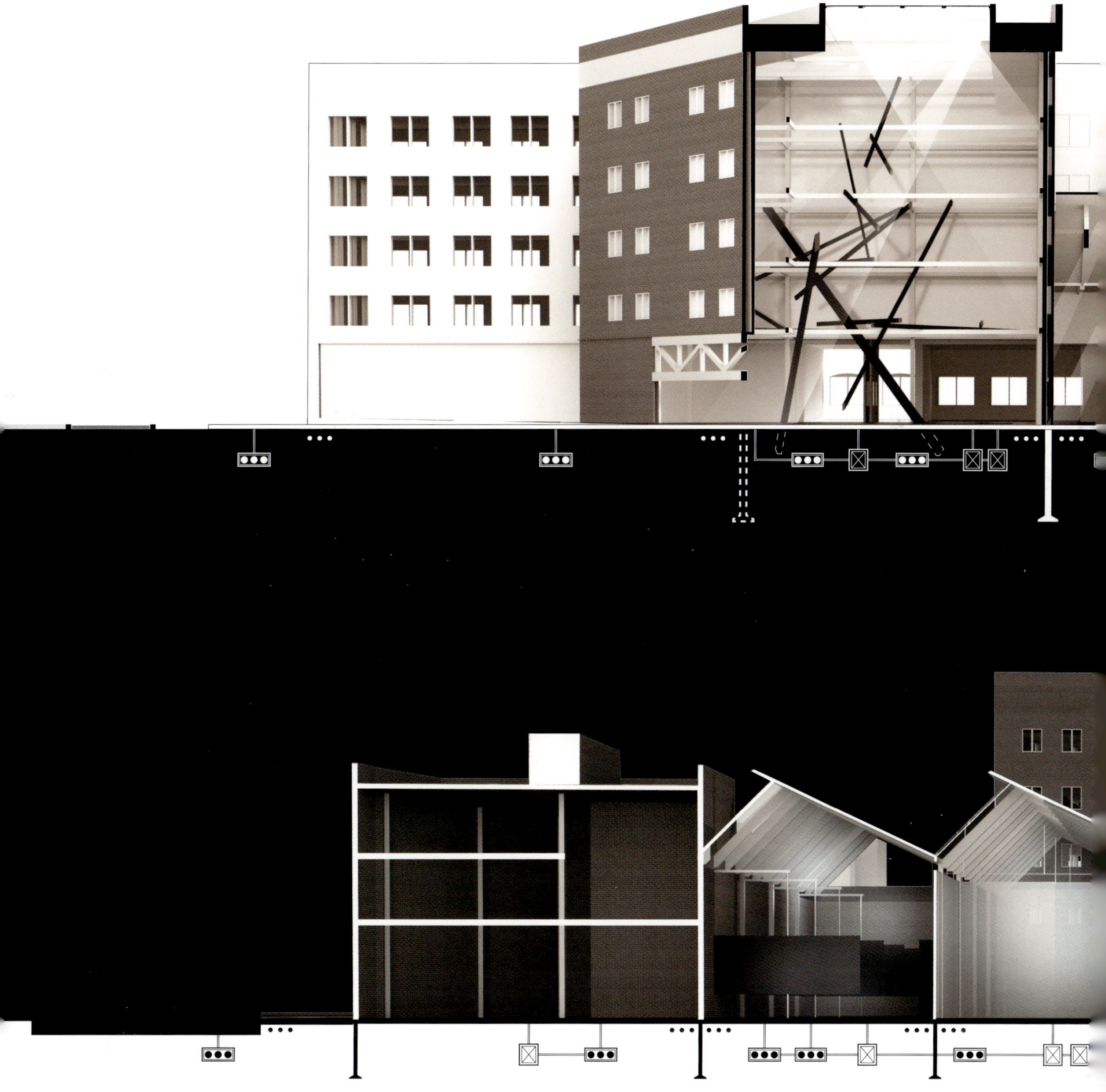

AVATAR | DEATH

GHOST TOWN: LIVING WITH THE PAST

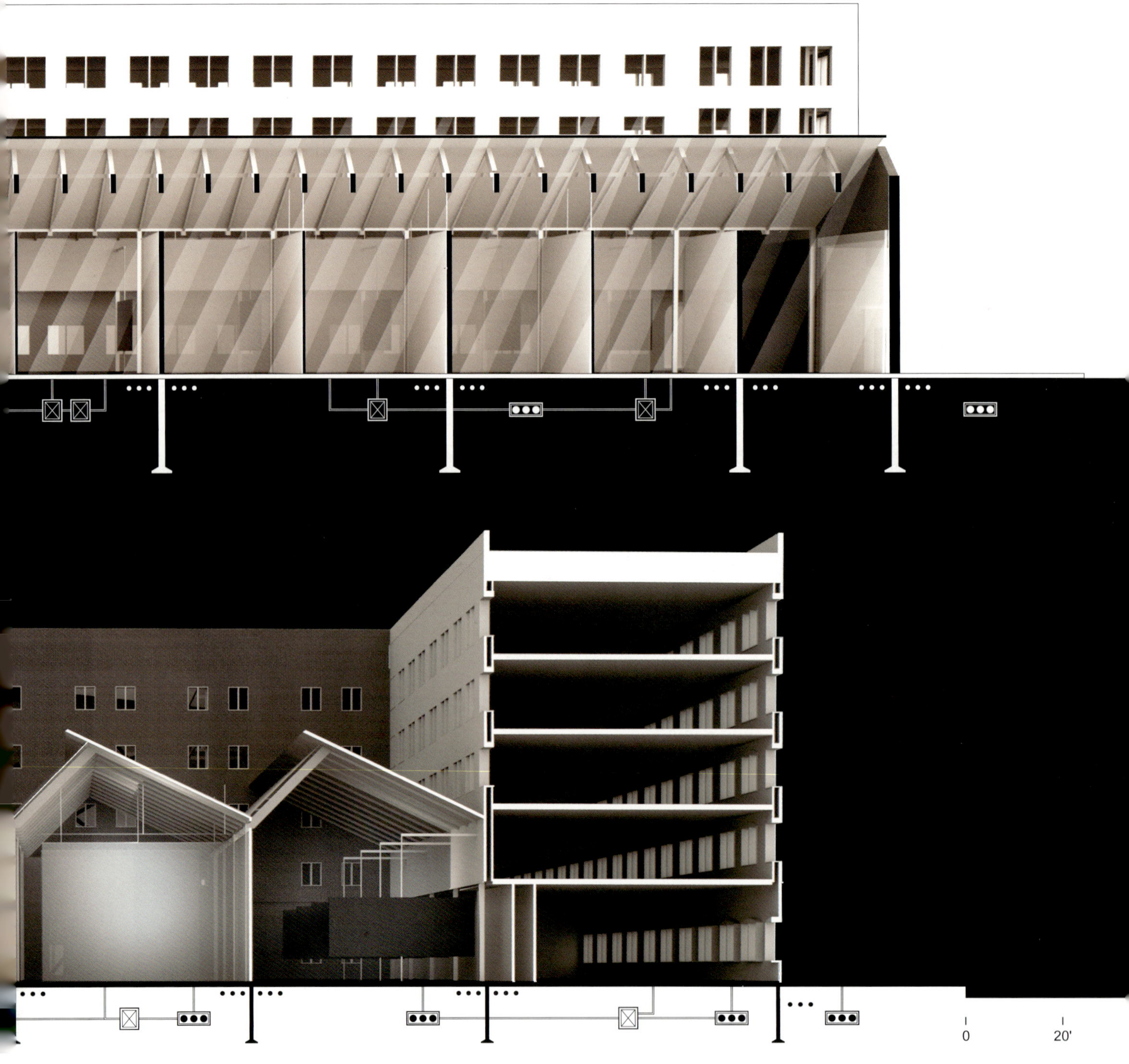

The existing fabric of the neighborhood is maintained by preserving the sectional structure of the factory. The interior walls are removed and recycled into free-form steel beams suspended from the steel structure of the open atrium, while the north facing skylights of the open factory space are repurposed with translucent panels creating ghost-like shapes with art and people.

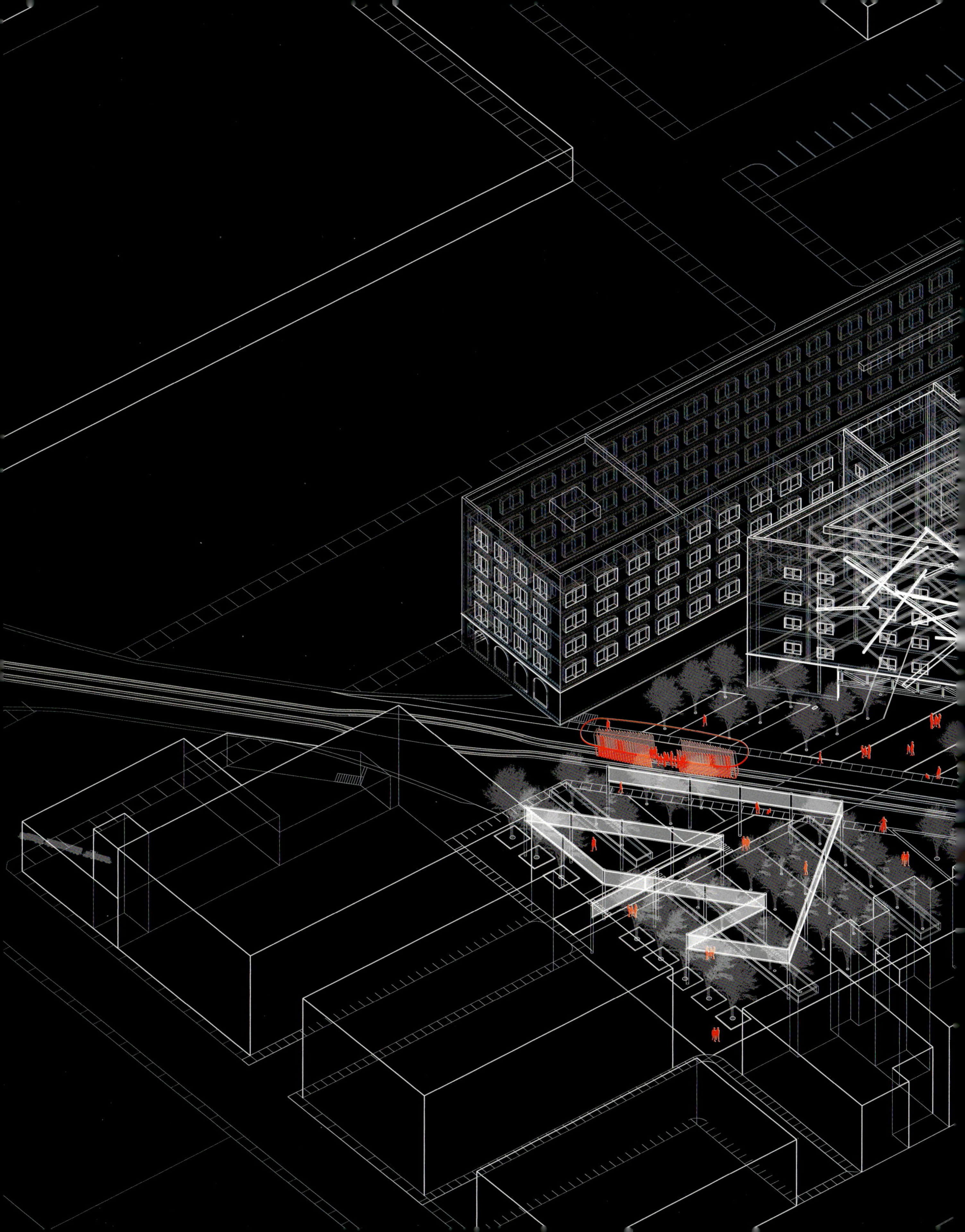

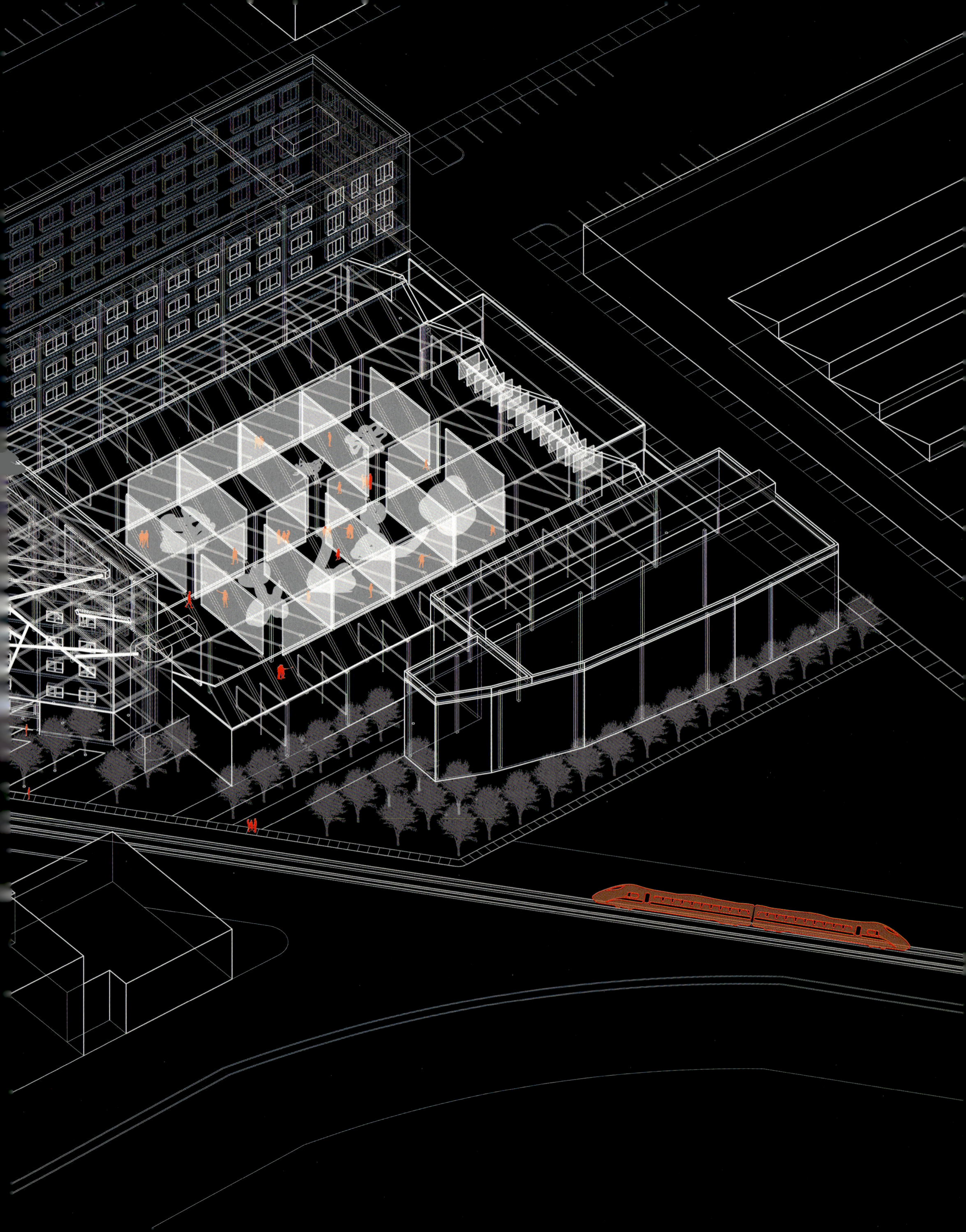

The natural process and rituals of dying vary in different cultures. The perspective of life after death also becomes a matter of perspective as different societies try to create memories of those who have passed. Perhaps buried, cremated, or preserved, each culture initiates different ceremonies, testimonies, shrines, and memorials.[206] With the new breakthrough technologies in AI and immersive projections, it is now possible to communicate with those who have passed into another realm through AI recordings.[207] This new concept of an ancestral shrine allows those deceased to create an avatar that can speak to their grandchildren and beyond. This project proposes a place that becomes a reconceived cemetery where descendants can communicate with the dead. The facility takes the visitor on a journey through a series of gardens that hyper activates the five senses and concludes in the sixth sense garden that becomes a reflection of one's life. The permanent shrines are based upon three

TALKING TO ANCESTORS IN AN AUGMENTED REALM

platonic solids. They house different immersive experiences for those visiting their ancestors. The Cerebro is a spherical design that places the mourner in the center with 360-degree views of past holographic and augmented reality projections. The Catacomb is a cubical space bathed in dark light that has a large monitor projecting a three-dimensional image of the deceased. This virtual image can answer any question from those in the space in real time. This brings the figure to life as emotional answers heighten the experience of talking with the dead. This reimagining of the ancestral shrine creates a new ritual that can change the perception of dying. Finally, the pyramidal chapel is a chamber of reflection that partially submerges itself into the ground and is open to the air while the living descend a ramp returning to the earth and isolating themselves in an INbetween space that is void of technologies.

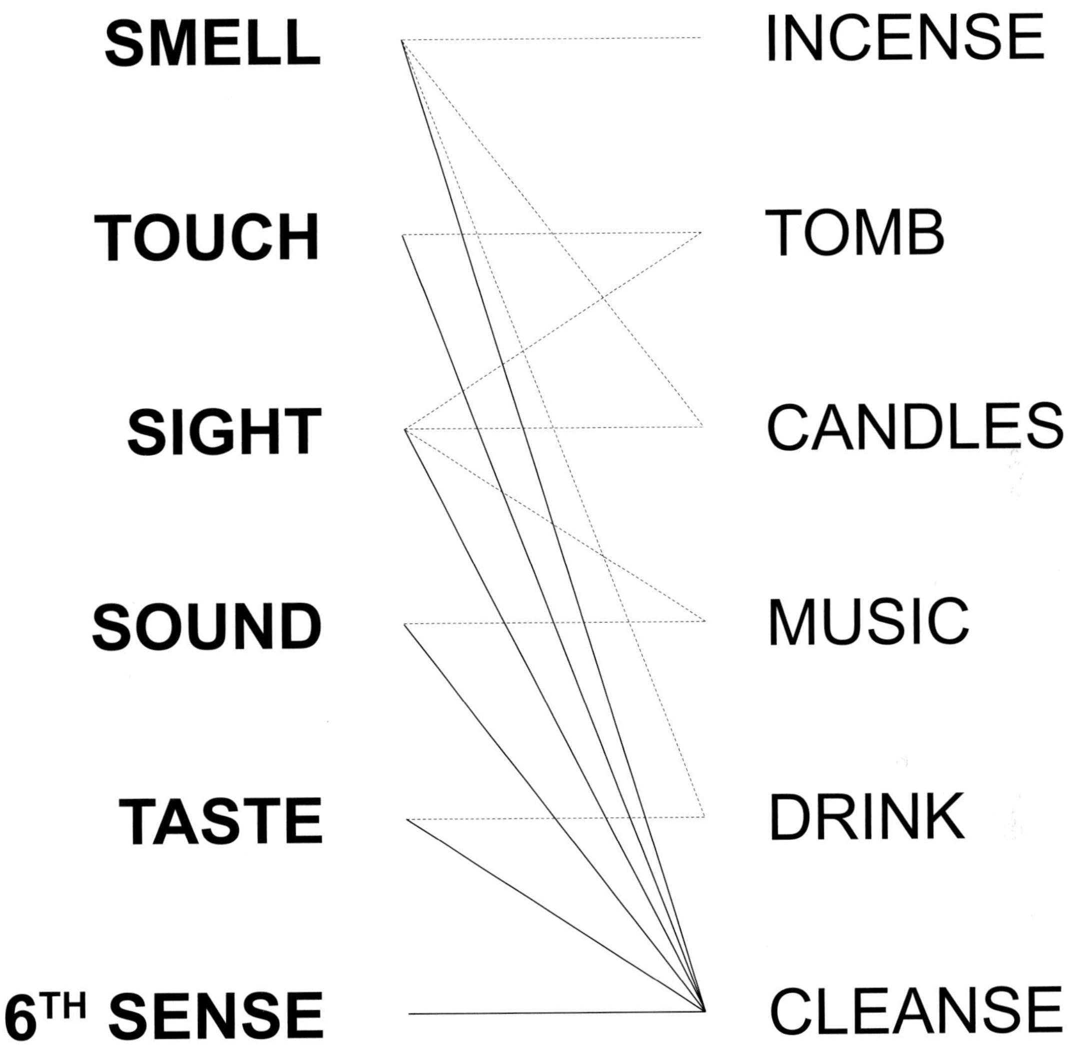
SMELL
INCENSE
TOUCH
TOMB
SIGHT
CANDLES
SOUND
MUSIC
TASTE
DRINK
6TH SENSE
CLEANSE

LIFE

BURY

**DEATH IN 4 MINUTES:
AFTER THE HEART STOPS**

CREMATE

**BODY IS A CONTAINER
FOR THE SOUL**

PRESERVE

**CRYOGENICS:
FREEZES THE BODY**

DEATH

Through burial, cremation, and preservation, each culture has a distinct concept of the afterlife. Burial believes that all organisms will return to the earth and become one again. Cremation is a ritual that liberates the dead from the physical world by burning their body to ashes. In contrast of the other two, Preservation honors the dead's physical manifestation by keeping it in state.

OBON LANTERNS

BUDDHISM: HONOR ANCESTRAL SPIRITS

SPIRIT HOUSE

EASTERN: TALK TO ANCESTORS

LIFE

DIA DE LOS MUERTOS

MAYAN: THE DAY ANCESTORS VISIT

PAPER HOUSE

TAOISM: SEND MATERIALS TO THE AFTERLIFE

DIVERSE CULTURES: COMMUNICATE WITH THE DEAD

SEANCE

WESTERN: TALK TO ANCESTORS

FET GEDE

HAITIAN: DANCE WITH ANCESTRAL SPIRITS

DEATH

INTERACTIVE TESTIMONIES

MODERN: CONVERSE WITH ANCESTORS IN SHOAH FOUNDATION

HOLOGRAM

FUTURE: CONVERSE WITH ANCESTORS

Eastern and Western cultures have different perspectives in communicating with the dead. Buddhism, Haitian, and Mayan cultures have in common their view of the afterlife being free of materiality and desires. Whereas, Taoism believes in sending materials to the dead in the afterlife. In the real world today, it is possible to converse with the dead through AI devices.[208]

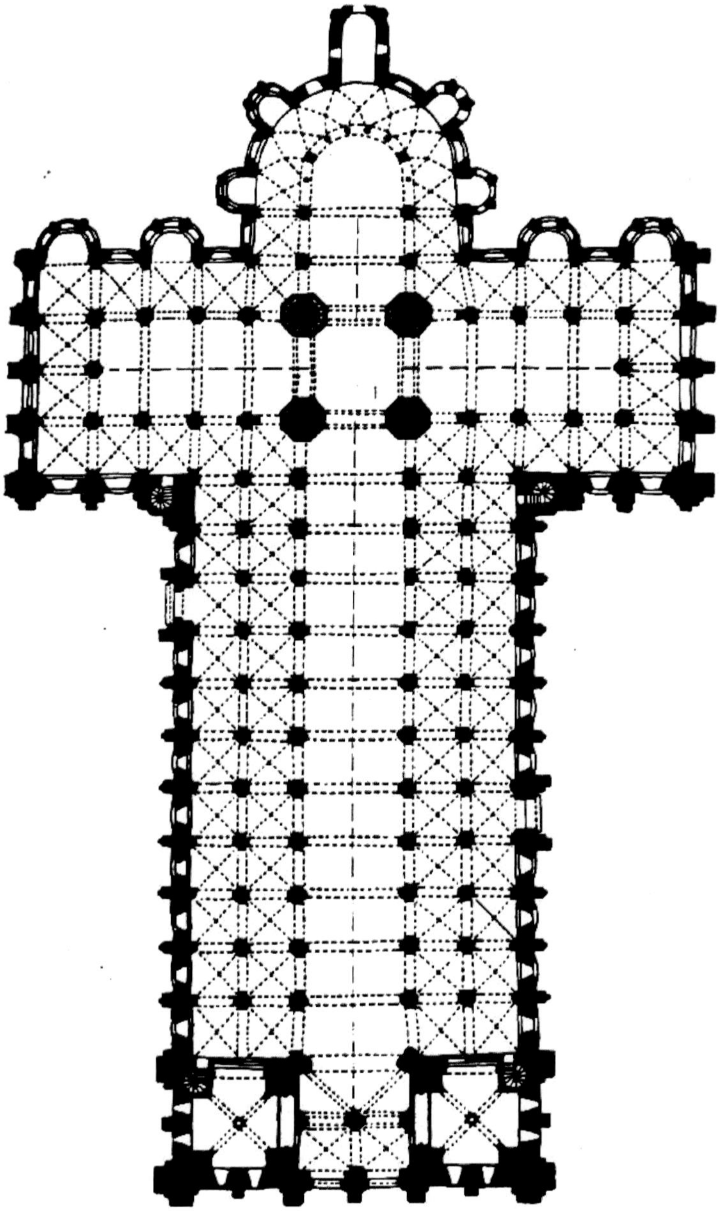

ST. SERNIN (1180)

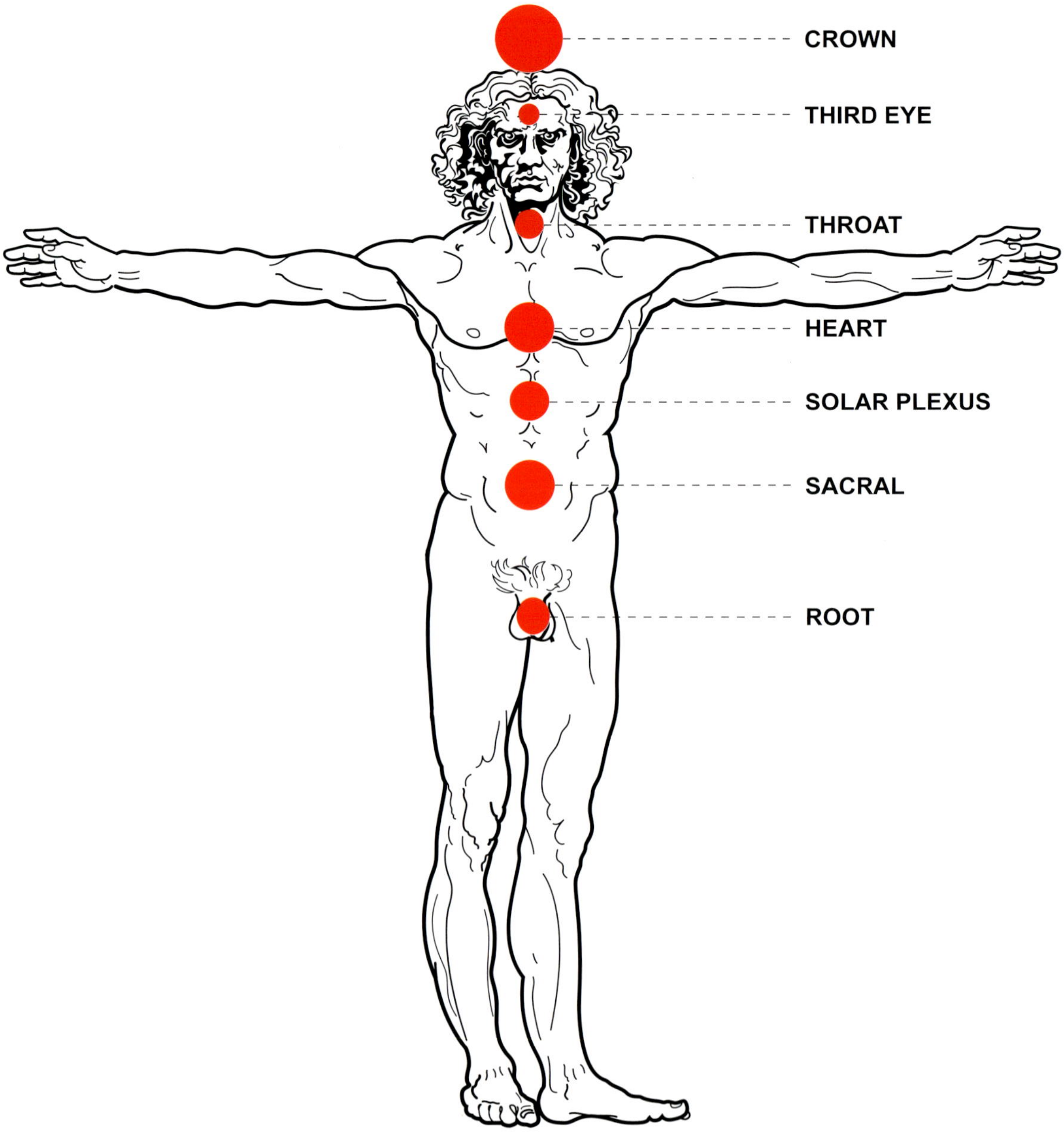

SEVEN CHAKRAS

EASTERN TRADITION: ENERGY POINTS IN THE BODY TAKE YOU FROM THE EARTH TO THE HEAVEN

This project poetically looks at the ceremonial procession of St. Sernin and the seven chakras of the body[209] as the crossroad of both the body and architecture. Fusing the two concepts together, the project will take the visitors on a procession through the seven gates of chakras and the six senses while arriving at a new way of communicating with the dead through technology.

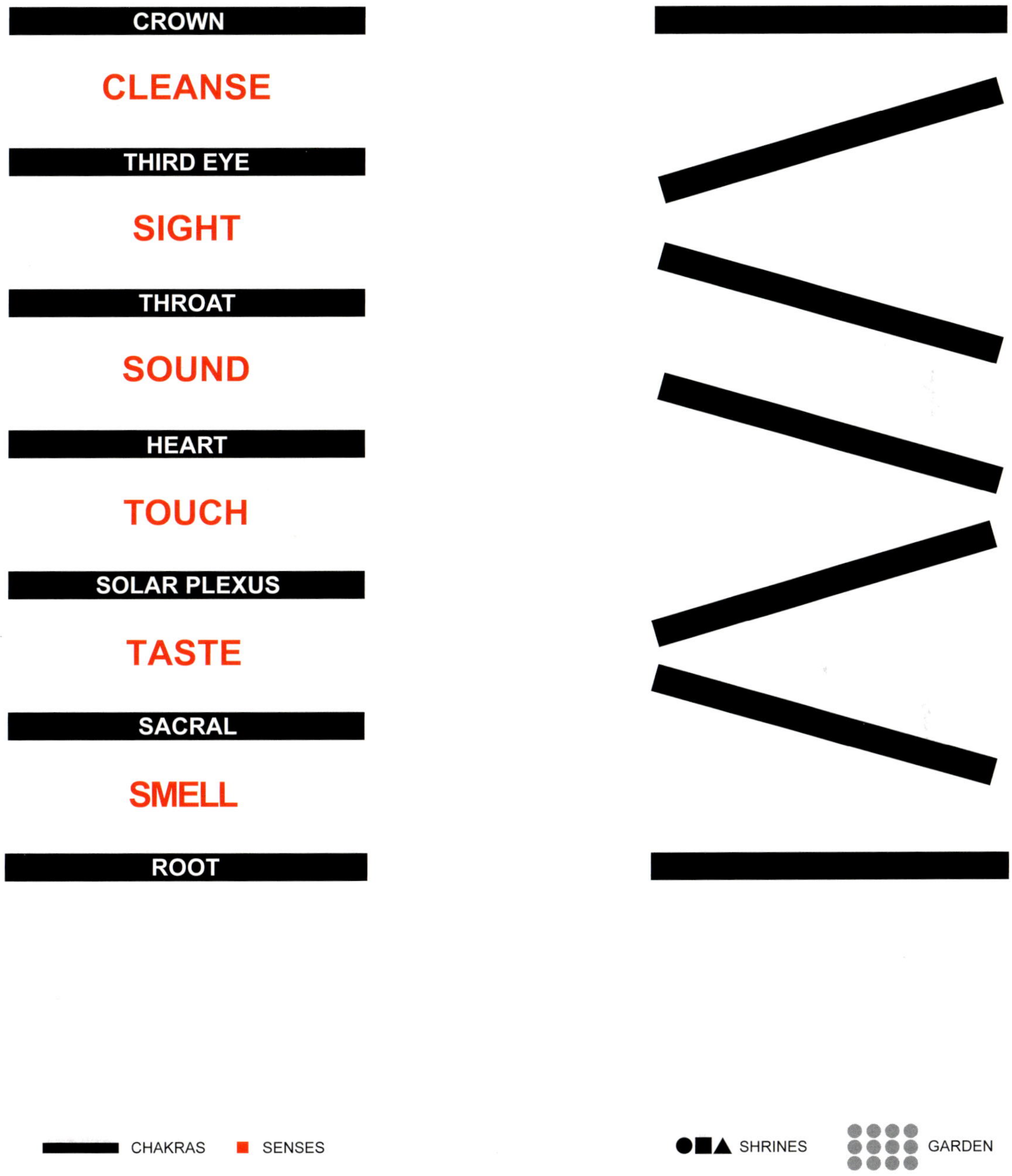

AVATAR | HERE AFTER

CHAKRAS + SENSES: A PATHWAY INBETWEEN

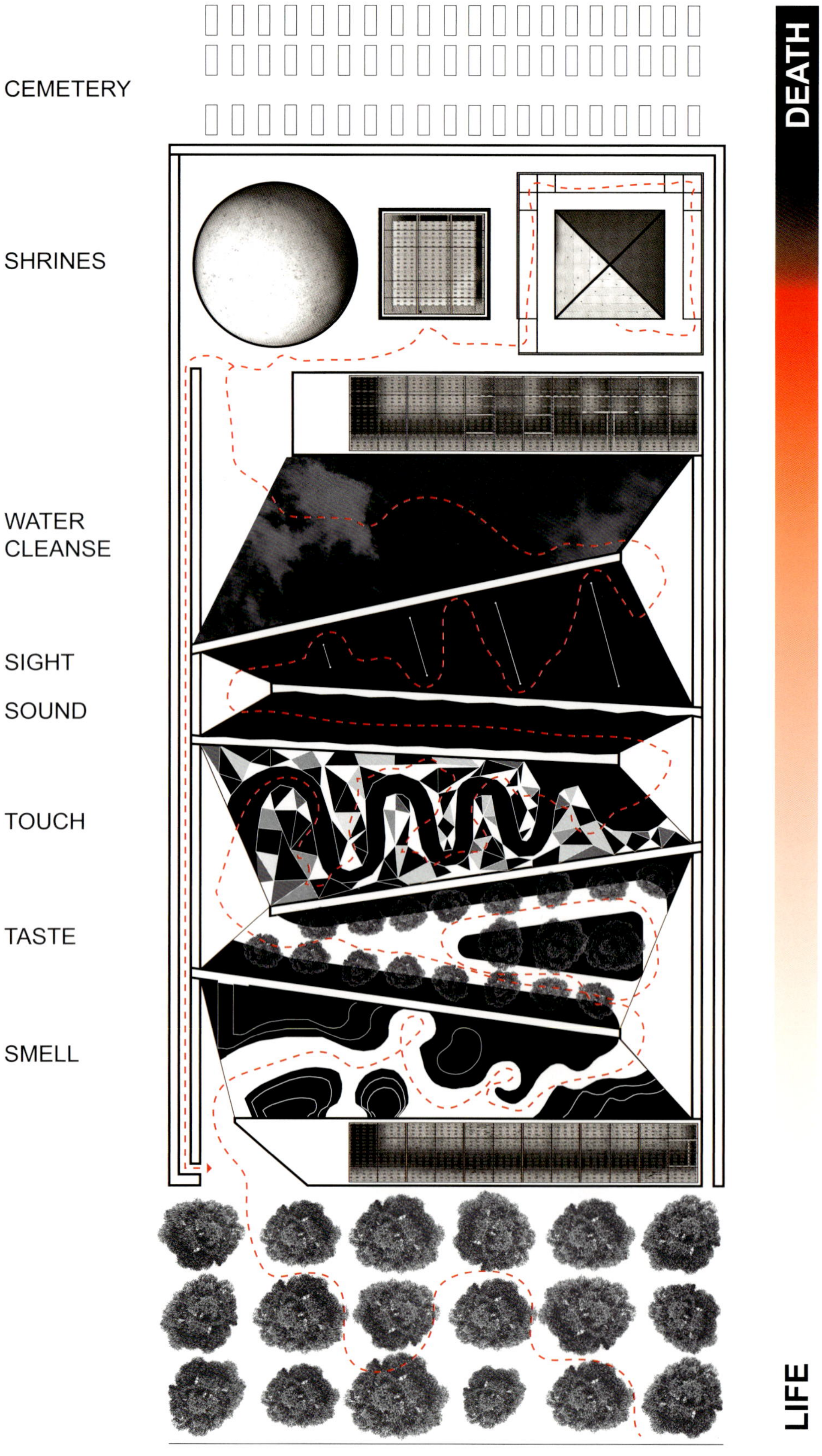
CEMETERY
SHRINES
WATER
CLEANSE
SIGHT
SOUND
TOUCH
TASTE
SMELL
DEATH
LIFE

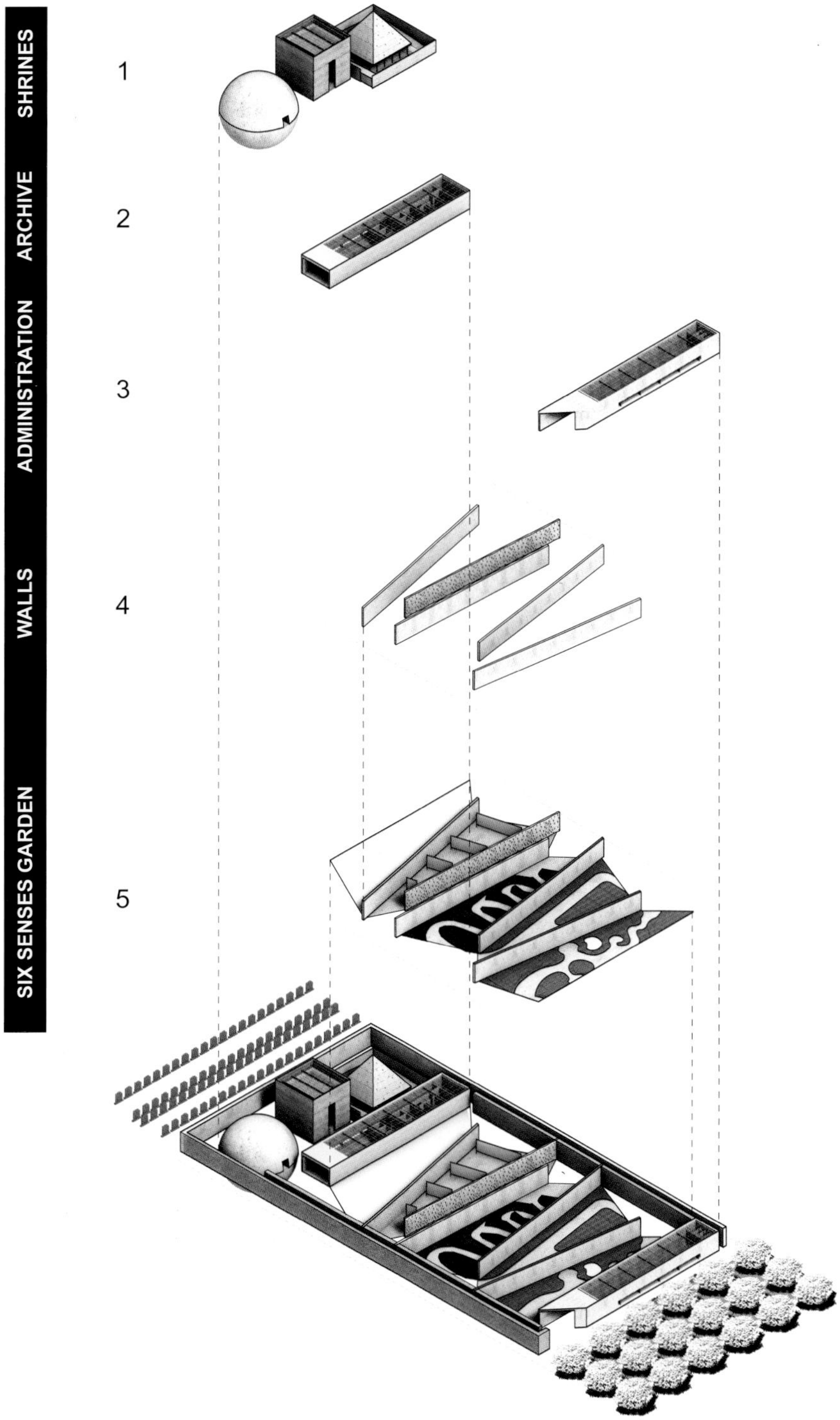

1 SHRINES 2 ARCHIVE 3 ADMINISTRATION 4 WALLS 5 SIX SENSES

i AVATAR | HERE AFTER

INBETWEEN: THE LIVING AND THE DEAD

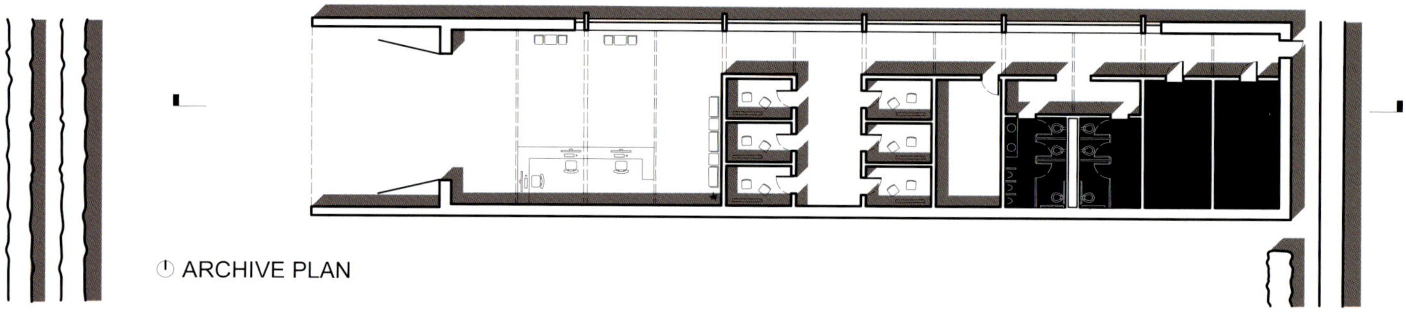

ARCHIVE PLAN

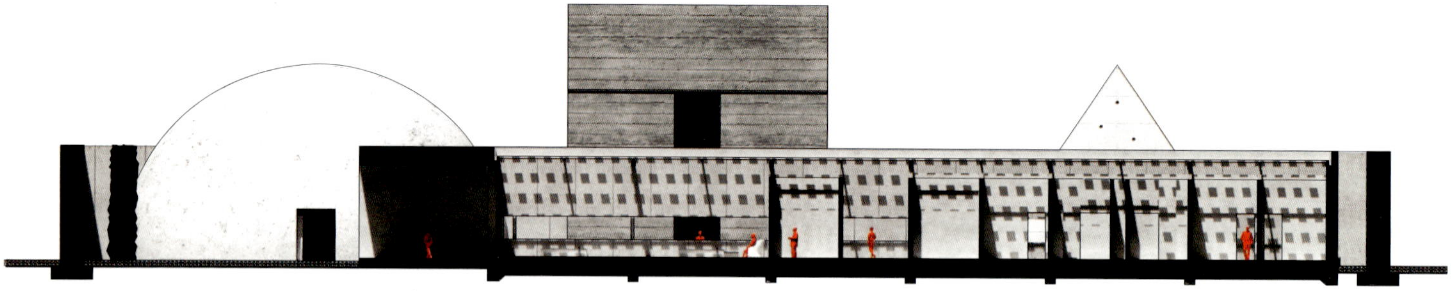

ARCHIVE SECTION

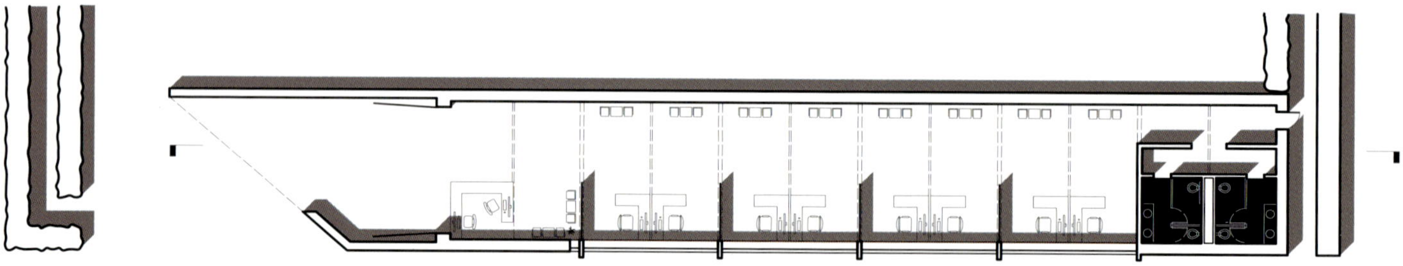

ADMINISTRATION PLAN

ADMINISTRATION SECTION

The ancestral shrine is designed as a prototype that can fit in various site contexts. In an urban setting, this project locates itself in a city block between two main streets. The programs are organized on a path that leads from life to death; the life of the city decompresses on a journey through the garden, sensory cemetery leading to shrines that reintroduce visitors to their ancestors.

CEREBRO

TESTIMONY WITH THE DEAD

CA

HOLOGR

AVATAR | HERE AFTER

SHRINES: COMMUNICATING WITH ANCESTORS

B

MBER

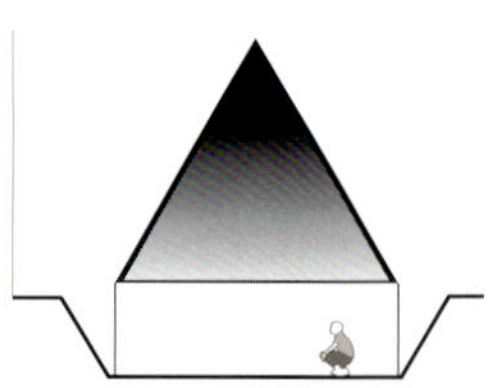

CHAPEL

REFLECTION HALL

The three shrines take on distinct mourning concepts: "'testimony with the dead,' 'talking to the dead,' and 'self-reflection.'" The spherical cerebro places the mourner in a 360-degree environment with augmented projections. The cubical catacomb allows for a holographic dialogue between the dead and the living, and the pyramidal chapel provides a spiritual place for mourning the dead.

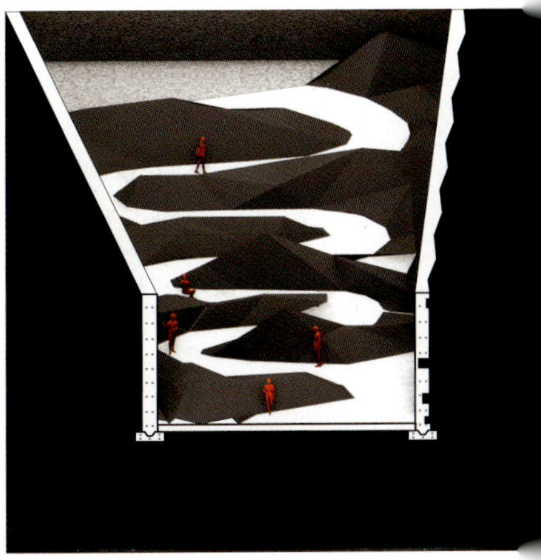

MOURNING JOURNEY: A PATH FOR THE SIX SENSES

The journey of the mourners go through a procession of seven gates of chakras and the six gardens of the senses with each garden engendering a stimulation that responds to the physical senses. As the mourners pass through the gardens, there are familiar triggers of past memories with their ancestors that use enhanced technology to conjure images of their own families.

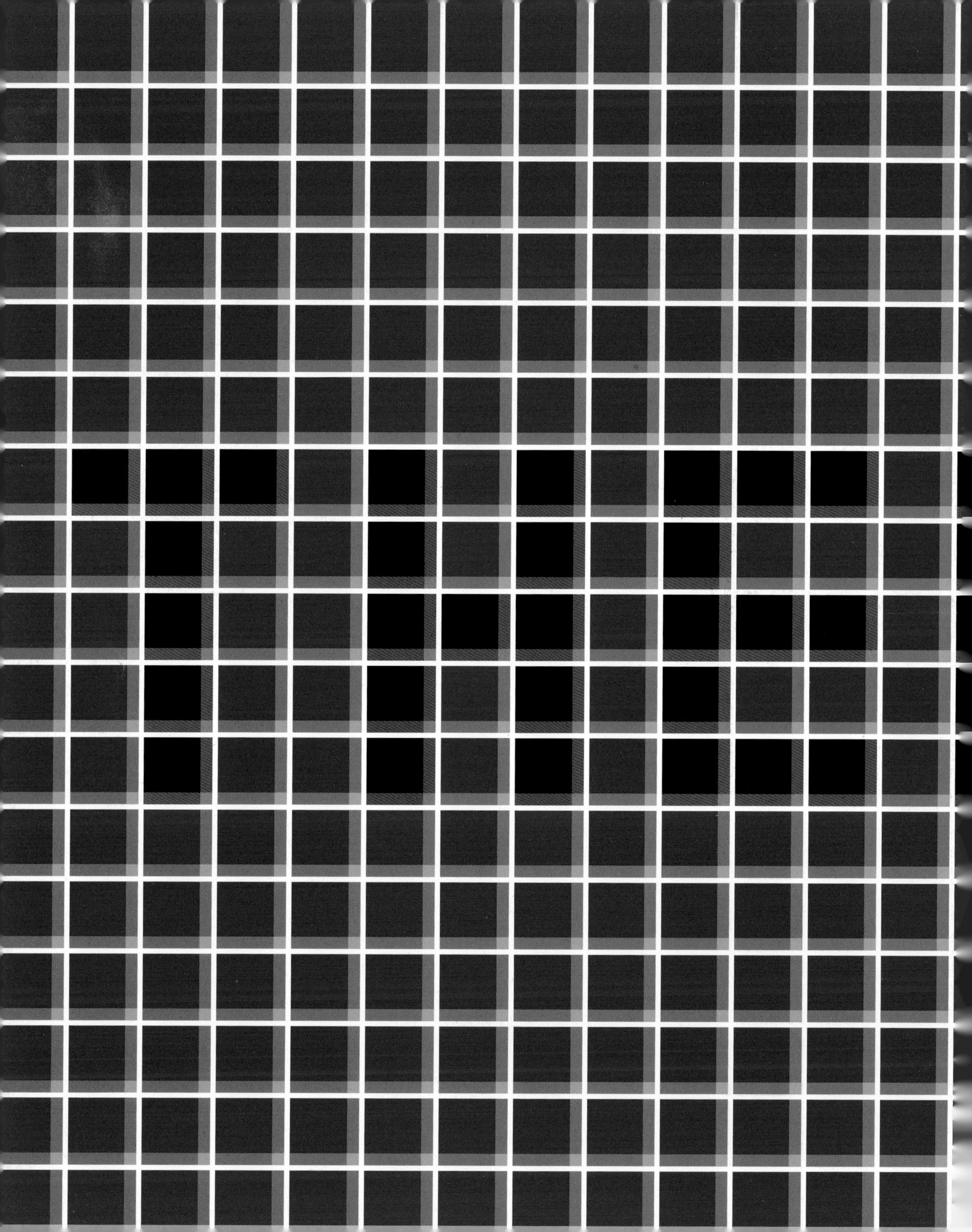
THE

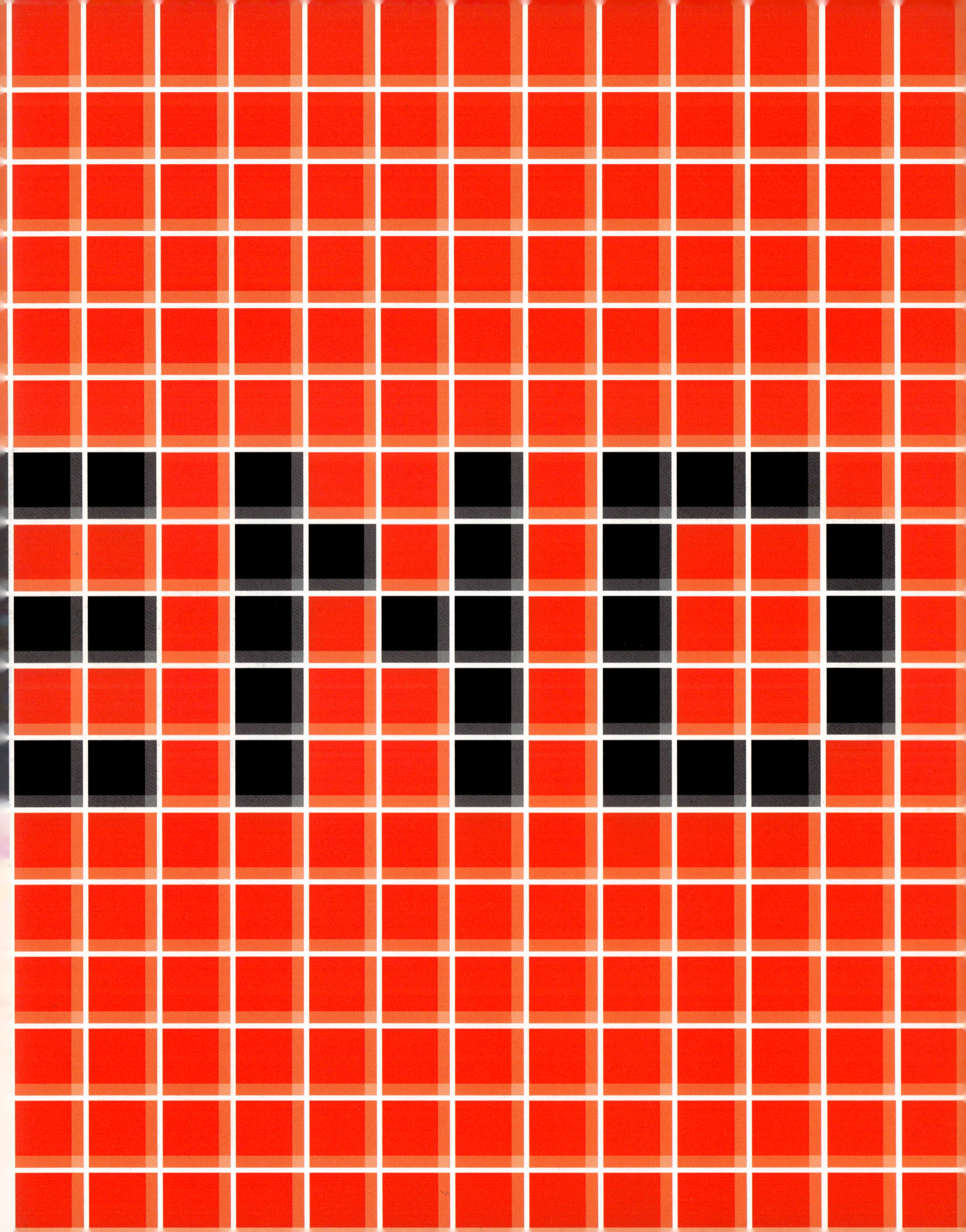
END

1969
EARTH ATMOSPHERE

On July 20, 1969 humans took the first photograph of the earth from outer space. The earth appeared untouched—a sphere floating in black with no borders, no countries, no boundaries. Astronauts have spoken about the inspiration from seeing a beautiful thin, blue band surrounding the globe that is the fragile atmosphere supporting life. Fifty years later, the impact of humans is undeniable, even from space. Today 27,000 fragments of debris, known as "space junk," circle the earth in perpetual orbit with city lights, bridges, and even humanmade islands visible from space. We must ask ourselves the same question that Jane Goodall once asked, "How come the most intellectual creature to ever walk Earth is destroying its only home?" There is no doubt that technology has made our lives easier, but at what cost? Through our own innovations, our species is pushing itself to the brink of extinction with warfare, climate change, and the proliferation of viruses. Technological advancements like artificial intelligence and genetic engineering provide exciting new capabilities but they also hold the power for people's total destruction. Humans' extraordinary capacity for communication made us the dominant species on

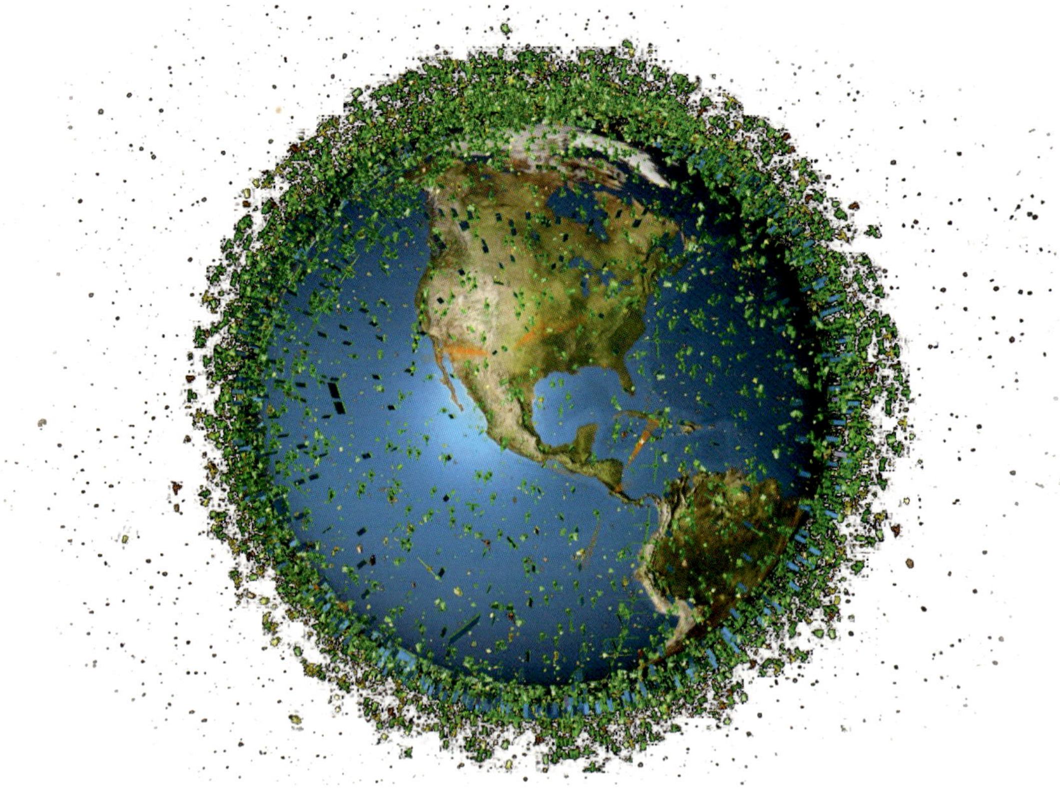

2022
27,000 SPACE DEBRIS

earth thousands of years ago[210], but now that we are communicating on a global scale, the risks are increasing too. Cyber warfare has the ability to bring businesses to a halt and cause political unrest. Misinformation is also becoming so common that it can be difficult to decipher the truth, and as a result, people are increasingly relying on faith for the answers. The technological evolution from real to unreal has made physical borders a thing of the past. The only barriers that remain are social constructs that, like people's faiths, are difficult to change. The agricultural, industrial, and information revolutions each caused a paradigm shift in society and, once again, humans are coming to a similar vortex. Before society changes, we must pause to reflect on what it is to be human. Do we go all-in and dive into the virtual world letting the chips fall where they may? Or do we transition into living + dying INbetween the real and the virtual? "I'll show these people what you don't want them to see. A world without rules and controls, without borders or boundaries. A world where anything is possible. Where we go from there is a choice I leave to you."

THE MATRIX, NEO, 1999[211]

The allure of technology creates a desire of humanity to take the next step toward the new. With each innovation there is the potential to evolve new possibilities for humankind, however, there is also the dark side. The migration from the real to the virtual can bring humans closer together, however, the loss of reality can become the ingredients for the ultimate extinction of humans.

PAGE 102

PAGE 166

PAGE 248

PAGE 326

AUGMENTED REALITY: TARGETS FOR VIEWING

INbetween has four pages that use Augmented Reality to enter the virtual world through targets illustrated on the left page. When viewed on an app entitled, "INBETWEEN the Real + the Virtual," you can enter the virtual world by using an iPhone, laptop, or iPad. Each AR environment encapsulates the possiblities of immersing the viewer into a three-dimensional world.

END NOTES

1 Eco, Umberto. *Travels in Hyperreality*. Translated by William Weaver, Harcourt Brace Jovanovich, 1986.

2 Cadet, L. B., Reynaud, E. & Chainay, H. Memory for a virtual reality experience in children and adults according to image quality, emotion, and sense of presence. Virtual Reality (2021). https://doi.org/10.1007/s10055-021-00537-y.

3 Stephenson, Neal. *Snow Crash*. Random House Publishing Group, 1992.

4 *The Matrix*. Directed by Lana Wachowski and Lilly Wachowski, Warner Bros., Village Roadshow Pictures, Silver Pictures, 1999.

5 https://elitecontentmarketer.com/screen-time-statistics. https://www.webmd.com/sleep-disorders/sleep-101.

6 Nagata J. M., Cortez C. A., Cattle C. J., et al. "Screen Time Use Among US Adolescents During the COVID-19 Pandemic: Findings From the Adolescent Brain Cognitive Development" (ABCD) Study. JAMA Pediatr. 2022;176(1):94–96. doi:10.1001/jamapediatrics.2021.4334.

7 https://www.statista.com/topics/7863/social-media-use-during-coronavirus-covid-19-worldwide/#dossierKeyfigures.

8 "Blue light has a dark side." Harvard Health Publishing, Harvard Medical School, July 7, 2020, https://www.health.harvard.edu/staying-healthy/blue-light-has-a-dark-side.

9 Dr. Giuseppe Riva, Fabrizia Mantovani, Claret Samantha Capideville, Alessandra Preziosa, Francesca Morganti, Daniela Villani, Andrea Gaggioli, Cristina Botella, and Mariano Alcañiz. "Cyber Psychology & Behavior." Feb 2007. 45–56. http://doi.org/10.1089/cpb.2006.9993.
Anna Felnhofer, Oswald D. Kothgassner, Mareike Schmidt, Anna-Katharina Heinzle, Leon Beutl, Helmut Hlavacs, Ilse Kryspin-Exner. "Is virtual reality emotionally arousing? Investigating five emotion inducing virtual park scenarios," *International Journal of Human-Computer Studies*, Volume 82, 2015, 48–56, ISSN 1071-5819, https://doi.org/10.1016/j.ijhcs.2015.05.004.

10 https://group.mercedes-benz.com/company/tradition/company-history/1885-1886.html?r=dai#:~:text=On%20January%2029%2C%201886%2C%20Carl,Patent%20Motor%20Car%2C%20model%20no.

11 https://www.britannica.com/technology/steam-engine.

12 https://www.eagletransmissionmesquite.com/history-of-the-automatic-transmission.

13 https://www.tomorrowsworldtoday.com/2021/08/09/history-of-autonomous-cars/.

14 https://highways.dot.gov/public-roads/summer-1996/three-states-claim-first-interstate-highway#:~:text=The%20Dwight%20D.,the%20new%20interstate%20construction%20funding.

15 https://www.aarp.org/politics-society/history/info-2018/first-cell-phone-call.html.

16 https://simpletexting.com/where-have-we-come-since-the-first-smartphone/.

17 https://www.history.com/news/who-invented-the-internet#:~:text=The%20first%20workable%20prototype%20of,communicate%20on%20a%20single%20network.

18 https://backlinko.com/zoom-users.

19 https://www.aps.org/publications/apsnews/200606/history.cfm

20 https://www.metroweekly.com/2014/07/to-the-moon-and-back-on-4kb-of-memory/.

21 https://www.nasa.gov/mission_pages/explorer/explorer-overview.html.

22 https://www.space.com/31420-spacex-rocket-landing-success.html.

23 https://www.statista.com/statistics/200002/international-car-sales-since-1990/#:~:text=Worldwide%20car%20sales%20grew%20to,pandemic%20in%20all%20key%20economie.

24 https://www.statista.com/statistics/264657/ranking-of-the-top-20-countries-by-length-of-railroad-network/.

25 https://www.rd.com/article/why-americans-drive-automatic/#:~:text=According%20to%20CarMax%2C%2096%20percent,percent%20of%20all%20vehicles%20sold.

26 https://policyadvice.net/insurance/insights/self-driving-car-statistics/#:~:text=There%20are%20over%201%2C400%20self,registered%20in%20California%20thus%20far.

27 https://www.census.gov/library/stories/2019/12/2020-census-count-guides-funding-of-new-roads-and-bridges.html.

28 https://www.statista.com/statistics/245501/multiple-mobile-device-ownership-worldwide/#:~:text=In%202021%2C%20the%20number%20of,devices%20compared%20to%202020%20levels.

29 https://www.statista.com/statistics/330695/number-of-smartphone-users-worldwide/.

30 https://www.statista.com/statistics/617136/digital-population-worldwide/.

31 https://backlinko.com/zoom-users.

32 https://www.history.com/topics/space-exploration/moon-landing-1969#:~:text=On%20July%2020%2C%201969%2C%20American,to%20walk%20on%20the%20moon.

33 https://www.gbmb.org/blog/apple-iphone-x-ram-and-internal-memory-1#:~:text=Apple%20iPhone%20X%2C%20also%20known,to%20256%20GB%20internal%20memory.

34 https://www.geospatialworld.net/blogs/how-many-satellites-are-orbiting-the-earth-in-2021/.

35 https://www.space.com/inspiration4-spacex.html.

36 https://www.statista.com/statistics/748044/number-video-gamers-world/.

37 https://www.statista.com/statistics/870620/china-number-of-game-users/.

38 https://www.statista.com/topics/3070/us-gamers/.

39 https://www.computerhistory.org/timeline/1941/.

40 https://www.bnl.gov/about/history/firstvideo.php#:~:text=Tennis%20for%20Two%20was%20first,an%20oscilloscope%20for%20a%20screen.

41 https://www.britannica.com/technology/online-gaming.

42 https://www.verdict.co.uk/history-virtual-reality-timeline/#:~:text=1968%20%E2%80%93%20The%20first%20VR%20HMD,virtual%20tour%20of%20Aspen%2C%20Colorado.

43 https://www.statista.com/statistics/268954/revenues-of-global-video-games-industry/.

44 https://www.statista.com/statistics/859275/china-revenue-of-leisure-mobile-games/.

45 https://www.statista.com/statistics/293659/revenue-of-video-games-in-the-us/.

46 https://www.statista.com/statistics/268954/revenues-of-global-video-games-industry/.

47 https://www.statista.com/statistics/859275/china-revenue-of-leisure-mobile-games/.

48 https://www.statista.com/statistics/293659/revenue-of-video-games-in-the-us/.

49 https://www.statista.com/topics/3070/us-gamers/.

50 https://www.statista.com/topics/3070/us-gamers/.

51 https://www.scmo.net/faq/2019/8/9/how-many-compaters-is-there-in-the-world#:~:text=In%202019%2C%20there%20were%20over,servers%2C%20desktops%2C%20and%20laptops.

52 https://gamingshift.com/how-many-video-games-exist/.

53 https://www.statista.com/topics/2290/mmo-gaming/.

54 https://www.statista.com/statistics/499714/global-virtual-reality-gaming-sales-revenue/.

55 https://newzoo.com/insights/articles/global-games-market-to-generate-175-8-billion-in-2021-despite-a-slight-decline-the-market-is-on-track-to-surpass-200-billion-in-2023/#:~:text=We%20now%20estimate%20that%20the,began%20tracking%20revenues%20in%202012.

56 https://www.statista.com/statistics/322200/video-game-revenue-in-china/.

57 https://www.statista.com/statistics/500070/gaming-revenue-north-america.

58 https://www.statista.com/statistics/511639/global-mobile-game-app-revenue/#:~:text=It%20is%20estimated%20that%20global,of%20the%20worldwide%20gaming%20market.

59 Harari, Yuval Noah. *Sapiens*. Harper, 2015.

60 Plato. *Plato's the Republic*. Books, Inc., 1943.

61 Harari, Yuval Noah. *Sapiens*. Harper, 2015.

62 https://www.statista.com/statistics/617136/digital-population-worldwide/

63 James Wright, ed., "*International encyclopedia of social and behavioral sciences*", Elsevier, 2014. https://sites.utexas.edu/dsb/tokens/the-evolution-of-writing/
Heath, F. G. "ORIGINS OF THE BINARY CODE." Scientific American, vol. 227, no.2, Scientific American, a division of Nature America, Inc., 1972, 76–83, http://www.jstor.org/stable/24927408.
Benko, Attila and Cecília Sik Lányi. "History of Artificial Intelligence." Encyclopedia of Information Science and Technology, Second Edition, edited by Mehdi Khosrow-Pour, D.B.A., IGI Global, 2009, 1,759–1,762. https://doi.org/10.4018/978-1-60566-026-4.ch276.

64 https://contextc.Com/chroma-key-green-and-blue-in-rgb-hex-values.

65 Jawed, Soyiba et al. "Classification of Visual and Non-visual Learners Using Electroencephalographic Alpha and Gamma Activities." "*Frontiers in Behavioral Neuroscience*" vol. 13, 86. 7 May. 2019, doi:10.3389/fnbeh.2019.00086.

66 Kalloniatis M., Luu C. The *Perception of Color*. May 1, 2005 [Updated Jul 9, 2007]. In: Kolb H, Fernandez E, Nelson R, editors. "Webvision: The Organization of the Retina and Visual System" [Internet]. Salt Lake City (UT): University of Utah Health Sciences Center; 1995. Available from: https://www.ncbi.nlm.nih.gov/books/NBK11538/.

67 https://news.yale.edu/2011/06/22/birds-eye-view-far-more-colorful-our-own.

68 https://www.pantone.com/articles/color-fundamentals/understanding-different-color-spaces.

69 https://pixelproductionssd.com/blog/a-brief-history-of-the-green-screen

70 https://psycnet.apa.org/record/2019-71491-001.

71 Albers, Josef. *Interaction of Color*. New Haven, Conn.: Yale University Press, 2006.

72 https://jov.arvojournals.org/article.aspx?articleid=2617976.

73 https://photo.stackexchange.com/questions/40118/is-16-7-million-colour-wide-gamut#:~:text=16.7%20millions%20colors%20is%20one,and%20256%20levels%20of%20blue.

74 https://spacenews.com/spacex-passes-2000-starlink-satellites-launched/#:~:text=SpaceX's%20current%20Starlink%20constellation%20is,orbits%20at%20around%20550%20kilometers.

75 https://www.open.edu/openlearn/history-the-arts/history/history-science-technology-and-medicine/history-science/latitude-and-longitude.

76 https://www.history.com/topics/inventions/printing-press.

77 https://www.aps.org/publications/apsnews/200606/history.cfm.

86 Deumlich, Fritz. *Surveying Instruments*. Walter de Gruyter, 2010.

80 https://www.encyclopedia.com/science/encyclopedias-almanacs-transcripts-and-maps/heliocentric-theory.

81 Gregersen, Erik. “How Albert Einstein Developed the Theory of General Relativity.”Encyclopedia Britannica, 2022, https://www.britannica.com/story/how-albert-einstein-developed-the-theory-of-general-relativity.

90 https://www.britannica.com/topic/Facebook.

82 https://www.jpl.nasa.gov/missions/explorer-1.

83 https://www1.grc.nasa.gov/beginners-guide-to-aeronautics/newtons-laws-of-motion/.

91 https://storymaps.arcgis.com/stories/43449bc48bbc4937b440e9ed3e2ea11c.

92 https://www.history.com/news/who-invented-the-internet.

84 https://www.esri.com/en-us/what-is-gis/history-of-gis.

85 https://www.aiaa.org/docs/default-source/uploadedfiles/about-aiaa/press-room/videos/iaf-60th-anniv-gps-nomination.pdf?sfvrsn=9bc64bfa_0.

87 https://earthobservatory.nasa.gov/features/BlueMarble/BlueMarble_history.php.

93 https://www.nasa.gov/mission_pages/hubble/about.

78 Howe, H.D. 1970. The “Story of Greenwich Time.” *Journal of the British Astronomical Association* 80, 208–211.

79 https://blog.google/products/maps/look-back-15-years-mapping-world/.

88 Grieves, Michael. (2016). *Origins of the Digital Twin Concept*. 10.13140/RG.2.2.26367.61609.

89 Benko, Attila and Cecília Sik Lányi. “History of Artificial Intelligence.” *Encyclopedia of Information Science and Technology*, Second Edition, edited by Mehdi Khosrow-Pour, D.B.A., IGI Global, 2009, pp. 1759-1762. https://doi.org/10.4018/978-1-60566-026-4.ch276.

94 Oestreicher, Christian. “A History of Chaos Theory.” *Dialogues in Clinical Neuroscience* vol. 9,3 (2007): 279-89. doi:10.31887/DCNS.2007.9.3/coestreicher.

95 https://www.amnh.org/exhibitions/einstein/legacy/quantum-theory.

96 *Ready Player One*. Directed by Steven Spielberg, Warner Bros., Village Road showPictures, Amblin Entertainment, 2018.

97 Zweig, Stefan. *The World of Yesterday: An Autobiography*. Lincoln, Neb: University of Nebraska Press, 1964.

98 https://courses.Lumenlearning.Com/boundless-physics/chapter/the-electromagnetic-spectrum/.

99 https://www.metroweekly.com/2014/07/to-the-moon-and-back-on-4kb-of-memory/.

100 Https://www.Nationalgeographic.Org/encyclopedia/prime-meridian/.

101 https://www.google.com/streetview/explore/.

102 creative commons cc-by-sa 3.0 License. “History of periodic table, main group elements.” *New World Encyclopedia*, Feb. 19, 2019.
“Timeline: The Discovery of Elements.” Chemistry explained.
Helmenstine, Anne. “When were the elements discovered? Timeline and periodicable.” Science notes and projects, Dec. 10, 2018.

103 Turkle, Sherry. Alone Together: *Why We Expect More from Technology and Less from Each Other*. New York :Basic Books, 2011.

104 https://courses.lumenlearning.com/ushistory2ay/chapter/the-rise-of-suburbs-2/.

105 https://www.un.org/development/desa/en/news/population/2018-revision-of-world-urbanization-prospects.html.

106 https://www.theesa.com/resource/2021-essential-facts-about-the-video-game-industry/.

107 Https://eagle.Northwestu.Edu/faculty/gary-gillespie/technology-of-communication-timeline/.

108 Harari, Yuval Noah. *Sapiens*. Harper, 2015.

109 “Controllers.” Valve index® - upgrade your experience - valve. Corporation,www.Valvesoftware.Com/en/index/controllers.

110 Moore’s Law: https://www.britannica.com/technology/Moores-law.

111 https://www.statista.com/statistics/245501/multiple-mobile-device-ownership-worldwide/#:~:text=In%202021%2C%20the%20number%20of,devices%20compared%20to%202020%20levels.

112 https://www.hp.com/us-en/shop/tech-takes/video-game-genres.

113 Minecraft wiki. Minecraft wiki, gamepedia, 4 sept. 2020, Minecraft.Gamepedia.Com/minecraft_wiki.

114 https://news.stanford.edu/2019/08/21/online-dating-popular-way-u-s-couples-meet/.

115 Kimberly M. Christopherson. “The Positive and Negative Implications of Anonymity in Internet Social Interactions: On the Internet, Nobody Knows You’re a Dog”, *Computers in Human Behavior*, Volume 23, Issue 6, 2007, 3,038–3,056, ISSN 0747-5632, https://doi.org/10.1016/j.chb.2006.09.001.

116 https://www.wsj.com/articles/could-ai-keep-people-alive-after-death-11625317200
https://www.reuters.com/article/us-global-tech-privacy-trfn/data-of-the-dead-virtual-immortality-exposes-holes-in-privacy-laws-idUSKBN21Z0NF.

117 https://www.washingtonpost.com/technology/2022/02/02/facebook-earnings-meta/.

118 “Timeline: the human genome project.” Facts, the public engagement team at the wellcome genome campus, 19 Jan. 2015, Www.Yourgenome.Org/facts/timeline-the-human-genome-project.
“File:human genome idiogram Svg.” File:human genome idiogram Svg - wiki media commons.
Genome research limited. “How do you map a genome?” Facts, the public engagement team at the wellcome genome campus, 2 June 2016.
Saey, tina hesman. “What I actually learned about my family after trying 5 DNA ancestry tests.” Science news, 13 May 2020.
“Common cloning applications and strategies.” Thermo fisher scientific - us. “Chromosome.” Wikipedia, wikimedia foundation, 5 Sept. 2020.

119 https://www.theverge.com/2016/6/2/11837854/neural-lace-cyborgs-elon-musk.

120 https://www.nigms.nih.gov/education/fact-sheets/Pages/circadian-rhythms.aspx.

121 https://mymodernmet.com/leonardo-da-vinci-vitruvian-man/.

122 https://www.iconeye.com/opinion/icon-of-the-month/modulor-man-by-le-corbusier.

123 Tilley, Alvin R. The Measure of Man and Woman: Human Factors in Design. New York:Wiley, 2002. Print.

124 Yee, Nick, et al. “The Proteus Effect: Implications of Transformed Digital Self-Representation on Online and Offline Behavior.” Communication Research, vol. 36, no. 2, Apr. 2009, pp. 285–312, doi:10.1177/0093650208330254.

125 https://www.newworldencyclopedia.org/entry/olympic_games.

126 https://daoinsights.com/news/china-launches-first-meta-human-virtual-influencer-ayayi/.

127 https://www.miamitimesonline.com/business/technology/sophia-the-robot-interviewed-during-tech-with-soul/article_3c70874c-7350-11ec-9faf-0bbf1f39882c.html.

128 Turkle, Sherry. Alone Together : Why We Expect More from Technology and Less from Each Other. New York :Basic Books, 2011.

129 Spielberg, Steven. Minority Report. Twentieth Century Fox, 2002.

130 https://www.theguardian.com/music/2021/oct/23/20-years-of-the-ipod-how-music-and-tech-new-era-steve-jobs.

131 https://www.fhwa.dot.gov/interstate/history.cfm.

132 Hawking, Stephen. "Is Everything Determined", Black Holes and Baby Universes and Other Essays, 1993.

133 https://blueprintforbetter.org/articles/architectures-carbon-problem/.

134 Turkle, Sherry. Alone Together : Why We Expect More from Technology and Less from Each Other. New York :Basic Books, 2011.

135 https://www.atlasobscura.com/places/cristo-de-la-concordia.

136 https://www.npr.org/sections/13.7/2013/12/02/248089436/the-truth-about-the-left-brain-right-brain-relationship.

137 https://www.grc.nasa.gov/www/k-12/rocket/state.html#:~:text=Any%20substance%20can%20occur%20in%20any%20phase.&text=But%20if%20we%20lower%20the,a%20gas%20called%20water%20vapor.

138 https://www.nasa.gov/vision/earth/everydaylife/jamestown-water-fs.html#:~:text=About%2070%20percent%20of%20the,match%20for%20life%20in%20general.

139 https://askabiologist.asu.edu/bird-hypothalamus.

140 https://www.kaenz.com/water-temperature-swimming/#:~:text=Movement%20in%20water%20increases%20heat,of%20body%20heat%20is%20precipitated.

141 https://www.nps.gov/articles/geothermal-systems-and-monitoring-hydrothermal-features.html.

142 Yahia, M.W., Johansson, E., Thorsson, S. et al. Effect of urban design on microclimate and thermal comfort outdoors in warm-humid Dar es Salaam, Tanzania. Int J Biometeorol 62, 373–385 (2018). https://doi.org/10.1007/s00484-017-1380-7.

143 https://www.fondation-vincentvangogh-arles.org/en/documentation-2/vincent-van-gogh/van-gogh-et-la-lumiere-provencale/.

144 https://solarsystem.nasa.gov/moons/earths-moon/lunar-phases-and-eclipses/#:~:text=These%20eight%20phases%20are%2C%20in,month%20(every%2029.5%20days).

145 https://www.britannica.com/place/Mount-Saint-Helens.

146 https://www.britannica.com/topic/mirage-optical-illusion.

147 https://www.etymonline.com/word/mirage.

148 https://psych.hanover.edu/research/vi-040.html.

149 Lauren Richey, Bailey Stewart, and Justin Peatross , "Creating and Analyzing a Mirage", The Physics Teacher 44, 460-464 (2006) https://doi.org/10.1119/1.2353594.

150 https://www.discovermagazine.com/mind/the-vision-thing-mainly-in-the-brain.

151 https://www.brainline.org/article/vision-our-dominant-sense#:~:text=Research%20estimates%20that%20eighty%20to,%2C%20and%2For%20cognitive%20response.

152 https://www.weather.gov/media/ind/seasons.pdf.

153 http://www.artandpopularculture.com/Music_is_liquid_architecture

154 Beranek, Leo L. "Music, acoustics, and architecture." Bulletin of

the American Academy of Arts and Sciences 45.8 (1992): 25-46.
155 https://www.npr.org/2021/02/26/971650046/climate-change-deforestation-threaten-monarch-butterfly-migration#:~:text=Monarch%20Migration%20Threatened%20By%20Climate%20Change%2C%20Deforestation%20%3A%20NPR&text=Monarch%20Migration%20Threatened%20By%20Climate%20Change%2C%20Deforestation%20The%20population%20of,according%20to%20a%20new%20report.
156 University Of Texas, Austin. "Extinction Rate Across The Globe Reaches Historical Proportions." ScienceDaily. ScienceDaily, 10 January 2002. <www.sciencedaily.com/releases/2002/01/020109074801.htm>.
157 Zweig, Stefan. The World of Yesterday: An Autobiography. Lincoln, Neb: University of Nebraska Press, 1964. Print.
158 https://www.google.com/maps/place/Sugar+Island/@46.4278308,-84.340491,11z/data=!3m1!4b1!4m5!3m4!1s0x4d36468ef5b46d3b:0x7decf19ba2efb365!8m2!3d46.5032439!4d-84.1769858.
159 https://www.vox.com/the-goods/22911159/new-space-age-billionaires-culture.
160 https://www.noaa.gov/education/resource-collections/climate/climate-change-impacts.
161 https://people.com/archive/edgar-mitchells-strange-voyage-vol-1-no-6/.
162 https://www.nde-ed.org/Physics/Magnetism/linesofforce.xhtml.
163 https://www.princeton.edu/~oa/manual/mapcompass2.shtml#:~:text=The%20angular%20difference%20between%20true,as%20shown%20in%20Figure%206.7.&text=The%20magnetic%20field%20lines%20of%20the%20earth%20are%20constantly%20changing,1%20degree%20every%20five%20years).
164 https://militarybase.net/defense-ready-condition-defcon-levels/.
165 https://www.nasa.gov/content/space-applications-of-hydrogen-and-fuel-cells.
166 https://www.cnet.com/news/harvesting-energy-with-space-solar-panels-could-power-the-earth-247/.
167 https://space.nss.org/wp-content/uploads/SSP-DOE-Brightness-Due-To-Reflected-Sunlight.pdf.
168 https://www.noaa.gov/education/resource-collections/climate/climate-change-impacts.
169 http://www.change-climate.com/Politics_Policies_Transparency/DEFCON_Defense_Warning_Conditions_Readiness_Codes/1DefCon_Code_Red_Nuclear_World_War_Imminent_Military_Maximum.html.
170 SABAH A. ABDUL-WAHAB & A. AL-ARAIRNI (2004) Environmental considerations in urban planning, International Journal of Environmental Studies, 61:5, 527-537, DOI: 10.1080/0020723041001688161.
171 Lubell, Jeffrey. "Preserving and Expanding Affordability in Neighborhoods Experiencing Rising Rents and Property Values." Cityscape, vol. 18, no. 3, US Department of Housing and Urban Development, 2016, pp. 131–50, http://www.jstor.org/stable/26328276.
172 https://www.mdpi.com/1996-1073/9/10/796.
173 "Natural Disasters and their Impact on Cities". In obo in Urban Studies. 8 Mar. 2022. <https://www.oxfordbibliographies.com/view/document/obo-9780190922481/obo-9780190922481-0014.xml>.
174 Rutz, C., Loretto, MC., Bates, A.E. et al. COVID-19 lockdown allows researchers to quantify the effects of human activity on wildlife. Nat Ecol Evol 4, 1156–1159 (2020). https://doi.org/10.1038/s41559-020-1237-z.
175 Frederica Perera, Alique Berberian, David Cooley, Elizabeth Shenaut, Hollie Olmstead, Zev Ross, Thomas Matte, Potential health benefits of sustained air quality improvements in New York City: A simulation based on air pollution levels during the COVID-19 shutdown, Environmental Research, Volume 193, 2021, 110555, ISSN 0013-9351, https://doi.org/10.1016/j.envres.2020.110555.
176 https://comptroller.nyc.gov/reports/the-pandemics-impact-on-nyc-migration-patterns/.
177 https://policyadvice.net/insurance/insights/electric-car-statistics/.
http://ti.org/antiplanner/?p=16539.
https://vancouver.ca/files/cov/transportation-2040-plan.pdf.
https://www.oberlo.com/blog/virtual-reality-statistics#:~:text=Nearly%20one%20in%20five%20(19,either%20extremely%20or%20moderately%20satisfied.
https://www.greenbiz.com/article/those-net-zero-building-commitments-are-adding.
https://www.mckinsey.com/~/media/McKinsey/Featured%20Insights/Digital%20Disruption/Harnessing%20automation%20for%20a%20future%20that%20works/MGI-A-future-that-works_In-brief.pdf.
https://loupfunds.com/auto-outlook-2040-the-rise-of-fully-autonomous-vehicles/
https://www.un.org/development/desa/en/news/population/2018-revision-of-world-urbanization-prospects.html
https://www.futuretimeline.net/21stcentury/2060-2069.htm
178 https://escholarship.org/content/qt43t4721n/qt43t4721n_noSplash_851827d8873af0c1dd3a591ed7a367a1.pdf?t=nrzndv.
179 https://www.latimes.com/archives/la-xpm-1988-06-21-fi-4744-story.html#:~:text=It%20Started%20With%20Levittown%20in%201947%20%3A%20Nation's%201st%20Planned%20Community%20Transformed%20Suburbia,-By%20JAMES%20F&text=It%20was%201947.,of%20ten%20in%20cramped%20city%20apartments.
180 https://www.merriam-webster.com/dictionary/suburb.
181 https://www.moma.org/explore/inside_out/2014/02/03/frank-lloyd-wrights-living-city-lives-onconserving-the-broadacre-city-model/
https://courses.lumenlearning.com/ushistory2ay/chapter/the-rise-of-suburbs-2/
https://www.nytimes.com/2001/04/17/us/90-s-suburbs-of-west-and-south-dens er-in-onesprawling-in-other.html
https://blog.nyhistory.org/the-shantytown-nineteenth-century-manhattans-straggling-suburbs/
https://www.newyorker.com/news/news-desk/what-puerto-rico-needs-after-hurricane-maria
https://nypost.com/2019/09/05/hurricane-dorian-causes-7b-in-property-damage-to-bahamas/
https://www.ft.com/content/052923d2-78c2-11ea-af44-daa3def9ae03
https://www.wsj.com/articles/how-covid-19-may-transform-urban-life-01593785900
https://romeonrome.com/2016/01/mussolinis-architectural-legacy-in-rome/
http://www.americanyawp.com/text/26-the-affluent-society/
182 https://www.thoughtco.com/overview-of-suburbs-1435799.
183 https://www.pewresearch.org/social-trends/2021/12/16/americans-are-less-likely-than-before-covid-19-to-want-to-live-in-cities-more-likely-to-prefer-suburbs/.
184 https://www.nbcnews.com/politics/national-security/climate-change-threatens-spark-instability-conflict-around-world-u-s-n1282078.
185 https://www.history.com/topics/cold-war/berlin-wall.
186 https://www.livescience.com/worlds-biggest-man-made-disasters
https://www.livescience.com/biggest-natural-disasters-throughout-history.
187 https://www.history.com/topics/cold-war/berlin-wall.
188 https://www.usatoday.com/story/news/world/2018/05/24/border-walls-berlin-wall-donald-trump-wall/553250002/.
189 Primack, Brian A et al. "Social Media Use and Perceived Social Isolation Among Young Adults in the US." American journal of preventive medicine vol. 53,1 (2017): 1-8. doi:10.1016/j.amepre.2017.01.010.
190 Keles, Betul, Niall McCrae, and Annmarie Grealish. "A systematic review: the influence of social media on depression, anxiety and psychological distress in adolescents." International Journal of Adolescence and Youth 25.1 (2020): 79-93.
191 Luciana, Runcan Patricia. "One minute more: Adolescent addiction for virtual world." Procedia-Social and Behavioral Sciences 2.2 (2010): 3706-3710.
192 Paschke, Kerstin, et al. "Adolescent gaming and social media usage before and during the COVID-19 pandemic." Sucht (2021).
193 Smahel, David, Lukas Blinka, and Ondrej Ledabyl. "Playing MMORPGs: Connections between addiction and identifying with a character." CyberPsychology & Behavior 11.6 (2008): 715-718.
194 https://www.nimh.nih.gov/health/statistics/major-depression.
195 https://www.nimh.nih.gov/health/statistics/major-depression.
196 https://www.un.org/press/en/2019/ga12131.doc.html.
197 https://www.britannica.com/technology/stepwell.
198 https://www.usatoday.com/story/news/nation/2020/02/14/climate-change-study-plant-animal-extinction/4760646002/.
https://mashable.com/article/climate-change-tipping-points-future/
https://www.nature.com/articles/s41467-019-12808-z.
199 https://ecos.fws.gov/ecp0/reports/ad-hoc-species-report?kingdom=P&status=E&status=T&status=EmE&status=EmT&status=EXPE&status=EXPN&status=SAE&status=SAT&mapstatus=3&fcrithab=on&fstatus=on&fspecrule=on&finvpop=on&fgroup=on&ffamily=on&header=Listed+Plants.
200 Randers, J., Goluke, U. An earth system model shows self-sustained thawing of permafrost even if all man-made GHG emissions stop in 2020. Sci Rep 10, 18456 (2020). https://doi.org/10.1038/s41598-020-75481-z
velvetchainsaw.com.
201 https://www.bbc.com/future/article/20160113-could-just-two-people-repopulate-earth.
202 https://climate.nasa.gov/faq/16/is-it-too-late-to-prevent-climate-change/.
203 https://medialibrary.climatecentral.org/resources/ipcc-6th-assessment-report-the-physical-science-basis
204 https://www.scientificamerican.com/article/theres-still-time-to-fix-climate-about-11-years/.
205 https://www.theactivetimes.com/travel/25-ghost-towns-around-us-and-history-behind-them/slide-7.
206 https://eterneva.com/blog/death-rituals/.
207 https://www.cnn.com/2021/01/27/tech/microsoft-chat-bot-patent/index.html.
208 https://www.cnn.com/2021/01/27/tech/microsoft-chat-bot-patent/index.html.
209 https://www.khanacademy.org/humanities/medieval-world/romanesque-art/romanesque-art-in-france/a/basilica-of-saint-sernin
https://www.healthline.com/health/fitness-exercise/7-chakras.
210 Harari, Yuval N. author. Sapiens : a Brief History of Humankind. New York :Harper, 2015.
211 Wachowski, Lana, and Lilly Wachowski. The Matrix. Warner Bros., 1999.

IMAGE CREDITS

P14 left	https://m.blog.naver.com/PostView.naver?isHttpsRedirect=true&blogId=thehury&logNo=221348072556.
P14 right	https://www.microsoft.com/en-gb/d/hololens-2/91pnzzznzwcp.
P15 left	https://www.buntefarben.com/.
P15 right	https://www.geekgirlauthority.com/behold-the-majesty-of-the-oasis-in-this-new-ready-player-one-trailer/.
P16-17	Google Image. Figure Cutouts.
P18-19	Ibid.
P20-23	Ibid.
P25	Google Image.
P27 right	Google Image. 3D Brain.
P34-35 top	https://www.pinterest.com/pin/414120128221531661/. https://fromtheparapet.wordpress.com/tag/ishtar-gate/. https://www.pinterest.com/pin/322077810850721757/. https://comicwow.tv/reviews/masters-of-venice-comicwow-review/. https://www.pinterest.com/pin/469711436134786699/. https://ageofrevolution.org/200-object/the-death-of-major-peirson-6-january-1781/the-death-of-major-peirson-6-january-1781-1783-by-john-singleton-copley-1738-1815/.
P34-35 bottom	https://www.conservativefriend.org/viewpoints.html. https://lib.unb.ca/news/new-collection-making-modern-world. https://www.wildernessarena.com/environment/signaling/how-to-communicate-with-morse-code-using-visual-audio-pressure-communication. https://daydaynews.cc/movie/406256.html. https://metroquest.com/citizenengagementatop5priorityfordigitalcities/. https://arinsider.co/2020/08/24/can-vr-save-virtual-events/.
P40	https://www.emage-plus.com/blog/2021/10/15/augmented-reality-and-virtual-reality-a-new-era-of-consumption.
P41	https://phys.org/news/2019-05-starlink-satellites-orbiting-altitude-space.html.
P44	Google Image. Physic diagrams.
P50	https://en.wikipedia.org/wiki/The_Blue_Marble.
P51	https://www.dreamstime.com/stock-illustration-earth-wireframe-d-model-panet-image59934143.
P55	Adapted for educational purposes. http://www.basedog.it/2016/09/19/the-history-of-space-exploration-on-a-single-space-map/.
P56 left	Google Image. View of Houston.
P56 right	Ibid.
P57 left	Ibid. Adapted for educational purpose.
P57 right	Google Earth Pro.
P61	Neri Oxman. MoMA Exhibition.
P64 - 71	Google Image. Images of technological inventions.
P74-75	Gameplay images of Limbo and The Last Night.
P76	Typical MineCraft cutouts.
P78 top	https://uploadvr.com/facebook-beat-saber-imu/.
P78 bottom	https://arthurbourgeais.artstation.com/projects/v1bEbD.
P79 top	Google Image. Family photo. https://www.ronenbekerman.com/200-free-2d-cutout-people-case3d/.
P79 bottom	The Sims 4. https://www.roadtovr.com/microsoft-altspacevr-update-event-hosting/.
P83	https://othersociologist.com/2014/02/20/visual-science-maps-representations-culture-history/. https://sites.nd.edu/biomechanics-in-the-wild/author/kschoolm/. Google Image. Sophia. Google Image. Ayayi.
P88-89 top	Google Image. Aggressive Game Characters. Google Image. Succubus Game Character.
P88-89 bottom	https://www.pinterest.com/pin/612137774327728728/. Google Image. Tall Game Character. https://www.pinterest.com/pin/806988826958841538/. https://www.pinterest.com/pin/391953973802640135/. https://www.pinterest.com/pin/540361655301308815/. https://www.pinterest.com/pin/336362665902317885/. https://www.pinterest.com/pin/815714551244837213/. https://www.pinterest.com/pin/820569994586101496/. https://www.pinterest.com/pin/540361655292880659/.
P90	Google Image.
P91	Google Image. ReadyPlayerMe Avatar. https://gameideas.fandom.com/wiki/Genjuro_Kibagami.
P93	https://www.subpng.com/png-cl7ood/. https://www.pinterest.com/pin/847591592340118948/.
P100	Left Column, Third Row: Google Image. Monarch Butterfly.
P101	Right Column, Third Row: https://www.pinterest.com/pin/493425702908361903/.
P101	Right Column, Fourth Row: https://therake.com/stories/code/impractical-choice-artwork-nick-veasey/.
P105	https://www.lostiempos.com/oh/actualidad/20200127/historias-que-guarda-cristo.
P107	Google Image.
P122-P123	Ibid.
P154-155	https://www.komar.org/faq/mount-saint-helens-volcanic-eruption/. https://yinalardanakta.medium.com/natures-fury-03-volcanoes-1a3719ec210f. https://www.mcchord.af.mil/News/Features/Display/Article/2190614/62nd-airlift-wing-recalls-mount-st-helens-eruption-on-40th-anniversary/. https://richflintphoto.blogspot.com/2010/05/blow-down.html. https://en.wikipedia.org/wiki/Mount_St._Helens.
P170	https://www.pinterest.com/pin/30962316174965006/.
P223	https://peacetour.org/blog/page/3/.
P260	Google Image. Avatars.
P286-287	Google Image. https://www.flickr.com/photos/markgregory/7651774934. https://www.pinterest.com/pin/564709240756867566/. https://en.wikipedia.org/wiki/Wikipedia:Recent_additions/2016/November. https://stellalabella.com/italy-images/benito-amilcare-andrea-mussolini-dictator/2/. Google Image. https://nypost.com/2019/09/05/hurricane-dorian-causes-7b-in-property-damage-to-bahamas/. https://www.newyorker.com/books/double-take/sunday-reading-responding-to-disaster. https://www.ft.com/content/052923d2-78c2-11ea-af44-daa3def9ae03. Google Image. https://www.e-ir.info/2020/09/06/covid-nationalism/. https://www.wsj.com/articles/is-another-exodus-ahead-for-u-s-cities-11592488220.
P290	http://thinkurban.org/blog/2012/07/24/city-growth-and-whats-a-city. Google Image.
P292	Ibid.
P296-297	Ibid.
P298 top	https://www.pinterest.com/pin/432134526739491547/.
P304-305	Bottom: Google Image.
P358-359	Google Image.
P394-395	Google Image.
P396	https://www.pinterest.com/pin/144044888057126588/.
P408	https://en.wikipedia.org/wiki/The_Blue_Marble.
P409	Google Image.

ACKNOWLEDGMENTS

Writing a book is a journey, a mental conversation that takes on a life of its own. In taking on a subject like the real and the virtual world, the inevitable question arises: What is reality? The one lifeline that kept me tethered to what is real is all the people that supported this monumental project. The original graphics, the architectural proposals to explore this new world, the writing and the research all came together to make this book possible.

The Gerald D. Hines College of Architecture + Design with Dean Patricia Oliver, FAIA were the inspiration for beginning this project. The 50 students in the 5th year and graduate school of the college worked with continued passion on the various stages of this three-year endeavor. It is this pursuit and curiosity of an architectural idea that I find the greatest source of creativity and trying to make sense of our own time.

Khoi Bui as the coordinator with Carlos Castellanos were the two alumni that have been the pillars that helped shape the graphic work and the research in achieving such success. Josh Stanton Smith has been collaborating for over a decade with me on multiple projects. His creativity and pursuit of augmented and virtual reality can be seen in this book with four pages that are experienced in AR. It has been enlightening to have had him to muse about future scenarios with. I also want to thank a past collaborator who is always unflinchingly there for critiques: Gail Peter Bordon, FAIA. I appreciate his time and his ability to scan a page and immediately see the improvements to be made given the concept and graphic layouts. I also want to thank the publishers AR+D, particularly Jake Anderson and Gordon Goff for being so supportive and professional.

One of the most difficult tasks in writing a book that attempts to connect very diverse dots is to make a cohesive narrative, a story that is undeniable not just about the present, but also offers a glimpse of what might come. Julie Anna Zweig has given me something that I will forever hold dear to my heart: the opportunity to talk about how I see the world with my daughter. What a rare moment to go into my inner thoughts and to share both my self-doubts and lofty ideas. Her contribution to the editing, to the story telling, and to conversing about the dystopian and visionary dreams of the future will stay with me forever (even into my virtual avatar).

A family is like a puzzle that comes together to create a picture of generations from the past, the present, and the future: a portrait that is passed through our DNA over time. Lea, my daughter, and with my grandchildren, Eli and Olivia, complete the family portrait. Behind the curtain my partner, my wife, my rock will and always has been Linda. She has been at my side for over four decades and we have lived through the real and the world of dreams together. She has kept our family as one and this has given me the strength to venture out into uncharted territory knowing I had a safe harbor to come home to when I returned.

Peter Jay Zweig
February 2022

CREDITS

LIVING + DYING INBETWEEN THE REAL + THE VIRTUAL

DIRECTOR + AUTHOR

Peter Jay Zweig, FAIA

SPONSORSHIP

Dean Patricia Belton Oliver, FAIA
University of Houston
Gerald D. Hines College of Architecture and Design

Gerald D. Hines + Barbara Hines

GRAPHIC DESIGN

Peter Jay Zweig, FAIA
Khoi Bui, Coordinator
Carlos Soto Castellanos
Julie Anna Zweig

AUGMENTED REALITY

Joshua Stanton Smith, Co-Director
Peter Jay Zweig, FAIA, Co-Director

Published by Applied Research and Design Publishing, an imprint of ORO Editions.
Gordon Goff: Publisher

www.appliedresearchanddesign.com
info@appliedresearchanddesign.com

Author: Peter Jay Zweig
Book Design: Peter Jay Zweig + Khoi Anh Bui + Julie Anna Zweig
Project Manager: Jake Anderson

10 9 8 7 6 5 4 3 2 1 First Edition

ISBN: 978-1-954081-78-9

Color Separations and Printing: ORO Group Ltd.
Printed in China.

AR+D Publishing makes a continuous effort to minimize the overall carbon footprint of its publications. As part of this goal, AR+D, in association with Global ReLeaf, arranges to plant trees to replace those used in the manufacturing of the paper produced for its books. Global ReLeaf is an international campaign run by American Forests, one of the world's oldest nonprofit conservation organizations. Global ReLeaf is American Forests' education and action program that helps individuals, organizations, agencies, and corporations improve the local and global environment by planting and caring for trees.

PROJECT TEAMS

TRUTH
Khoi Bui
Andrea Cortes
Rachel Williams

GAME ON
Khoi Bui
Ashley Nicole Coffy
Carlos Soto Castellanos
Phong Dung
Madeleine Garcia
Jonathan Guerrero
Kyle Swisher

NO BORDERS
Sebastian Andrade
Khoi Bui
Carlos Soto Castellanos
Aileen Mendoza
Miguel Lopez
Christopher Zhu
Lap Nguyen
Mariana Galvan
Nicholas Le
Nastassia Lite Chua
Xuchuan Liu
Francisco Chan
Aileen Mendoza
Gabriela A Martinez

AVATAR
Khoi Bui
Carlos Soto Castellanos
Eduardo Cano Garcia
Grace Ehmling
Jonathan De La Rosa
Alejandro Reyez
Hector Jimenez
Aileen Mendoza
David Espinoza
Joselyn Nguyen

AUGMENTED REALITY
Joshua Stanton Smith
Hind Alkhaffaf
Eduardo Garcia Cano
Lucio Alexander Chavez
Elvia K Delgado
Luis Alexandro Hinojosa
Simone Ashley Holmes
Megan D Julian
Jasleen Kaur
Miguel Angel Lopez
Pierce Mercer
Maisie Munoz
Jacob Ramos
Martin Rodriguez
Joel J Samuel
Zehra Z Shenwari
Gabriel De La Torre

RESEARCH
Julie Anna Zweig
Linhan Chen
Grace Ehmling
Michael Fajtl
Valeria Laverde
Stefan Nino
Ekaterina Bondarenko Ondarza
Hector Rodriguez
Martin Rodriguez
Shyam Kumar Thapa

LIVING + DYING

INBETWEEN

THE REAL + THE VIRTUAL

WHAT IT MEANS TO BE A HUMAN

0 1 1 1 0 1 0 1 0 1 1 0 1 1 1 0 0 1 1 1 0 0 1 0 0 1 1 0 0 1 1 0 0 1 0 1 0 1 1 0 0 0 0 1 0 1 1 0 1 1 0 0

0 1 1 1 0 1 0 1 0 1 1 0 1 1 1 0 0 1 1 1 0 0 1 0 0 1 1 0 0 1 1 0 0 1 0 1 0 1 1 0 0 0 0 1 0 1 1 0 1 1 0 0

0 1 1 1 0 1 0 1 0 1 1 0 1 1 1 0 0 1 1 1 0 0 1 0 0 1 1 0 0 1 1 0 0 1 0 1 0 1 1 0 0 0 0 1 0 1 1 0 1 1 0 0

0 1 1 1 0 1 0 1 0 1 1 0 1 1 1 0 0 1 1 1 0 0 1 0 0 1 1 0 0 1 1 0 0 1 0 1 0 1 1 0 0 0 0 1 0 1 1 0 1 1 0 0

0 1 1 1 0 1 0 1 0 1 1 0 1 1 1 0 0 1 1 1 0 0 1 0 0 1 1 0 0 1 1 0 0 1 0 1 0 1 1 0 0 0 0 1 0 1 1 0 1 1 0 0

LIVING INBETWEEN

UNREALITY

REAL

INBETWEEN

PETER JAY ZWEIG, FAIA
UNIVERSITY OF HOUSTON, GERALD D. HINES COLLEGE OF ARCHITECTURE + DESIGN

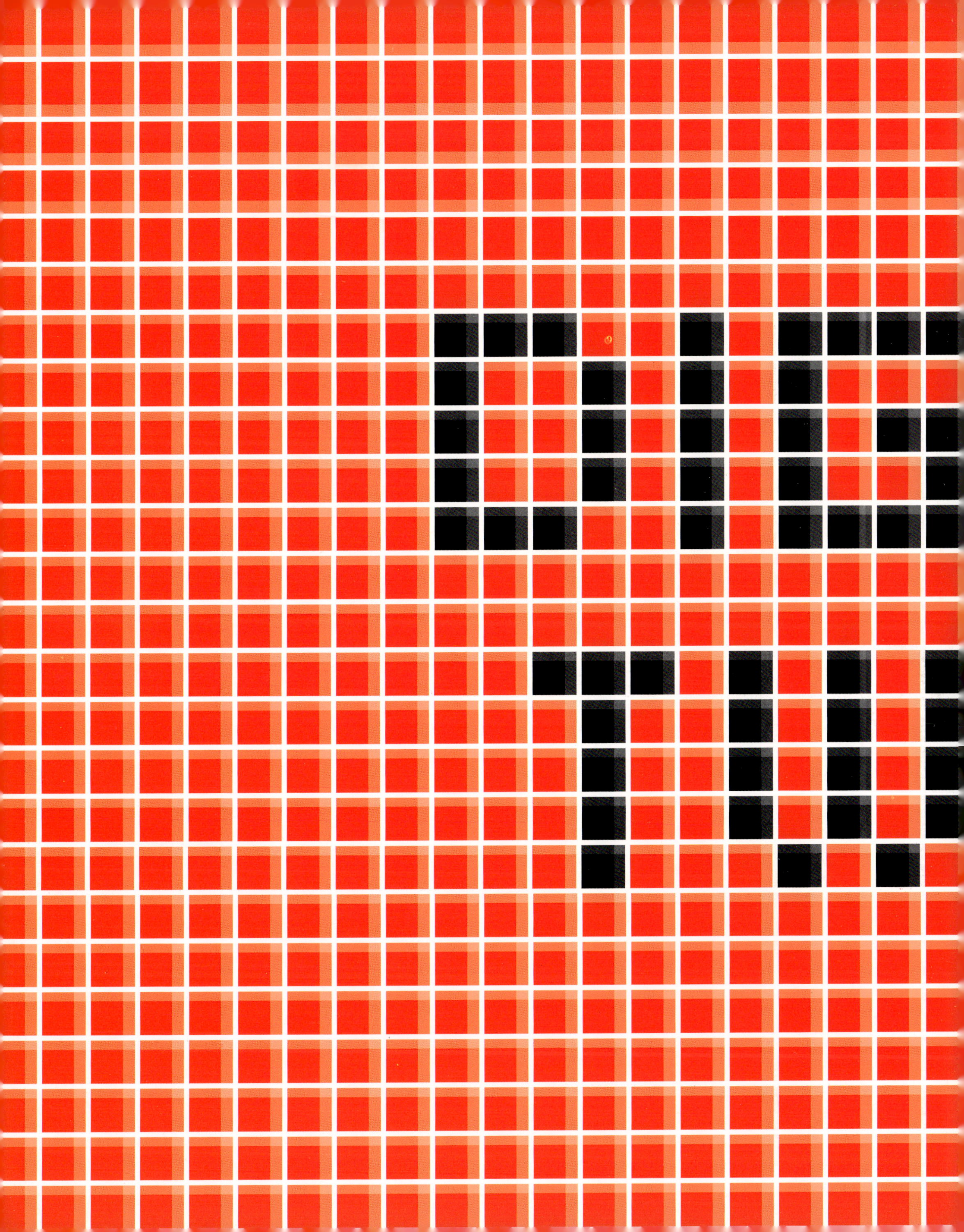